Lecture Notes in Artificial Intelligence   2317

Subseries of Lecture Notes in Computer Science
Edited by J. G. Carbonell and J. Siekmann

Lecture Notes in Computer Science

Edited by G. Goos, J. Hartmanis, and J. van Leeuwen

Springer
*Berlin
Heidelberg
New York
Barcelona
Hong Kong
London
Milan
Paris
Tokyo*

Mary Hegarty   Bernd Meyer
N. Hari Narayanan (Eds.)

Diagrammatic Representation and Inference

Second International Conference, Diagrams 2002
Callaway Gardens, GA, USA, April 18-20, 2002
Proceedings

Springer

Series Editors

Jaime G. Carbonell, Carnegie Mellon University, Pittsburgh, PA, USA
Jörg Siekmann, University of Saarland, Saarbrücken, Germany

Volume Editors

Mary Hegarty
University of California, Department of Psychology
Santa Barbara, CA 93106, USA
E-mail: hegarty@psych.ucsb.edu

Bernd Meyer
Monash University, School of Computer Science and Software Engineering
Clayton Campus, Wellington Road, Victoria 3800, Australia
E-mail: bernd.meyer@acm.org

N. Hari Narayanan
Auburn University, Department of Computer Science and Software Engineering
107 Dunstan Hall, Auburn, AL 36849, USA
E-mail: narayan@eng.auburn.edu

Cataloging-in-Publication Data applied for

Die Deutsche Bibliothek - CIP-Einheitsaufnahme

Diagrammatic representation and inference : second international conference,
diagrams 2002, Callaway Gardens, GA, USA, April 18 - 20, 2002 ; proceedings
/ Mary Hegarty ... (ed.). - Berlin ; Heidelberg ; New York ; Barcelona;
Hong Kong ; London ; Milan ; Paris ; Tokyo : Springer, 2002
    (Lecture notes in computer science ; Vol. 2317 : Lecture notes in
    artificial intelligence)
    ISBN 3-540-43561-1

CR Subject Classification (1998): I.2, D.1.7, G.2, H.5, J.4, J.5

ISSN 0302-9743
ISBN 3-540-43561-1 Springer-Verlag Berlin Heidelberg New York

Springer-Verlag Berlin Heidelberg New York
a member of BertelsmannSpringer Science+Business Media GmbH

http://www.springer.de

© Springer-Verlag Berlin Heidelberg 2002
Printed in Germany

Typesetting: Camera-ready by author, data conversion by PTP-Berlin, Stefan Sossna e.K.
Printed on acid-free paper   SPIN: 10846636   06/3142   5 4 3 2 1 0

Preface

Beginning with prehistoric cave drawings, diagrams have been a common means of representing and communicating information throughout history. Humans are skilled at creating, understanding, and making inferences from diagrams. In recent years, with advances in graphic technologies, innovations such as animations and interactive visualizations have made diagrammatic representations even more important in scientific and technical discourse and in everyday life. There is increased interest in fields such as artificial intelligence, computer vision, and visual programming languages to endow computers with human-like diagrammatic reasoning capacities. These developments have triggered a new surge of interest in the study of diagrammatic notations, which is driven by several different scientific disciplines concerned with cognition, computation, and communication.

"Diagrams" is an international and interdisciplinary conference series on the theory and application of diagrams in all scientific fields of inquiry. It grew out of a series of workshops during the 1990s: Thinking with Diagrams (TWD), Theory of Visual Languages (TVL), and Reasoning with Diagrammatic Representations (DR). The conference series was successfully launched in Edinburgh in September 2000. It attracts researchers from a variety of academic disciplines who are studying the nature of diagrammatic representations, their use in human communication, and cognitive or computational mechanisms for processing diagrams. Thus, it reflects the realization that the study of diagrammatic representation and communication must be pursued as an interdisciplinary endeavor. "Diagrams 2002" was the second event in this series. It took place at Callaway Gardens, Georgia, USA, April 18-20, 2002.

The call for contributions to Diagrams 2002 attracted 77 submissions from disciplines such as architecture, artificial intelligence, cognitive science, computer science, education, human-computer interaction, logic, philosophy, and psychology. The conference program was determined by a distinguished Program Committee that brought both interdisciplinary expertise and international flavor to the endeavor. Each submission was thoroughly peer-reviewed by three members of the Program Committee or additional referees they nominated. This labor-intensive process was intended to equitably identify the highest quality scientific and technical contributions, effectively communicated, that provided the balanced multidisciplinary intellectual record of research appearing in these proceedings. The acceptance rate was about 30% with 21 full papers accepted for presentation at the conference. In addition, 19 submissions were accepted as posters.

Besides paper and poster presentations, Diagrams 2002 included two invited talks. One was by B. Chandrasekaran, a respected researcher in artificial intelligence who played a key role in the very first meeting on this topic (1992 AAAI Spring Symposium on Reasoning with Diagrammatic Representations) and the subsequent development of this field. The second invited talk was presented by James A. Landay, an emerging researcher in human-computer interaction, who has studied how designers use sketches in the early stages of user interface design for the web and has leveraged his findings to build novel computational tools that support design by sketching.

We gratefully acknowledge financial support from the Office of Naval Research, the American Association for Artificial Intelligence, and the Cognitive Science Society. Their support enabled us to provide scholarships to all student first authors of papers and posters presented at the conference, and present a best paper award which was announced at the conference. The generosity of our sponsors is very much appreciated. In addition, the conference was held in cooperation with the Japanese Cognitive Science Society and the Japanese Society for Artificial Intelligence. We thank Hiroshi Motoda, Atsushi Shimojima, and Masaki Suwa for their efforts in securing this cooperation.

We thank members of the program and organizing committees for making the meeting and this volume a success. We are grateful for the continued support of Springer-Verlag. The staff of Callaway Gardens provided a pleasant setting for our intellectual exchanges. Finally, the core of any such enterprise is the participants and contributors. Their effort and enthusiasm made this a worthwhile endeavor.

March 2002

Mary Hegarty
Bernd Meyer
N. Hari Narayanan

Organization

General Chair

N. Hari Narayanan   Auburn University, USA

Program Chairs

Mary Hegarty   University of California at Santa Barbara, USA
Bernd Meyer   Monash University, Australia

Administration

Finance & Local
Organization Chair   Roland Hübscher, Auburn University, USA
Publicity Chair   Volker Haarslev, University of Hamburg, Germany

Sponsorship

Office of Naval Research
American Association for Artificial Intelligence
Cognitive Science Society

In Cooperation with

Japanese Cognitive Science Society
Japanese Society for Artificial Intelligence

Program Committee

Michael Anderson	Fordham University, USA
Dave Barker-Plummer	Stanford University, USA
Alan Blackwell	Cambridge University, UK
Dorothea Blostein	Queen's University, Canada
Paolo Bottoni	University of Rome, Italy
Jo Calder	Edinburgh University, UK
B. Chandrasekaran	Ohio State University, USA
Peter Cheng	University of Nottingham, UK
Richard Cox	Sussex University, UK
Max J. Egenhofer	University of Maine, USA
Norman Foo	University of Sydney, Australia
Ken Forbus	Northwestern University, USA
George Furnas	University of Michigan, USA
Meredith Gattis	University of Sheffield, UK
Helen Gigley	Office of Naval Research, USA
Mark Gross	University of Washington, USA
Corin Gurr	Edinburgh University, UK
Volker Haarslev	University of Hamburg, Germany
Patrick Healey	University of London, UK
Mary Hegarty	University of California at Santa Barbara, USA
John Howse	University of Brighton, UK
Roland Hübscher	Auburn University, USA
Maria Kozhevnikov	Rutgers University, USA
Zenon Kulpa	Institute of Fundamental Technological Research, Poland
Stefano Levialdi	University of Rome, Italy
Robert Lindsay	University of Michigan, USA
Ric Lowe	Curtin University, Australia
Bernd Meyer	Monash University, Australia
Richard E. Mayer	University of California at Santa Barbara, USA
Mark Minas	University of Erlangen, Germany
N. Hari Narayanan	Auburn University, USA
Kim Marriott	Monash University, Australia
Nancy Nersessian	Georgia Institute of Technology, USA
Daniel L. Schwartz	Stanford University, USA
Priti Shah	University of Michigan, USA
Atsushi Shimojima	Advanced Institute of Science and Technology, Japan
Sun-Joo Shin	University of Notre Dame, USA
Masaki Suwa	Chukyo University, Japan
Barbara Tversky	Stanford University, USA
Yvonne Waern	Linkoeping University, Sweden

Additional Referees

D. Jacobson
M. Jamnik
S. Kriz
Truong Lan Le
D. Waller
M. Wessel

Table of Contents

Logic and Diagrams

Diagrams in Human-Computer Interaction

Tracing the Processes of Diagrammatic Reasoning

Invited Talk

Author Index

What Does It Mean for a Computer to Do Diagrammatic Reasoning? A Functional Characterization of Diagrammatic Reasoning and Its Implications[1]

B. Chandrasekaran

Laboratory for Artificial Intelligence Research,
Department of Computer and Information Science,
The Ohio State University,
2015 Neil Avenue, Columbus, OH 43210 USA
chandra@cis.ohio-state.edu

Abstract. One might have thought that the issue of use of mental images in reasoning has been put to rest after years of debate, but one would be wrong. The journal Behavioral and Brain Sciences will soon publish a paper by Zenon Pylyshyn, which restates his earlier thesis - this time as a null hypothesis that cannot yet be rejected - that images do not play any role in reasoning, and that information which people often think they get from images when reasoning is really tacit knowledge that they already have. He does not compare the case of use of mental images with the case when the reasoner uses an external diagram for the same problem, so it is not clear if in those cases also he would claim that the diagram does not play a role. In any case, one might think that this is an issue for psychology, but not for artificial intelligence. However, conceptually, unless one has a robot with a vision system and the robot draws a diagram on a surface and uses its visual system to extract information as it reasons about a situation, any claim that a computer program performs diagrammatic reasoning in some situation has a status similar to that of the claims of human use of mental imagery during reasoning. In parallel with skeptics such as Pylyshyn of the role of images in reasoning, there are those in artificial intelligence who claim that the idea of diagrammatic reasoning by a computer is incoherent: what it means for the internal representation to be an image or a depiction is not clear, and the objections to "seeing" or perceiving information in such an image apply equally to the human and the computer cases. I do not think that the idea of computers performing diagrammatic reasoning is incoherent. I will argue my position on this by characterizing diagrammatic reasoning functionally. Then, I will argue that a computer program can be said to be performing this kind of reasoning if there exists a coherent and non-vacuous level of description of the operations of the program that satisfies the functional properties. Along the way, I will identify diagrammatic reasoning as an instance of a larger class of rea-

---

[1] Prepared through participation in the Advanced Decision Architectures Collaborative Technology Alliance sponsored by the U.S. Army Research Laboratory under Cooperative Agreement DAAD19-01-2-0009.

M. Hegarty, B. Meyer, and N. Hari Narayanan (Eds.): Diagrams 2002, LNAI 2317, pp. 1–2, 2002.

soning in which a model of some kind is used to get some information about a situation. Psychology has many more constraints about what sorts of models are available than artificial intelligence, which may be, at least in theory, freer in this regard.

Biography. B. Chandrasekaran is Professor Emeritus, Senior Research Scientist and Director of the Laboratory for Artificial Intelligence Research in the Department of Computer and Information Science at Ohio State University. His major research activities have been in knowledge-based reasoning (especially causal understanding as applied to design and diagnosis), image-based reasoning, architecture of the mind, and cognitive science. His work on generic tasks in knowledge-based systems is among the most heavily cited in recent research in the area of knowledge-based systems, as is his work on engineering design and functional reasoning. He co-authored the book Design Problem Solving (Morgan Kaufmann, 1989). His current focus is on causal understanding and use of images in problem solving. His research has been supported over the years by NSF, NIH, DOE, DARPA, and AFOSR in addition to industrial firms such as IBM, DEC, and Boeing. He is currently a participant in the Army Research Laboratories Collaborative Technology Alliance on Advanced Decision Architectures. Chandrasekaran was Editor-in-Chief of IEEE Expert from 1990 to 1994, and he currently serves on the editorial boards of International Journal of Human-Computer Studies, Journal of Experimental and Theoretical Artificial Intelligence, and Artificial Intelligence in Engineering, among others. He has been elected Fellow of IEEE, AAAI and ACM. He and Herbert Simon chaired the 1992 AAAI Spring Symposium on Diagrammatic Reasoning. He co-edited Diagrammatic Reasoning: Cognitive and Computational Perspectives (AAAI Press/The MIT Press, 1995).

Movement Conceptualizations in Graphical Communication

Ichiro Umata[1], Yasuhiro Katagiri[1], and Atsushi Shimojima[2]

[1] ATR Media Information Science Laboratories, Seika Soraku Kyoto, 619-0288, Japan
[2] Japan Advanced Institute of Science and Technology, 1–1 Asahi Tatsunokuchi
Nomi Ishikawa 923-1292, Japan

Abstract. Graphical representations such as maps and diagrams play an important role in everyday communication settings by serving as an effective means of exchanging information. In such communication, graphical representations work not only as "windows" through which we can see the target situations the graphics describe, but also serve as information processing "sites" because we can take advantage of their handiness. Though the final aim of speakers is to exchange information about the target situation, graphical representations are so deeply a part of information processing that they may affect the way people grasp the target situations. This paper presents an empirical investigation of language usage in graphical communication. Drawing on the HCRC Map Task Corpus data, we demonstrate that the existence of graphics affects linguistic expressions of motion when people collaboratively work on a task. This effect demonstrates that the use of graphical representations has an influence on movement event conceptualizations.

1 Introduction

Daily conversation often involves graphical representations that serve as strong visual aids for the exchange of information and for use in collaborations. People often give or ask directions by referring to maps, or they might draw a floor plan when describing where furniture should be placed in a room. People communicate with each other effectively by integrating information from linguistic and graphical sources in such communications (Neilson and Lee (1994); Lee and Zeevat (1990); Umata, Shimojima, and Katagiri (2000)).

It has been proposed that the use of graphics not only serves as a memory aide, but also affects the strategies of problem-solving and understanding. For example, Schwartz (1995) observed the effect that diagrammatic fidelity has on inference. Suwa and Tversky (1997) examined focus shifts and successful exploration of related thought by conducting protocol analyses of architectural designers' reflections on their own sketching behavior. However, little is known about the effect graphical representations have on language usage in conversational situations.

In our research we studied how the availability and configuration of graphics affect language usage in communication and problem-solving tasks. We focused

M. Hegarty, B. Meyer, and N. Hari Narayanan (Eds.): Diagrams 2002, LNAI 2317, pp. 3–17, 2002.
© Springer-Verlag Berlin Heidelberg 2002

on the influence of graphical representations on the perspectives from which people conceptualize motion events. Suppose that John and Mary are at Goodge Street tube station in London talking about their friend Paul's trip. Mary might say as in (1) below, but she would not as in (2):

(1) He's going down to Waterloo Station via the Northern Line.
(2) He's coming down to Waterloo Station via the Northern Line.

The current position where the two people are located becomes the reference point of the movement in this case, and the movement can only be conceptualized as a movement away from the reference point, and hence the use of "go." Suppose, on the other hand, that John and Mary are talking over a map of the London Underground shown in Figure 1. Mary could use, in this case, either (1) or (2). The availability of the map and the configuration of the icons on the map affect the conceptualization of the movement here: the location of the Waterloo Station icon in relation to them makes it possible for her to conceptualize the movement, in addition to the previous distal movement conceptualization, as a movement *in the map-world* toward the reference point, their current position as an observer to the map. Thus, graphical representation can have an influence on language usage.

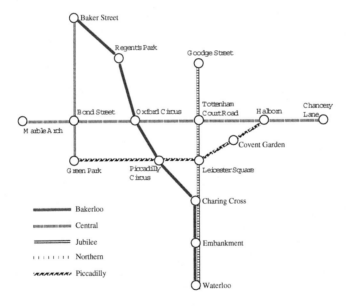

Fig. 1. Route Map of London's Underground System

The use of "come" in (2) is possible because the map and the graphical objects contained in it are readily available to the speaker as a resource to formulate messages to be communicated and problems to be resolved. The locations and arrangements of objects can be expressed in terms of the relationships between the graphical objects and the speakers, as well as those between the objects themselves and the speakers. This availability, or the ease of accessibility, of graphical representations should work to amplify our communicative and reasoning capabilities by providing us with a novel set of possibilities to construct perspectival event conceptualizations.

We investigated the effect of graphical representations on perspectival event conceptualizations through the empirical analysis of the use of motion verbs in the HCRC Map Task Corpus. The next section is a brief description of the Edinburgh HCRC Map Task Corpus. In the following section, we outline the classification of the types of perspectival conceptualizations available in communications that involve graphical representations. The subsequent section deals with the correlation between the perspectives and the alignment of directional axes. We argue that there is a trade-off between the cognitive cost and the ease of alignment. We then examine, in the final section, the distribution of the deictic motion verbs "*come*" and "*go*," and show that the configuration of graphics affects conceptualization of motion events.

2   Data

The data analyzed here is from the HCRC Map Task Corpus. This map task is a cooperative one involving two participants. The two speakers sit opposite one another, and one speaker gives instructions for a route to the other. Each has a map that the other cannot see, and a route is marked on the Instruction Giver's map while no route is marked on that of the Instruction Follower. The speakers are told that their goal is to reproduce the Giver's route on the Follower's map. Their maps are not exactly identical and the speakers are told this explicitly at the beginning of their first session. It is, however, up to them to discover how the two maps differ. The maps describe fictitious areas. We selected and analyzed 32 conversations (non-eye-contact, unfamiliar pair condition) from the entire corpus.

3   Perspectives in Graphical Communication

When we expand the domain of discourse to include maps and graphical objects so as to encompass a whole set of communicative behaviors in graphical communications, the possibility arises of setting up a novel set of perspectives involving graphics in addition to the existing perspectives concerned solely with the world. In graphical communication, theoretically possible categories of perspectives on motion events are as follows:

(a) Observer-to-World Perspective

   A movement is taken as a movement in the real-world and conceptualized from the viewpoint of the observer within the real-world. This perspective is concerned solely with the real-world.

(b) Agent Perspective

A movement is taken as a movement in the real-world and conceptualized from the viewpoint of the agent of motion. This perspective is concerned solely with the real-world.

(c) Observer-to-Graphic Perspective

A movement is taken as a movement in the map space and conceptualized from the viewpoint of the observer relative to the map. This perspective concerns both the real-world and the graphic space and creates the bridge between the two.

(d) Protagonist Perspective

A movement is conceptualized from the viewpoint of an imaginary agent in a narrative world. In graphical communication situations, a graphic provides the narrative domain for this perspective. The agent can be identified with either the speaker or the listener. This perspective belongs solely to the narrative world.

Of the four perspectives above, the first two are easy to recognize. When you say "John come to my place from Goodge Street" without a map, you are taking the Observer-to-World Perspective. If you are actually driving to somebody's place, you might say "I'm now going to the right-hand side of Piccadilly Circus," taking the Agent Perspective. If you are showing somebody the way via a cellular phone, you might say, "Go south, turn left at Leicester Square," taking the Agent Perspective of the person whom you are talking to.

In graphical communication, the latter two perspectives are available in addition to (a) and (b). You can say, "Mary is coming down to Waterloo Station from Goodge Street Station," using the route map shown in Figure 1, taking the Observer-to-Graphic Perspective. The Protagonist Perspective typically works in fictitious settings. When used in graphical communication, the map space becomes the narrative space and makes a graphics-oriented sub-type by providing concrete and tangible graphical objects upon which the perspectival conception is laid out. You can say, "Suppose you're actually moving around on the map. Go south from Goodge Street, and turn right at Leicester Square. Then you'll be at Piccadilly Circus," referring to the map shown in Figure 1. In this case, you are taking the Protagonist Perspective. These two perspectives constitute a set of perspectival event conceptualizations specific to graphical communication behavior.

In the HCRC Map Task, the maps described not real-world situations, but fictitious ones, and the speakers did not have direct access to information on the target situations. Thus, the only perspectives available were (c) and (d). We will examine those perspectives with examples of language use drawn from the HCRC Map Task Corpus in the following subsections.

3.1 Observer-to-Graphic Perspective

First, consider the following utterances:

(3) (The Giver is showing the Follower the movement shown in Figure 2.)
     Giver: Okay? Now you need to drop straight down towards the gazelles.
  Follower: Right, coming in at the top of them.
     Giver: That's right, and then go round the bottom of the gazelles.
  Follower: On the left-hand side?
     Giver: And head off to the right-hand side ... So you go under the gazelles

Fig. 2. Movement described in (3)

The expression "drop straight down to" would not have been suitable without a map. It describes the motion from the Observer-to-Graphic perspective, making use of the spatial relation on the map. Other spatial expressions ("at the top of," "bottom of," "right-hand side," and "under") also describe the spatial relation of the target situation via the graphical relations. For example, "go round the bottom of the gazelles" does not mean going actually under the gazelles in the target situation. Also, notice that the deictic spatial expression "to the right of" is based on the Observer-to-Graphic Perspective. It would have been "to the left-hand side" if the giver was taking the Protagonist Perspective.

3.2 Protagonist Perspective

Now we will examine the examples of the Protagonist Perspective shown in (4):

(4) (The Giver is showing the Follower the movement shown in Figure 3.)
     Giver: Now, keep the buffaloes to your right-hand side.
  Follower: Uh-huh
     Giver: And turn right immediately past them.
  Follower: Right.

The deictic spatial expressions "to your right-hand side" and "right" are based on the Protagonist Perspective in these utterances. If the speakers were talking based on the Observer-to-Graphic Perspective, they would have been "to your-left-hand side" and "left" respectively.

There are also more explicit Protagonist utterances. One such example is shown in (5):

(5) (The Giver is showing the Follower the movement shown in Figure 4.)

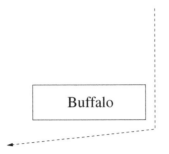

Fig. 3. Movement described in (4)

Giver: Then down.
Follower: What do you mean down? Towards the bottom of the paper?
Giver: Uh-huh
Follower: Uh-huh. Do I ... do I go by the collapsed shelter?
Giver: Uh-huh
Follower: Uh-huh
Giver: And then ... so that ... until you've got ...
Follower: The collapsed shelter's on my right?
Giver: Uh-huh
Follower: Right
Giver: And then go round to your left
Follower: My left?
Giver: As you're the wee guy

Fig. 4. Movement described in (5)

The spatial expressions "on my right" "to your left" and "my left" are clearly based on the Protagonist viewpoint. The giver confirmed it with the utterance "as you're the wee guy," introducing an imaginary agent explicitly. Thus, we can find plenty of such expressions based on those two perspectives involving graphical representations in graphical communication.

4   Perspectives and Alignment of Coordinates

The external graphical representations provide people with a new resource for information processing and communication. The existence of graphics introduces a set of novel perspectives, namely the Observer-to-Graphic Perspective and the Protagonist Perspective, for speakers to adopt in describing and reasoning about situations and events depicted by graphics. Now, we will turn to the problem of identifying the factors that determine or affect the choice of perspectives. We will explore the correlation between the alignment of directional coordinates and the choice of perspectives.

In graphical communication, the alignment of coordinates between conversation participants is usually quite important. Aligning the coordinates is comparatively easy when people are talking to each other face-to-face, sharing the same graphical representation. People can see where their partners stand in relation to the graphical representation. Even in such cases, however, people often try to make alignment easier by standing on the same side of the graphical representation to share their viewpoints. Aligning the coordinates is much more difficult when people do not know where their partners stand with respect to the graphics (e.g. talking over the phone with the same map in each person's hand, etc.). Alignment often fails in such cases, causing miscommunication.

Aligning the coordinates is especially important when people are taking the Observer-to-Graphic Perspective in which the spatial relation between the speaker and the graphical representation plays a key role. In this perspective, describing the target situation is basically easy once the coordinates are aligned; all the speakers have to do is just report the situation on the graphics as they see it. However, this should be based on a firm alignment of coordinates, and alignment failures are fatal when people are taking the Observer-to-Graphic Perspective.

On the contrary, aligning the coordinates in the Protagonist Perspective is not as important as in the Observer-to-Graphic Perspective. Conversation participants "go into the graphical representation" in this perspective, and the spatial relations between the actual speakers and their graphics are not so crucial as long as they can describe the relations between the graphical objects and fictitious agents. Describing the target situation from this perspective is not always easy, however. The fictitious agent often has to "move" around in various directions on a graphical representation and often ends up orienting differently from the speakers themselves. Hence, it requires an operation such as either a mental or physical rotation of the graphical representation to arrive at adequate descriptions from the perspective of the fictitious agent.

4.1   Alignment Failures and the Protagonist Perspective

Thus, there seems to be a trade-off between the cost of description and the robustness against alignment failure.

Description cost/misalignment robustness trade-off.

- The Observer-to-Graphic Perspective is *low cost* for descriptions, as speakers can describe directions and movements as they perceive them, but *fragile* against misalignment, as it requires perfect alignment of coordinates between speakers for successful communication.
- The Protagonist Perspective is *robust* against misalignment, as it doesn't require alignment of coordinates between speakers at all, but it is *high cost* for descriptions, as it requires mental or physical rotations to obtain adequate descriptions of directions and movements.

Because of this trade-off, it is likely that people tend to stay in the Observer-to-Graphic Perspective otherwise they risk running into the danger of miscommunication due to alignment failure. They would adopt the Protagonist Perspective only when they encounter or predict difficulties in communication.

Hypothesis: perspective/communication difficulty correlation.

- The Observer-to-Graphic Perspective tends to be assumed by the speakers when the information exchange goes smoothly.
- The Protagonist Perspective is adopted by the speakers when they encounter difficulties in communication.

To verify this hypothesis, we examined the HCRC Map Task Corpus Data again. First we classified all 32 dialogues we analyzed into two groups: one contained the dialogues that were based solely on the Observer-to-Graphic Perspective, and the other contained those that had Protagonist-perspective-based utterances. Then, we compared the rate of the dialogues which contained miscommunications concerning the direction the followers are to take in the initial stage of the task. The speakers were expected to adopt the Observer-to-Graphic Perspective if they did not have any problems in aligning the coordinates in the initial stage. The distribution is shown in Figure 5.

Out of all 32 dialogues, 28 were solely based on the Observer-to-Graphic Perspective, and 4 dialogues included utterances based on the Protagonist Perspective. The Observer-to-Graphic Perspective was dominant in this task setting. The Protagonist Perspective was persistent, however, once it was adopted. The speakers almost constantly took the Protagonist Perspective when they described either rightward or leftward motions in those dialogues after they adopted the perspective. We could find only one exceptional utterance throughout four dialogues.

The frequencies of initial directional misalignment exhibit significantly different distributions between the dialogues based on the Observer-to-Graphic Perspective and those based on the Protagonist Perspective ($\chi^2_{(2)} = 5.991, p < 0.05$). The Protagonist-perspective-based dialogues had more initial misalignments (adjusted residual: Protagonist = 2.74, Observer = -2.74). Thus, our hypothesis that the speakers would tend to adopt the Observer-to-Graphic Perspective if their information exchange went smoothly, and that they would adopt the Protagonist Perspective when they encountered communication difficulties was supported. It is likely that people adopt the Protagonist Perspective to avoid the danger of miscommunication resulting from misalignment of coordinates.

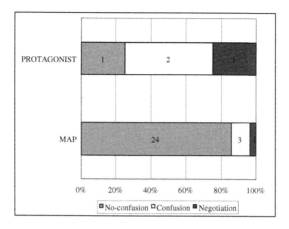

Fig. 5. Distribution of dialogues containing directional confusion for the two perspectives

4.2   Incoherence in the Protagonist Perspective

So far we have argued that there is a trade-off between the cost of description and the robustness against misalignment, and observed that this trade-off reveals itself in the form of the correlation between the speakers' choice of perspective and the communication difficulties they encounter. Upon closer examination, however, we found a systematic directional incoherence in the Protagonist Perspective. In contrast to the Observer-to-Graphic Perspective, which people tend to adopt if they do not have any problems in alignment, people adopt the Protagonist Perspective to avoid misalignments. They persistently take the Protagonist Perspective when they describe horizontal motions. There is, however, some incoherence found when they describe vertical motions. Consider (5) again, cited as (6) below with emphasis added:

(5) (The Giver is showing the Follower the movement shown in Figure 4.)
  Giver: Then down.
 Follower: What do you mean down? Towards the bottom of the paper?
  Giver: Uh-huh
 Follower: Uh-huh. Do I ... do I go by the collapsed shelter?
  Giver: Uh-huh
 Follower: Uh-huh
  Giver: And then ... so that ... until you've got ...
 Follower: The collapsed shelter's on my right?
  Giver: Uh-huh
 Follower: Right
  Giver: And then go round to your left
 Follower: My left?
  Giver: As you're the wee guy

The word "down" is describing a spatial relation on the map from the Observer-to-Graphic Perspective. Although the speakers describe horizontal motions constantly from the Protagonist viewpoints, they describe the vertical motions sometimes from the Observer-to-Graphic Perspective. It would seem that the high accessibility afforded by the external graphical representations is so strong that people tend to be attracted to the graphic relative conceptualizations of movements even at the cost of violating coherence among different instances of movement conceptualizations.

An even clearer example is shown in (7):

(7) Giver: You keep going.
   Follower: Uh-huh.
     Giver: For a couple of inches then turn left again so you're going back
           up the map
   Follower: Uh-huh, right.

The two perspectives coexist in one utterance in this case. Examples like this suggest that people can somehow reserve a viewpoint as an observer even when they are taking the Protagonist Perspective. The mechanism of this coexistence and the reason why people are more sensitive to horizontal motions than to vertical motions are still unclear, and will be left for future research.

5 Motion Verbs and Accessibility of Graphical Representations

We have observed that the existence of graphics provides two novel perspectives, and that those perspectives are used in an interesting trade-off between description cost and misalignment robustness. The Observer-to-Graphic Perspective is less robust against misalignment, but takes less description cost as it makes use of the relative spatial relation between the speaker and graphical objects once the alignment is fixed. It is likely that the way people conceptualize the target situations is affected by spatial relations between speakers and graphical objects under the Observer-to-Graphic Perspective if we consider its dependence on such relations. We will examine the effect of the spatial relations between the speaker and graphical objects on the use of the motion verbs *come* and *go* in this section.

5.1 Motion Verbs: *Come* and *Go*

Verbs like *come* and *go* reflect a speaker's reference point, as is shown by (1) and (2). *Go* indicates motion to a location which is distinct from the reference point. *Come* indicates motion toward the reference point[1]. If we classify motion in relation to the reference point, the possible categories are as follows:

---

[1] Actually, *come* and *go* have more complicated semantics, as is shown in Fillmore (1997). The scheme presented here is a rather simplified version, but serves well enough for the present purpose.

(i) Motion from the reference point
(ii) Motion toward the reference point
(iii) Motion neither from nor toward the reference point

Go can cover (i) and (iii), because the only restriction it has is that the goal of the motion should not be the reference point. The restriction in the use of come is stronger, in the sense that it indicates only motion toward the reference point and can cover only (ii). Thus, go is more frequently and widely used than come, although they are usually considered as antonyms.

The reference point tends to be set to the speaker's current position under the Observer-to-World Perspective, so come often indicates motion toward the speaker. Go does not have such a tendency because of its general semantic nature compared to that of come. Thus, the distributions of go and come indicating actual motion in the real world are summarized as follows:

Usage of motion verbs in the real world domain.

– Go is used more frequently and generally than come.
– Compared with go, come is used more frequently in indicating motion toward the speaker than in indicating motion away from the speaker.

In graphical communication, a reference point is set based on one of the perspectives shown in section three. For example, the utterance "go round to your left" in (5) was based on the Protagonist Perspective, and the reference point was likely set at the position the follower is looking at on the map at that moment. The speaker is "going into the map" in this perspective, and the reference point is likely to be set from an agent viewpoint in the fictitious map world. The utterance "coming in at the top of them" in (3) is from the Observer-to-Graphic Perspective. In the former case, it is likely that the reference point was at the position of the imaginary agent rather than at that of the speaker at that moment. On the contrary, in the latter case the speaker grasped the movement on the map as an observer in the real world, reserving the possibility that his actual position in the real world could be the reference point in relation to the graphical objects. Thus the reference point setting can be made between the graphical objects and the speaker in a parallel manner to that for the real-world motion description. Thus, our hypothesis on motion description from the Observer-to-Graphic Perspective is as follows:

Hypothesis: usage of "come" under the Observer-to-Graphic Perspective.

– Go will be used more frequently and generally than come in graphical communication.
– Compared with go, come will be used more frequently in indicating motion in the graphics domain toward the speaker than in indicating motion in the graphics domain away from the speaker.

In conversation involving a graphical representation, people describe the target-world information through the graphical representation (see Umata, Shimojima, and Katagiri (2000)). The target world is captured via its representation in such cases, and our prediction is that people tend to conceptualize the target event based on the relations between the graphical objects and themselves. If this is right, the configuration of a graphical representation will affect the conceptualization of its target world events. To verify this assumption, we examined the HCRC Map Task corpus by focusing on the usage of the verbs *come* and *go* under the Observer-to-Graphic Perspective. Our hypothesis is that the spatial configuration between graphical objects and the speaker affects the usage of these motion verbs under the Observer-to-Graphic Perspective just as if those graphical objects were actual ones.

5.2 Analysis

We analyzed the usage of *come* and *go* in 28 out of 32 conversations under the Observer-to-Graphic Perspective. Of all the occurrences of *come* and *go*, only those that describe motion were analyzed here. The occurrences of "fictive motion" expressions such as "the bay goes like that," were not considered. The direction of the motion they describe was analyzed for all the occurrences.

The maps used in this task were fictitious, and the subjects had no direct access to the target world of the map. The Observer-to-World Perspective was not available in this task setting. Almost all of the Instruction Givers adopted the strategy of giving their Followers local instructions step-by-step along their route on the map, which finally led the Followers to their goals. Each step in this case is a motion from the current position of the Follower to some landmark.

If we assume that the configuration of graphics affects the conceptualization of motion events under the Observer-to-Graphic Perspective, then it is likely that the actual distance between the speaker and objects in the graphics plays the key role. The occurrence of *go* would be prominent anyway because of its neutral semantic nature, as described above. Therefore, *go* would be used widely to express either motion toward the speaker (*toward motion*) or motion away from the speaker (*away motion*) in the graphics. The usage of *come* is more restricted, and its frequency is expected to be smaller. However, the distribution of motion directions of "*come*" should be similar to that of "*go*" if the spatial relations between the speaker and the graphical objects do not have any effect. The positions of the start and the goal were balanced among the dialogues (e.g. one map has its start at the top right corner and its goal at the bottom left, and another has its start and goal the other way round, etc), so the directions of motion were also balanced. Thus, there is no correlation between orientation of motion and the beginning or the end of each motion step. If the ratio of *toward motion* is larger for *come* than *go* in this setting, then it will be because of the effect of the spatial relations between the actual speaker as an observer and the graphical objects.

5.3   Results

As was expected, *go* was used much more frequently than *come* in graphical communication also. There were 487 occurrences of the verb *go* and 88 occurrences of the verb *come* used to describe motion. The distribution of motion described is shown in Table 1.

Table 1. Distribution of *come* and *go*

	away	toward	round	horizontal	sum
come	17	50	2	19	88
go	123	165	17	182	487
sum	140	215	19	201	575

A *round motion* involves both up and down motions in one sequence. *Horizontal motion* means either rightward or leftward horizontal movement. The frequency of *go* is higher than that of *come*, as was expected.

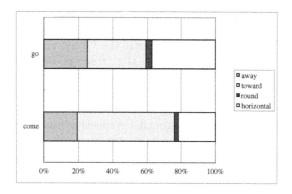

Fig. 6. Distributions of the direction of motion indicated by *come* and *go*

Figure 6 shows the distribution of the direction of motion indicated by *come* and *go*. The directions of motion exhibit significantly different distributions between *come* and *go* ($\chi^2_{(3)} = 17.21, p < 0.01$). More concretely, the ratio of *toward motion* is significantly larger in *come* occurrences compared to *go* occurrences (adjusted residual: *come* = 4.09, *go* = -4.09), as is shown in Figure 7. Thus, the assumption that the usage of *come* is affected by the configuration of graphics under the Observer-to-Graphic Perspective is also supported. The distribution of *come* shows that the configuration of graphical representation affects the conceptualization of motion events in the same way as the real-world configuration affects the conceptualization of an actual motion event.

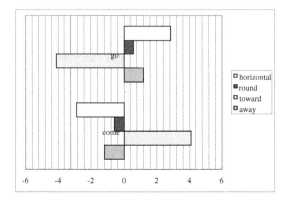

Fig. 7. Adjusted Residuals

6   Conclusions

We analyzed the use of movement expressions in the HCRC Map Task Corpus to elucidate the functions and effects that external graphical representations have for communications in conversational problem solving situations. We have found that (1) the use of graphical representations induces two new types of perspectival conceptualizations of events: the Observer-to-Graphic Perspective and the Protagonist Perspective; (2) the Observer-to-Graphic Perspective and the Protagonist Perspective have different trade-off patterns between description cost and misalignment robustness, and hence, they are employed in different situations according to the degree of communication difficulties; (3) a systematic deviation from the Protagonist Perspective was observed for specific types of movement (up/down movement) conceptualizations; and (4) the spatial configurations of graphical elements with respect to the speaker can affect the conceptualizations of, and hence the choice of descriptions of, motion events.

These results strongly indicate that people utilize graphical representations at hand not only as representational tools, but also as a resource for information processing and communication by taking advantage of their accessibility. This accessibility of graphics makes it possible for people to perform efficient communication and reasoning on the one hand, but on the other it also complicates the interaction of graphical and linguistic information. Analysis of graphic/linguistic information integration in terms of the expanded notions of perspectives we developed in this paper contributes to the elucidation of the interaction and hence, we believe, to establishing the basis for the effective design of graphical tools that support smooth collaboration and communication between people.

References

Fillmore, C. J. (1997). *Lectures on Deixis*. Stanford, CA.: CSLI Publications.

Lee, J., and H. Zeevat (1990). Integrating Natural Language and Graphics in Dialogue. In Diaper, D., Gilmore, D., Cockton, G., and B. Schackel. (Eds.) *Human Computer Interaction—INTERACT'90*, 211–234. Amsterdam.

Neilson, I., and J. Lee (1994). Conversations with Graphics: Implications for the Design of Natural Language/Graphics Interfaces. *International Journal of Human-Computer Studies 40*, 509–541.

Schwartz, D. L. (1995). Reasoning about the Referent of a Picture versus Reasoning about the Picture as the Referent: An Effect of Visual Realism. *Memory & Cognition 23*, 709–722.

Suwa, M., and B. Tversky (1997). How Do Designers Shift Their Focus of Attention in their Own Sketches? *Papers from the AAAI-97 Fall Symposium*, 102–108. AAAI Press.

Tversky, B. (1981). Distortions in Memory for Maps. *Cognitive Psychology 13*, 407–433.

Umata, I., Shimojima, A., and Y. Katagiri (2000). Talking through Graphics: An Empirical Study of the Sequential Integration of Modalities. *Proceedings of the 22nd Annual Conference of the Cognitive Science Society*, 529–534.

Toward a Model of Knowledge-Based Graph Comprehension

Eric G. Freedman[1] and Priti Shah[2]

[1] Department of Psychology, 411 Murchie Science Building, University of Michigan-Flint, Flint, MI 48502
(freedman@umich.edu)
[2] Department of Psychology, 525 East University, University of Michigan, Ann Arbor, MI 48109-1109
(priti@umich.edu)

Abstract. Research on graph comprehension has been concerned with relatively low-level information extraction. However, laboratory studies often produce conflicting findings because real-world graph interpretation requires going beyond the data presentation to make inferences and solve problems. Furthermore, in real-world settings, graphical information is presented in the context of relevant prior knowledge. According to our model, knowledge-based graph comprehension involves an interaction of top-down and bottom up processes. Several types of knowledge are brought to bear on graphs: domain knowledge, graphical skills, and explanatory skills. During the initial processing, people chunk the visual features in the graphs. Nevertheless, prior knowledge guides the processing of visual features. We outline the key assumptions of this model and show how this model explains the extant data and generates testable predictions.

1 Introduction

In today's Information Age, decision-makers have a large amount of complex, visually-presented data available to them. Frequently, visual displays make relevant information easy to understand. However, the poor design of a graph or a user's difficulty in using data presented in a graph can lead to tragic consequences. Indeed, the decision to launch the space shuttle Challenger has been blamed on the use of displays that obscured the relationship between temperature and o-ring failures (Oestermeier & Hesse, 2000; Tufte, 1997). In this case, engineers who decided to launch the Challenger were not convinced by the data that the cold weather was dangerous. This tragic decision, in retrospect, might have been avoided if either the graphs depicting the data were designed to highlight the relevant trend (Oestermeier & Hesse, 2000; Tufte, 1997), or if the engineers appropriately brought to bear their prior knowledge in interpreting the data (Hirokawa, Gouran, & Martz, 1988).

As the example above illustrates, comprehension of graphs can be effortful and error-prone (e.g., Maichle, 1994; Shah & Carpenter, 1995). At the same time, graphs are so pervasive because they seem to make quantitative information easy to understand (Tversky, 2001). But what factors influence whether graph comprehension is successful or not? Much of the previous research on graph interpre-

M. Hegarty, B. Meyer, and N. Hari Narayanan (Eds.): Diagrams 2002, LNAI 2317, pp. 18–30, 2002.
© Springer-Verlag Berlin Heidelberg 2002

tation focused on visual features of data, such as a graph's format, and considered the effect of such factors on the speed and accuracy of retrieving simple facts from graphs. This research provided some guidelines for presenting data to facilitate fact-retrieval (e.g. Kosslyn, 1989; Shah & Hoeffner, in press). However, it is not clear how one might apply such guidelines in real-life contexts, such as making a decision based on complex data, developing a scientific theory to account for a number of quantitative relationships, or making a judgement about whether the theory and the data are consistent. In such real-life contexts, factors such as prior knowledge and a viewer's biases play a major role. In addition, accurate and fast fact-retrieval is less critical than a deep understanding of relationships and trends.

A fundamental question in graph research, then, is the characterization of the comprehension processes and the principled identification of the characteristics of graphic formats, data sets, and interpretation tasks that shape the interpretation process. Fortunately, research on graph comprehension has made a number of advances in the last few years. There are a number of studies that consider not just the visual features of the graph, but also viewers' prior knowledge and skills in the context of relatively complex comprehension, scientific reasoning, and decision-making tasks. Given this new body of research, it is now possible to develop models of graph comprehension that takes into account these different factors.

We propose a model of graph comprehension that incorporates the effects of display characteristics and prior knowledge. We build on previous models of graph comprehension as well as other models of comprehension more generally. The Construction-Integration (CI) model of text and discourse comprehension (Kintsch, 1988) provides a useful framework for understanding graph comprehension because it explains how the reader's mental model of text is influenced by both the surface features of the text as well as his or her prior knowledge and expectations. In this paper, we describe how this model may be applied to graph comprehension.

2 A Construction-Integration Model

In this section, we briefly outline characteristics of the CI model of text comprehension and the corresponding features in graph comprehension. According to the CI model, comprehension takes place in two phases, a construction phase and a comprehension phase. During the construction phase, a reader or listener first activates textual information as well as a large set of prior knowledge associated with that text. During this stage, the reader attempts to construct a coherent representation of the available information. The superficial aspects of the text (such as the sentence structure and complexity) influence the kinds of representations that are formed because only information explicitly stated in the text, but not that which requires inferences, is initially activated. In addition, prior knowledge and goals guide the initial processing of the text and what information is encoded. Similarly, during the initial processing of graphical information, we argue that there is an automatic activation of perceptual features that guide processing of data.

During the second, integration phase of the CI model, disparate knowledge is combined into a coherent representation. When information is explicitly represented in a text so that no inferences are required to form a coherent representation, less effort is required in this phase. However, when the reader must draw some inferences in order to form a coherent representation or relate it to the task, then the integration process is effortful. Graph comprehension shares this characteristic with text comprehension. When relevant information is explicitly represented in the visual features that have been identified (on the basis of the display characteristics) and can be easily linked to prior knowledge, the comprehension is effortless. Consider, for example, the case of an expert in cognitive psychology viewing a line graph representing mental rotation data. A cognitive psychologist knows something about the cognitive phenomenon: when people identify a non-upright figure, they mentally rotate that figure. He or she also knows that studies of mental rotation involve presenting figures at different orientations and plotting the time to recognize objects at these orientations in a line graph. Thus, the expert automatically forms a link between the visual features (the shape of the line) and the theoretical interpretation of the data. When a perceiver lacks the relevant prior knowledge, or the display does not explicitly represent information that must then be inferred, comprehension is effortful. In the section on display characteristics we consider how the format of a graph might influence what inferences must be made.

The third characteristic of the CI model that is consistent with research on graph interpretation is that the construction and integration phases take place in alternating cycles. Readers or listeners construct a representation of one sentence or small set of propositions at a time, carry out the integration process, then move on to the next stage. Previous studies of eye fixations as viewers comprehend graphs are consistent with such a serial and incremental process in the context of graphs (Carpenter & Shah, 1997). This study examined the pattern and duration viewers' gazes on line graphs as they described and answered questions about graphs. Viewers identified individual quantitative facts and relations, based on the component visual chunks of a display, and related them to their graphic referents (i.e., the content of the graph serially. This iterative model can predict the distribution of viewers' gazes across different parts of a graph as well as the total number of gazes required to interpret graphs that vary in complexity.

We implemented a preliminary model using the CI framework which accounted for the interactive influence of graphical display characteristics, domain knowledge, and graph reading skill that are discussed in more detail in the next sections (Shah et al, 2000). Consistent with CI models of text comprehension, this model was a constraint-satisfaction model. However, the types of representations were to accommodate graphical information. Specifically, the model had three pools of units: visual features, domain knowledge, and interpretation propositions (see Figure 1). Visual feature units represented properties of the graph such as "the red line is increasing." When different graphic formats were "inputted" to the model, different visual features were activated. Domain knowledge units represented viewers' prior knowledge about quantitative relationships, such as "in general, rich people have expensive cars." Differences in domain knowledge could then be modeled by activation of different knowledge elements. Finally, proposition units represented possible interpretations of the information in a graph, such as "as income increased, the average value of cars owned increased." In the model visual feature and domain knowledge units were connected to the propositions by weighted links that corresponded to a viewer's graph reading skills. Thus, a model of a "good" good graph viewer incorporated strong weights between visual features and interpretations, whereas a model of a "poor" graph viewer did not have strong direct links between visual features and interpretations. Input to the model was a pattern of activation across the visual feature and domain knowledge units (representing the type of graph and the prior knowledge of the graph viewer). The model computed activation values for the interpretation proposition units based on this input. The steady state activation of the proposition units represented the model's interpretation of the graph.

Overall, the model was able to mimic the data from a study in which high and low graph-skilled readers described line and bar graphs depicting meaningful data (about which they had prior knowledge) and arbitrary data (about which they had little knowledge; Shah & Shellhammer, 1999). This preliminary implementation of the CI model suggests that it can provide a useful framework for describing the nature of graph comprehension and accounting for the influence of display characteristics and prior knowledge on viewers' comprehension of data. In the next sections, we provide more detailed accounts of exactly how display characteristics and prior knowledge influence the processing of graphs.

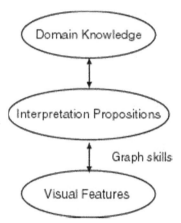

Fig. 1. Schematic illustration of CI model of graph comprehension in which visual features, domain knowledge, and graph skills influence the interpretation of data (Shah et al, 2001).

2.1 Display Characteristics

According to a CI model of graph comprehension, visual display characteristics, such as format and color, should have an effect on both the low-level perceptual aspects of graph comprehension as well as the high-level cognitive processes. Because there is a large amount of research on visual characteristics of a display and its influence on viewers' interpretations, we outline only briefly here some key data relevant to our model. More detailed reviews may be found in a number of papers, including (Lewandowsky & Behrens, 1999; Shah, Freedman, & Vekiri, forthcoming; Shah & Hoeffner, in press).

For the purposes of our model, the most important feature of display characteristics is those that influence how information is mentally grouped (into what we have referred to as *visual chunks*, such as individual lines in a line graph or groups of bars in a bar graph). When relevant quantitative information is directly represented in the visual chunks, then pattern perception and association processes are sufficient to interpret relevant quantitative information, and viewers are likely to be able to interpret that information accurately and quickly (Shah & Carpenter, 1995). However, when relevant information must be derived by mentally transforming data to make inferences about relationships or facts, then viewers are often unable to accurately comprehend that information (e.g. Casner & Larkin, 1989; Larkin & Simon, 1987; Shah & Carpenter, 1995). Indeed, they may not even consider those relationships. The general principle is that graphs should be designed so that the relevant quantitative facts and relations are explicitly represented in the main visual chunks of the display. Graphical displays are most useful when they make quantitative information perceptually obvious (Casner, 1990; Casner & Larkin, 1989; Larkin & Simon, 1987; Lohse, 1993; Pinker, 1990). In previous studies, we have shown that we can redesign graphs presented in middle school history textbooks according to this main principle and dramatically increase the number of students who identify the relevant trends in the data (Shah, Hegarty, & Mayer, 1999). By redesigning the graphical information, one can decrease the processing load by increasing the automatic construction and decrease the effortful integrative processing.

However, as Shah, Freedman, and Vekiri (forthcoming) review, the selection of the most effective display depends on the task demands determined by the situation or the viewer's goals. For example, viewers are more likely to describe x-y trends (e.g., as x increases, y decreases) when viewing line graphs (Carswell, Emery, & Lonan, 1993; Shah et al., 1999; Zacks & Tversky, 1999), and are more accurate in retrieving x-y trend information from line graphs than from bar graphs (Carswell & Wickens, 1987). Carswell & Wickens (1988) found that bar graphs and other "separable" displays were more appropriate for identification of individual facts. By contrast, line graphs, and other displays that integrated two or more variables, were better suited for tasks that required synthesis.

As suggested by the CI model, visual features are likely to interact with a graph viewer's prior knowledge. For example, information should be presented in a line graph if the viewer's goal is to identify a quantitative trend. However, if the goal is to find the relative differences between data points, then information might be best presented in a bar graph (Shah, Hegarty, & Mayer, 1997; Zacks & Tversky, 1999). Additionally, an individual with graphical skills will more readily chunk specific types of lines (e.g., dashed vs. unbroken) together in a line graph depicting multiple variables when compared with individuals with poor graph skills. In the next section, we discuss how prior knowledge also influences the interpretation of data.

2.2 Role of Prior Knowledge

According to the CI model, readers' prior knowledge has an influence on both the construction and integration phases of discourse comprehension. Specifically, a reader's skills in reading as well as his or her prior knowledge about the content of the text influences their interpretation of text. Just as in the case of text comprehension, several types of knowledge are brought to bear on graphs, including domain knowledge, graphical literacy skills, and explanatory skills.

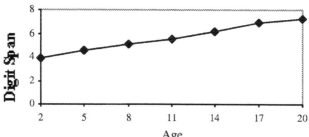

Figure 2. Sample stimulus from Freedman & Shah (2001).

Domain knowledge includes any mental representation of the content of the graphical information. This knowledge may include underlying theories and model, previous findings, and methodological techniques. The likelihood that domain knowledge will be brought to bear on the processing of a graph depends on several factors. First, the likelihood that this knowledge will be activated depends on the accessibility. It is presumed that the more often prior knowledge is employed the more accessible it will be. In experts, this knowledge may be automatically activated. In terms of the CI model, the automatic activation of prior knowledge during the construction phase increases the likelihood that this can be incorporated into a coherent representation of the graphical display during the integration phase. A second dimension is the salience, which refers to the meaningfulness or relevance of a particular explanation in a specific context.

One role of domain knowledge is to make the numerical relationships more salient as well as facilitate the inferential processing of graphs. Gattis and Holyoak (1996) believe that expert viewers make inferences about the graphical displays in order to understand the corresponding conceptual relations. These data are consistent with the CI model of language, which predicts that prior knowledge increases the likelihood that the inferences will be generated.

Another related role of domain knowledge is to guide processing at the level of meaningful quantitative functions. Freedman and Shah (2001) recently examined the domain knowledge differences between experts and novices. Doctoral students in cognitive psychology and cognitive aging and undergraduates described to graphics depicting cognitive aging data (e.g., RT and age) and non-cognitive aging data (e.g., political participation and age). Figure 2 shows a typical stimulus from this study. Each graph was accompanied with a short written description of the independent and dependent variables. In these graphs, the appropriate level of description was the underlying function. In their descriptions of the data, novices were more like to describe the main trends (e.g., 'acuity decrease as one gets older') while experts were more likely to describe particular patterns that, in the case of these graphs, were meaningful (e.g., exponential functions). Figure 3a provides the percentage of main effect descriptions and Figure 3b provides the percentage of responses that referred to particular functions (e.g., increasing slope). This difference was not due to the level of detail that they provided, but instead to the meaning that they were able to attach to their description. Novices were likely to describe irrelevant or extraneous details as frequently as experts (but experts distinguished between details that were possibly relevant and those that were not. Thus, the present study supports the idea that novices tend to rely on lower-level perceptual processing of the graphical display, whereas even in their description of the display characteristics, experts were able to use their graph skills to detect specific mathematical trends in the data.

In a related study, Shah (1995) compared viewers interpretations of familiar data for which they had expectations about trends (number of car accidents, number of drunk drivers, and traffic density) and unfamiliar, to data for which viewers did not have any expectations (ice cream sales, fat content, and sugar content). The results suggest that when viewers had particular expectations, they were likely to express those relationships (for example, as drunk driving increases, car accidents increase), ignoring "idiosyncratic" data points inconsistent with the general expected trends (e.g., local maxima and minima). By contrast, when viewers did not have expectations, they were less likely to describe general trends, and more likely to describe local maxima and minima. These results suggest that viewers' familiarity with quantitative trends influenced whether or not they would describe those trends. In a second study, novice viewers described expected relationships even when the graph did not explicitly depict those relationships. For example, viewers said that a graph showed that increased levels of drunk driving led to greater numbers of car accidents, even when those two variables were depicted in the graph as orthogonal and not related. Thus, domain knowledge can lead to systematic biases and errors during the interpretation of graphs.

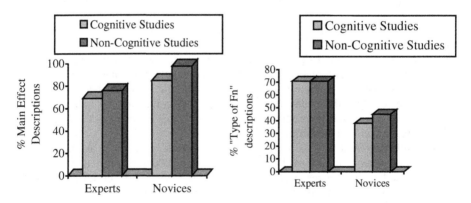

Figure 3a. Percentage of main effect descriptions

Figure 3b. Percentage of description of line functions

Another body of research on possible influences of prior knowledge have examined the effects of prior beliefs on covariation assessment (Broniarczyk & Alba, 1994; Jennings, Amabile, and Ross, 1982; Trolier & Hamilton, 1986; Wright & Murphy, 1984). Although this earlier work does not use graphically presented data, it likely has implications for the interpretation of graphs. Jennings, Amabile, and Ross (1982), for example, found that people who had prior theories about covariation made estimates of covariation that were unrelated to the actual correlations. In general, subjects typically overestimated the actual relationships.

In follow-up research, Wright and Murphy (1984) found the presence of prior beliefs increased participants' sensitivity to the actual covariation in data presented in tabular format. Covariation ratings were higher when participants had expectations about the relations than when subjects had no expectations. They interpreted their results to mean that prior knowledge may make people more resistant to noise.

Freedman and Smith (1996) extended this work to graph comprehension. We asked viewers to estimate the covariation in scatterplots. When no prior knowledge was available, viewers' estimates were relatively conservative. Conversely, when subjects estimated the degree of association without scatterplots, they perceived a relatively high degree of relation among the traits. When subjects had prior knowledge and were presented with data, we found that perceivers struck a compromise between their prior beliefs and the available data. However, Freedman and Smith found that prior beliefs did not influence the judgments about the scatterplots unless the subjects had 30 seconds to think about their theories. Thus, with novice participants it is possible that it takes time before prior knowledge may be integrated with display information.

Although prior beliefs influence quantitative judgments, making participants aware of the characteristics of the stimuli reduced the influence of prior beliefs (Sá, West, and Stanovich, 1999). At the same time, participants' beliefs are also affected by viewing data inconsistent with their beliefs. These studies suggest that it is possible to overcome biases based on prior beliefs and that perhaps the influence of prior beliefs may be dependent on the strength of those beliefs and on the data. When data are unambiguous, weak prior beliefs do not influence their interpretation (Anderson & Kellum, 1992). However, when prior beliefs were strong, viewers were biased even in the interpretation of unambiguous data (Anderson, 1995).

Together, the research on prior knowledge in data interpretation suggests that when prior knowledge is activated it guides the processing and encoding of graphical information. The CI model provides a mechanism for accounting for the effects of prior knowledge in terms of the top-down influence on the processing of display characteristics and in the types of inferences that are made.

Graph Skills. According to the CI model, viewers differ in their knowledge about the association between different visual features and their interpretation. This aspect of our model was originally proposed by Pinker (1990) in the context of graphs. According to his model, *graph schemata* mediate the translation of visual descriptions into a conceptual representation. The graph schemata includes viewers' prior knowledge about graphs, such as the fact that there's an x-axis and a y-axis, that straight lines means linear relationships, our knowledge of the properties of particular types of graph formats, and so on.

The CI model assumes that graphical knowledge is activated early during the construction phase and it is used to facilitate the chunking of the elements. Furthermore, highly skilled graph viewers are less influenced by both the visual characteristics of the graphs because they have associations between the formats and inferences that can be made (Shah, 2000). In one study that illustrates this point, we compared interpretations of line graphs and bar graphs provided by skilled and unskilled graph viewers (Shah & Shellhammer, 1999). To compare individual differences in domain knowledge, half the participants saw line graphs and bar graphs that included familiar semantic content (e.g., Age of owner vs. value of car). The other half of the participants saw graphs that had no semantic content, only letters such as "A vs. B". Overall, graph reading skill and domain knowledge both interacted with graph format. Unskilled graph viewers, as well as viewers with no domain knowledge, typically provided surface-level descriptions of the data (for example, describing each line individually in a line graph). By contrast, skilled graph viewers and those with domain knowledge made inferences about the data (e.g., describe a general trend or main effect) regardless of format. Similarly, Maichle (1994) found that whereas participants with poor graph comprehension skills relied on the trends in the data where as the good graph skill subjects emphasized comparisons among trends.

Like domain knowledge, graphical knowledge can lead to biases in the interpretation of graphs. Viewers' schemas for graphs and maps can distort their representations of them (Schiano & Tversky, 1992; Tversky & Schiano, 1989). Participants who drew line "graphs" from memory tended to distort the lines and draw them as being closer to 45° than the lines originally were. When they were told that the same display depicted a map, however, they distorted the lines so that they were closer to 0° or 90°.

Explanatory Skills. One of the main reasons that graphical, information is often used is to provide an account of the underlying mechanisms or processes. For instance, a cognitive psychologist might interpret a linear increase in response time as a function of increasing items as indicating a serial search mechanism. While explanation may be a viewer's central goal, far less is known about the explanatory processes employed during graph comprehension. Few studies have systematically examined the role of explanatory processes during graph comprehension. Oestermeier and Hesse (2000) suggest that use of graphics in providing causal explanations presupposes that the view has prior knowledge of the domain. However, Oestermeier and Hesse also suggest that the ability to provide explanation of visual data depends on abstract knowledge about the concept of causation. Tabachneck-Schijf, Leonardo, and Simon (1997) assume that experts engage in visual-graphical thinking during explanatory processes. Chinn and Brewer (1992) examined the role of prior knowledge in dealing with anomalies. Chinn and Brewer found that undergraduates tended to ignore the anomalous data. Indeed, Trafton and Trickett (2001) have found that unlike novices who may ignore anomalous data, experts focus their attention on anomalies. One explanation consistent with our CI model is that novices may ignore anomalies while scientists focus on them because explanations are activated in experts but not in novices. While these studies establish the empirical link between explanatory processes and evidence interpretation and the theoretical link between explanation and graph interpretation, further research is needed to determine whether explanatory processes influence the comprehension of graphical information.

In a recent study, we explicitly considered the effect of expertise on scientific reasoning skills, including explanatory skills, in viewers' interpretations of data (Freedman & Shah, 2001). Experts were more likely than novices to generate spontaneous explanations of the data (see Figure 4a). Furthermore, experts were more likely to provide evaluations of the study that indicated possible reinterpretation of the data and suggested additional possible analyses or studies (see Figure 4b). Thus, experts used their prior knowledge to go beyond a description of the available data. These results suggest that explanatory skills, perhaps a specific component of

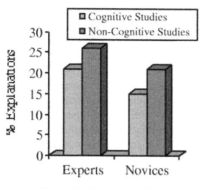

Figure 4a. Percentage of explanations of data

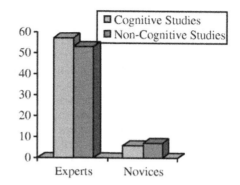

Figure 4b. Percentage of evaluations of graphs.

scientific reasoning skills, influence the types of interpretations viewers give to data. Again, experts may offer explanations because this knowledge is automatically activated during the construction phase.

3 Conclusions

In summary, the cognitive research suggests that graph comprehension is influenced not only by the display characteristics of a graph, but also a viewer's domain knowledge, graphical literacy skills, and explanatory and other scientific reasoning skills. These factors are analogous to factors that influence text comprehension. We suggest therefore that models of text comprehension, and in particular the CI model (Kintsch, 1988) may provide a useful modeling framework for considering the precise mechanisms through which prior knowledge and perceptual features interact with each other. Specifically, the CI model assumes that graph comprehension involves an interaction between display characteristics and prior knowledge. In addition, further research on the interaction of prior knowledge and display characteristics provides a fruitful avenue for future research.

The relative ease or difficulty in interpreting graph occurs because two kinds of processes are involved in comprehending different kinds of information from a graph. If the graph supports visual chunks that the viewer can map to relevant quantitative information, then it may be automatically retrieved. In terms of the CI model, if an individual possesses domain knowledge, graph skills or explanatory skills, this information is retrieved during the construction phase. If not, quantitative information must be computed by inferential processes that consist of a number of retrieval and comparison substeps. Thus, the CI model can account for novice's difficulty with the comprehension of graphics. To what extent does prior knowledge influence the comprehension of graphs? Consistent with the CI model of discourse comprehension, we believe that prior knowledge is activated during the early processing of graphs. Thus, the model proposes that viewers' knowledge about graphic formats, and expectations about the relationships between particular variables, have a top-down influence on the kinds of interpretations that viewers give to graphs. However, novice viewer's may not be able to activate their prior knowledge automatically so that their prior knowledge may be applied during a relatively effortful integration stage.

The studies reviewed in this paper also support our claim that viewer's bring several types of knowledge to bear on the processing of graphical information. Each of these types of knowledge must be studied separately as well as demonstrating how they interact with the other types of knowledge and the display characteristics. We (Shah, Freedman, & Vekiri, forthcoming) have assumed that the comprehension of graphs coordinating propositional and spatial representations of the data display. Clearly, propositional knowledge interacts with spatial representations (Tabachneck-Schijf, Leonardo, & Simon, 1997). Consequently, the precise mechanisms by which propositional and nonpropositional knowledge are integrated needs to be articulated.

The major challenge for knowledge-based approaches to graph comprehension is the need to specify the precise mechanisms through which prior knowledge and perceptual features interact with each other. To investigate this interaction systematically, studies need to be conducted in which both the display characteristics and the viewer's prior knowledge are manipulated. For instance, viewers without prior knowledge may focus on superficial aspects of the data rather than the relevant relationship, at least when the graphs are not well designed. Another way to disentangle the interaction between prior knowledge and data is to present graphical displays that are incongruent with viewer's prior knowledge. With inconsistent or incongruous data, prior knowledge will lead to biased decision making, especially when the format of the graphical display is non-optimal. Finally, the CI model must be used to make testable predictions that can differentiate this model from other possible models. Thus, recognizing this interaction between prior knowledge and data provides a fruitful avenue for future research.

References

1. Alloy, L. B., Tabachnik, N. (1984). Assessment of covariation by humans and animals: The joint influence of prior expectations and current situational information. Psychological Review, 91, 112-149.
2. Anderson, C. A. (1983). Abstract and concrete data in the theory perseverance of social beliefs: When weak data lead to unshakable beliefs. Journal of Experimental Social Psychology, 19, 93-108.
3. Anderson, C. A., Lepper, M. R., Ross, L. (1980). Perseverance of social theories: The role of explanation in the persistence of discredited information. Journal of Personality and Social Psychology, 39, 1037-1049.
4. Broniarczyk, S. M., Alba, J. W. (1994). Theory versus data in prediction and correlation tasks. Organization Behavior and Human Decision Processes, 57, 117-139.
5. Carpenter, P. A., Shah, P. (1998). A model of the perceptual and conceptual processes in graph comprehension. Journal of Experimental Psychology: Applied, 4, 75-100.
6. Carswell, C. M., Wickens, C. D. (1987). Information integration and the object display: An interaction of task demands and display superiority. Ergonomics, 30, 511-527.

7. Carswell, C. M., Emery, C., Lonon, A. M. (1993). Stimulus complexity and information integration in the spontaneous interpretation of line graphs. Applied Cognitive Psychology, 7, 341-357.
8. Casner, S. M. (1990). Task-analytic design of graphic presentations. Unpublished doctoral dissertation, University of Pittsburgh, Pittsburgh, PA.
9. Casner, S. M., Larkin, J. H. (1989). Cognitive efficiency considerations for good graphic design. Proceedings of the Cognitive Science Society. Hillsdale, NJ: Erlbaum.
10. Chapman, L. J., Chapman, J. P. (1969). Illusory correlation as an obstacle to the use of valid psychodiagnostic signs. Journal of Abnormal Psychology, 74, 271-280.
11. Chinn, C. A., Brewer, W. F. (1992). Psychological responses to anomalous data. Proceedings of the 14th Annual Conference of the Cognitive Science Society, 165-170.
12. Culbertson, H. M., Powers, R. D. (1959). A study of graph comprehension difficulties, Audio Visual Communication Review, 7, 97-100.
13. Freedman, E. G., Shah, P. S. (November, 2001). Individual differences in domain knowledge, graph reading skills, and explanatory skills during graph comprehension. Paper presented at the 42nd Annual Meeting of the Psychonomic Society, Orlando, FL.
14. Freedman, E. G., Smith, L. D. (1996). The role of theory and data in covariation assessment: Implications for the theory-ladenness of observation. Journal of Mind and Behavior, 17, 321-343.
15. Gattis, M., Holyoak, K. J. (1996). Mapping conceptual to spatial relations in visual reasoning. Journal of Experimental Psychology: Learning, Memory, & Cognition, 22, 231-239.
16. Hirokawa, R.Y., Gouran, D.S., Martz, A.E. (1988). Understanding the sources of faulty group decision making: A lesson from the Challenger disaster. Small Group Behavior,19, 411-433.
17. Jennings, D. L., Amabile, T., Ross, L. (1982). Informal covariation assessment: Data-based versus theory-based judgments. In D. Kahneman, P. Slovic, and A. Tversky (Eds.), Judgment under uncertainty: Heuristics and biases. Cambridge: Cambridge University Press.
18. Kintsch, W. (1988). The role of knowledge in discourse comprehension. A construction-integration model. Psychological Review, 95, 163-182.
19. Kosslyn, S. (1989). Understanding charts and graphs. Applied Cognitive Psychology, 3, 185-225.
20. Larkin, J. H., Simon, H. A. (1987). Why a diagram is (sometimes) worth ten thousand words. Cognitive Science, 11, 65-99.
21. Lewandowsky, S., Behrens, J. T. (1999). Statistical graphs and maps. In F. T. Durson, R. S. Nickerson, R. W. Schvaneveldt, S. T. Dumais, D. S. Lindsay, and M. T. H. Chi (Eds.) Handbook of Applied Cognition (pp. 513-549). Chichester, England: John Wiley and Sons, Ltd.
22. Lohse, G. L. (1993). A cognitive model of understanding graphical perception. Human-Computer Interaction, 8, 353-388.
23. Maichle, U. (1994). Cognitive processes in understanding line graphs. In W. Schnotz and R. W. Kulhavy (Eds.), Comprehension of Graphs (pp 207-226). Amsterdam, Netherlands: Elsevier Science.
24. Pinker, S. (1990). A theory of graph comprehension. In R. Freedle, (Ed.), Artificial Intelligence and the Future of Testing, (pp. 73-126). Hillsdale, NJ: Lawrence Erlbaum Associates.
25. Oestermeier, U., Hesse, F. W. (2000). Verbal and visual causal arguments. Cognition, 75, 65-104.
26. Sá, W. C., West, R. F., Stanovich, K. E. (1999). The domain specificity and generality of belief bias: Searching for a generalizable critical skill. Journal of Educational Psychology, 91, 497-510.
27. Schiano, J. D., & Tversky, B. (1992). Structure and strategy in encoding simplified graphs. Memory and Cognition, 20, 12-20.

28. Shah, P. (2000). Graph comprehension: The role of format, content, and individual differences. In M. Anderson, M., B. Meyer, & P. Olivier. (Ed) Diagrammatic Representation and Reasoning. Springer Verlag.

29. Shah, P., Carpenter, P. A. (1995). Conceptual limitations in comprehending line graphs. Journal of Experimental Psychology: General, 124, 43-61.

30. Shah, P., Hoeffner, J. (in press). Review of Graph Comprehension Research: Implications for Instruction. Educational Psychology Review.

31. Shah, P., & Shellhammer, D. (1999). The Role of Domain Knowledge and Graph Reading Skills in Graph Comprehension. Presented at the 1999 Meeting of the Society for Applied Research in Memory and Cognition, Boulder, CO.

32. Shah, P., Freedman, E. G., Vekiri, I. (forthcoming). Graph Comprehension. In A. Miyake and P. Shah (Eds.). Handbook of Visuospatial Cognition. New York, NY: Cambridge University Press.

33. Shah, P., Mayer, R. E., & Hegarty, M. (1999). Graphs as aids to knowledge construction: Signaling techniques for guiding the process of graph comprehension. Journal of Educational Psychology, 91, 690-702.

34. Shah, P., Hoeffner, J., Gergle, D., Shellhammer, D., & Anderson, N. (2000). A construction-integration approach to graph comprehension. Poster presented at the 2000 annual meeting of the psychonomics society, New Orleans, LA

35. Tabachneck-Schijf, H. J. M., Leonardo, A. M., Simon, H. A. (1997). CaMeRa: A Computational Model of Multiple Representations. Cognitive Science, 21, 305-350.

36. Trafton, J. G., Trickett, S. B. (2001). A new model of graph and visualization usage. Unpublished manuscript.

37. Trolier, T. K., & Hamilton, D. L. (1986). Variables influencing judgments of correlational relations. Journal of Personality and Social Psychology, 50, 879-888.

38. Tufte, E. R. (1983). The visual display of quantitative information, Cheshire, CT: Graphics Press.

39. Tversky, B., Schiano, D. J. (1989). Perceptual and conceptual factors in distortions in memory for graphs and maps. Journal of Experimental Psychology: General, 118, 387-398.

40. Wright, J. C., Murphy, G. L. (1984). The utility of theories in intuitive statistics: The robustness of theory-based judgments. Journal of Experimental Psychology: General, 113, 301-322.

41. Zacks, J., Tversky, B. (1999). Bars and lines: A study of graphic communication. Memory & Cognition, 27, 1073-1079.

Learning on Paper: Diagrams and Discovery in Game Playing

Susan L. Epstein[1] and J.-Holger Keibel[2]

[1]Department of Computer Science, Hunter College and The Graduate School of
The City University of New York, New York, NY 10021, USA
`susan.epstein@hunter.cuny.edu`
[2]Germanics Department and Graduate Programme in Human and Machine Intelligence
University of Freiburg, 79098 Freiburg, Germany
`keibel@uni-freiburg.de`

Abstract. Diagrams play an important role in human problem solving. In response to a challenging assignment, three students produced diagrams and subsequent verbal protocols that offer insight into human cognition. The diversity and richness of their response, and their ability to address the task via diagrams, provide an incisive look at the role diagrams play in the development of expertise. This paper recounts how their diagrams led and misled them, and how the diagrams both explained and drove explanation. It also considers how this process might be adapted for a computer program.

1 The Problem

This paper recounts how three people generated and explored diagrams as they learned about a simple game. Despite identical instructions, their verbal and graphic protocols reveal a surprising diversity of process and of final perspectives, albeit a similar level of skill at the game itself. Our thesis is that visual cognition interleaves with high-level reasoning in complex ways. The primary results of this paper are a preference for one-dimensional representation in thinking about a two-dimensional space, the power of the diagram to displace the original reasoning context, and a demonstration of the synergy between visual and high-level processes during abstract reasoning.

Pong hau k'i is an old Korean game, played between black and white on the board shown in Figure 1 [1]. The *mover* (the contestant whose turn it is to move) slides her piece along a line to the single empty location on the board. By game-playing stan-

Pong hau k'i

- Two-player game
- 5 locations
- First to move plays black
- Second to move plays white
- Move = slide to the next empty location
- Initially, the board is as shown
- Mover who cannot slide loses

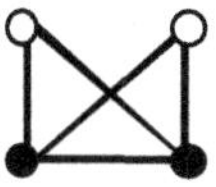

Fig. 1. As presented to the students, the game of pong hau k'i.

M. Hegarty, B. Meyer, and N. Hari Narayanan (Eds.): Diagrams 2002, LNAI 2317, pp. 31–45, 2002.

dards, this is a very simple game: there are only 60 possible *states* (locations of the markers plus the mover). A sequence of states is a *contest*. The state in Figure 1 is the *start state*, from which every contest begins. Contestants take turns until a mover loses because she is *trapped* (cannot move either of her pieces) or until a draw is declared because a state has repeated more than 3 times. Indeed, pong hau k'i is a *draw game*, that is, if both sides play correctly, every contest ends in a draw.

This work addresses cognition about movement in an abstract, rather than a realistic, two-dimensional space. Although there are studies of how individuals respond to a variety of representations for the same two-person game, to the best of our knowledge there are no studies of an individual's diagrammatic analysis of a game. This paper is in the tradition pioneered in [2], particularly the protocol analysis on problem reformulation from Subject 2. The first section describes the experiment that exposed a group of students to pong hau k'i. Subsequent sections describe the students' solutions and, more importantly, the processes they went through to reach them. The final sections discuss the results and the potential for automation of these processes.

2 The Experiment

Pong hau k'i was an informal assignment for students at The Cognitive Science Summer School at the New Bulgarian University in July, 2001. During a lecture they were given the rules with the slide in Figure 1, and told that it was a really simple, even boring, game. None of them had seen it before. They were asked to "play it, take protocols on the development of your own expertise, and capture any representational shifts," defined as "a different way of seeing things." (No example was provided.) They were encouraged to obsess over it and to report back on their experience.

That lecture included an introduction to artificial intelligence, with some discussion of the search space paradigm. One slide depicted part of the search space for another game, lose tic-tac-toe, as the *game tree* in Figure 2. In such a diagram, each *node* (picture of the board) represents a state and each *edge* (line joining two nodes) represents a move transition from one state to the next. The topmost node is called the *root*. A node that ends play is *terminal*. Later in the lecture, the students were asked to "draw out the state space for pong hau k'i, label [it] with outcomes, produce a rule set to play it, and produce a good heuristic for it." These directions appeared on slides.

This task began as a teaching device, not as a planned experiment. As a result, it lacks the timed photographs, videos, and recordings that lend rigor to some other work in this field, (e.g. [3-6]). Once several students began to respond in such great depth, however, documentation was as extensive as possible. The students of the Cognitive Science Summer School are a diverse lot, drawn from many countries and many disciplines. The Summer School is conducted in English, but the students' written materials were idiosyncratic, and included notation in both German and Polish. For uniformity, in what follows the first author has made minor spelling corrections, inserted clearer or more grammatical language [in brackets like these], and substituted B for "black," W for "white," and "-" for the empty position. Any italics are the subjects'

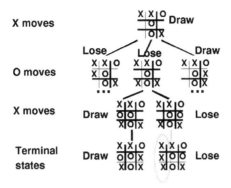

Fig. 2. An example of a game tree as presented to the subjects.

own, and all names have been omitted. Prior to this paper, none of the three subjects described here had seen the protocols of the others.

3 Subject 1 — A (Relatively) Quick Response

Subject 1 is a Ph.D. candidate in cognitive psychology. Originally, he offered no verbal protocol with his analysis, only an elaborate, tree-like diagram on a single sheet of paper, with some small annotations described below. Subject 1 drew his diagram in a few hours, stopping "when I thought it was complete…." He did not use a board, rather, "At the beginning I tried to imagine the situation on it. Then I just manipulated the letters and numbers. I drew a sketch when I was afraid I was lost to check it."

Subject 1's diagram is difficult to read (and therefore not included here). It is, however, possible to deduce a good deal of his process from it. Figure 3 reproduces the top portion. In the upper left corner is the start state from Figure 1. Below it are two sketches symbolizing how white could win. (Line segments were indeed omitted.) To the right of the start state is an encoding of it (WW–BB), from which one may deduce the numbering: 1 = upper left, 2 = upper right, 3 = center, 4 = lower left, 5 = lower right. "12345" lists the labels assigned to the vertices. Immediately below it is a list of the 7 (undirected) edges along which a move may occur.

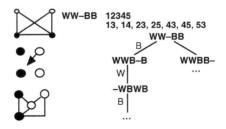

Fig. 3. The top portion of Subject 1's protocol. Ellipses are the authors'.

On the entire page there are only two other sketches of the board, both to the right. The first, Figure 4(a), appears near the two losing positions highest in the diagram, about 8 levels down in the tree. It represents the state just before a possible win for either mover. The second, Figure 4(b), appears next to a subtree placed far to the right but connected to the diagram (the *subtree-to-the-right*). It represents a state from which white, if the mover, can win; if black is the mover the contest will merely continue. These are the only sketches Subject 1 drew to check his reasoning.

Most of the page is occupied by a game tree, where each node is a 5-letter string representing the contents of the 5 locations on the board. Nodes at the same depth in the tree are horizontally aligned on the unlined paper. In most of the tree, a W or B to the left on each level indicates the mover in the previous state; in the subtree-to-the-right these labels were to the right. The two alternative first moves for black are the neighbors of the start state. Subject 1 chose to follow only the move from 4 to 3, shown below and to the left of WW-BB in Figure 3. At most levels there are only two or three nodes, but occasionally there are as many as six. The diagram reaches all margins of the paper, and includes 80 nodes in all. One node, on the eleventh level, points with a long arrow to its expansion, the subtree-to-the-right. Otherwise, exploration appears to have had a strong downward tendency, and to have moved from left to right at every level.

Most tree edges are undirected lines, but at some point Subject 1 noticed that some nodes repeated. When he found a repeated node (encoding and mover), he drew an unlabeled edge that was *directed* (had an arrow on at least one endpoint), as in Figure 4(c). A *backward edge* begins at one node and runs to another *higher* (closer to the root) in the tree; there are 11 backward edges. One backward edge in particular, from the leftmost portion of the subtree-to-the-right, returned to the root and presumably eliminated the need to expand WWBB–. A *forward edge* runs between two nodes and has arrows on both ends. The diagram includes two forward edges; the only difference apparent between backward edges and forward ones is their length — the repeating pairs joined by a forward edge are those farthest apart on the page. There is also one directed edge between two nodes on the same level, and one dashed forward edge between identical nodes associated with different movers. In the latter case, Subject 1 presumably observed the similarity, noted it with an edge, and judged it not worth pursuing.

Every terminal node is either the source of a backward edge (indicating repetition in play) or is surrounded by a rectangle. There are eight such nodes, each with a W or

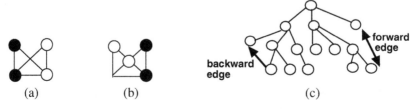

Fig. 4. The other two board sketches in Subject 1's protocol: (a) the state penultimate to a loss; either contestant can win from here, (b) a state from which white, but not black, can win, and (c) orientation of forward and backward edges in Subject 1's tree.

B label indicating the winner in that state. Only four of these are distinct: two wins for white and two for black. There is one significant error: the forward edge from the un-explored first move WWBB– is not to a later copy of itself. There is, however, another copy of WWBB– (one unremarked upon by Subject 1) that is fully expanded.

4   Subject 2 — An Obsession

Subject 2 is a Ph.D. student in linguistics and cognitive science with a strong mathematical background. His exploration of the game was remarkable. The day after the assignment, Subject 2 proudly displayed a diagram he had drawn of the search space. He seemed exhausted, and questioning revealed that he had devoted a good deal of time and energy to the task. Although he thought little of them, he had not yet discarded his first five diagrams, which chronicled his path to the sixth and final one. He remembered a great deal of the process he had been through in the past 24 hours, and we numbered and annotated his early diagrams as we discussed them. From these, Subject 2 began to write what eventually became a 13,000 word protocol, covering about 6 hours of intensive exploration and many more of retrospective reconstruction. The material in this section is based on observation of the diagrams, the debriefing session the day after the assignment, and the original and later versions of the written protocols. Subject 2 identifies 9 stages in his path to expertise at pong hau k'i.

Stage 1. (15 minutes) Subject 2 attempted to "get a feeling for the game and its state space." At lunch after the lecture, several students made pieces from bread (the crusts against the bread centers) and began to play. Initially, Subject 2 played "very much at random without even knowing how a [losing] state might look…." Indeed, he hypothesized one state as a loss that was not a loss at all. Nonetheless, "After a few more (random) moves my opponent pointed out that his [pieces] were stuck and that he had lost. I looked and realized that indeed, by accident, I had won this first game. Both his [pieces] were on the [right side] of the board and I was blocking him from moving…. From seeing this I figured that *one could lose if (and only if) one ended up having both figures on the same (left/right) side of the game board*." In a second contest, Subject 2 lost. Then, as the others watched and commented, he began to experiment alone, moving "… according to the rules, without consciously following any strategy. … I

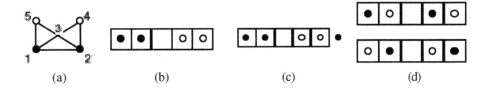

Fig. 5. (a) Subject 2's numbering of the board. (b) The starting board and (c) the start state represented in Subject 2's rectangular coding. (d) The "seeds" for Diagram 4.

got the feeling that the dynamics of the game contained several symmetries…. by a few straightforward moves I could invert the initial configuration (interchange the black and white pieces). And I also got the feeling that I would always end up in these same two [states] (the initial one and its inverse)…. I had the sensation of running into 'attractor states.' Then I tried to … [force either player to lose]. It only occurred occasionally and rather by accident than by controlled force. I could not really find out which moves one had to do/avoid in order to win."

Stage 2. (20 minutes) Subject 2 began to draw Diagram 1. He computed 30 possible *configurations*, ways to place the pieces on the board; with the label for the mover, he arrived at 60 possible states. "However, I was not yet sure that all of these states could actually be reached by legal moves from the initial configuration" For compactness, Subject 2 now shifted to a rectangular representation for a node. (He was drawing on graph paper, where a sequence of squares would be easy to draw clearly.) Because black moves first, Subject 2 labeled the 2 lower locations 1 and 2 (left to right, respectively). He labeled the center 3 because he felt it retained "some of the symmetry of the original game board." Originally the top locations were (left to right) 4 and 5, but he interchanged them (5 and then 4) so that 2 to 4 and 1 to 5 were moves. This produced the numbering in Figure 5(a). Using this numbering he wrote sample nodes, such as Figure 5(b), to confirm his calculation of the possible states.

Still at lunch with his friends, Subject 2 now began to draw Diagram 2, his first full state space diagram. All nodes were in the format of Figure 5(c), where the additional circle to the right denotes the mover. Although he initially checked his diagram against the pong hau k'i board, Subject 2 eventually relied solely on his new representation. He set up the two opening moves at the top, beneath the start state, just as Subject 1 had. "Then I followed only one of the two now opened paths …depicting white's and black's possible moves in alternating order. At a certain point I would interrupt myself and work a bit on the right path hoping for some cycle to close. The whole diagram became very confusing: with almost every new state that I drew I would get the feeling of having already encountered this particular one." As he searched his diagram looking for cycles, Subject 2 became "rather annoyed by the fact that I had to distinguish the mover with the state." He decided not to extend "paths starting from those states that occurred for the second time, but with the opposite mover. I believed I could just refer to some sort of symmetry with the paths starting at the corresponding state that I had visited first." At some point he realized that he had forgotten about the edge from 1 to 2: "… indeed, I found two new 'twigs' to be grown in my graph." Returning to it hours later, however, "Everything just became more and more messy and confusing…." He abandoned the task.

Stage 3. (45 minutes) Somewhat later, Subject 2 began alone, with new determination, to draw Diagram 3 in color. The playing pieces became green and red, edges were colored green or red to denote the mover transforming the state, and the external circle for the mover was eliminated. Edges now had arrows on both ends, indicating that moves were invertible. Subject 2 referred back to Diagram 2 regularly, checking for oversights, and frequently found them. Eventually, he extended the tree methodically, keeping both sides at about the same depth, and soon reached four winning nodes, two for red and two for green. These were highlighted in orange to denote termination of search. Then he expanded the non-terminal states. "I started to count the states. There were 30 of them. I took this correspondence with my earlier calculation as a strong confirmation for finally having covered the entire state space. … But my representation was just unbearably entangled. (And in fact, I was surprised by the apparent complexity of the state space.) I wanted to shape it up towards a comprehensible new diagram." Subject 2 considered the diagram crowded, and it did not support the symmetry he expected; in particular, he "had an expectation that the opposite of the start state would be on the very bottom…naturally."

Stage 4. (50 minutes) Subject 2 drew Diagram 4, beginning not from the initial state but from "another state of maximal connectivity." He chose the states in Figure 5(d), and positioned them first, horizontally, in the center of his diagram, with the upper one on the left, and the lower one on the right. Subject 2 then traced backward through the previous diagram to position the start state appropriately. He checked off states from Diagram 3 as he reached and reproduced them in the left half of Diagram 4, moving outward from the upper state in Figure 5(d). At the same time, he maintained elaborate tests for symmetry and edge color, for states he regarded as "opposites," and for edges that were incorrect but somehow "missing. Although he "was expecting some asymmetries," intersecting edges drove him to redraw it. After the 50 minutes of this stage, he writes that he now considered himself "obsessed."

Stage 5. (50 minutes) Diagram 5 positioned the start state at the top, surrounded by empty rectangular nodes, that is, states without any playing pieces. Subject 2 was concentrating on the shape of the space now, not the moves in the game. At some point he filled in the nodes, and then focused on introducing the losing states so that the ar-

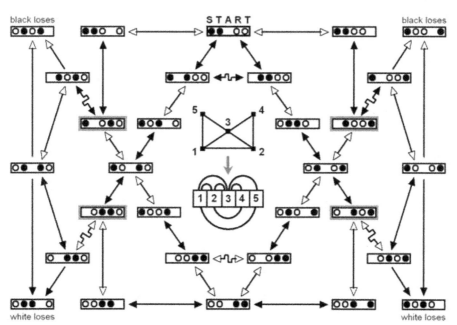

Fig. 6. Subject 2's Diagram 6, the product of 6 hours of intensive work. The four delicate states have gray borders.

rangement was spatially pleasing. To his surprise, Subject 2 found that his desire for symmetry now led to the discovery of edges that he had completely overlooked in earlier diagrams. Moreover, the angle of the edges now became significant to him. Subject 2 initially drew the right side of the diagram with empty nodes and colored edges, and re-derived the nodes before confirming them with Diagram 4.

Stage 6. (duration unknown) Concerned about an asymmetry on the left in Diagram 5, Subject 2 explained the picture to his roommate, pointing out four *delicate states* located between "the secure central area" and the areas on either side where one can lose. "The discovery of the *delicate states* and all my following considerations about strategies were entirely based on the arrow pattern and the state frames – which in my mind appeared almost deprived of their contents." Subject 2 now correctly recognized that pong hau k'i depends entirely on play in the delicate states: "…to prevent losing, I … simply have to avoid the two bad moves… I only need to memorize two specific configurations… whenever my opponent allows me to win, I … simply have to make the one good move…. For this I only need to memorize two further specific configurations…." He was startled, however, by how irrelevant his full diagram was to actual play: "I am surprised that … one would not have to know anything about the global pattern of the state space. … My two strategies … are local rules – local in terms of time (which translates to space on my diagram). … I also noticed that the … strategy …for making the winning move …also serves for …preventing my defeat…."

Stage 7. (20 minutes) The next morning, driven by "global symmetry," Subject 2 drew Diagram 6, shown in Figure 6. Although he also wanted to renumber the nodes, Subject 2 was pressed for time, and "wanted to focus on the intended rearrangements on

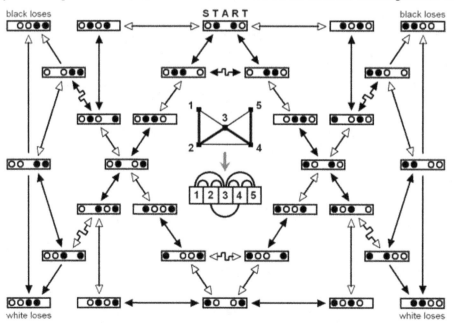

Fig. 7. Subject 2's "W-shaped encoding," Diagram 7, with the locations renumbered as in the center of the diagram.

the left-hand side of the preceding draft." Therefore he "… drew all 30 empty state frames starting on the right half where I did not have to change much. Then I added all the arrows until finally I copied the states themselves." He carefully aligned states vertically and horizontally, highlighting terminal states and delicate ones. The result was satisfying — he felt like an expert player and remarked on the symmetry of his diagram. "The stunning thing that was new here is the 'reflection through the center' symmetry and the fact that it operates not only on the state frames and arrow pattern but as well on the state configurations!! I liked this."

Stages 8 and 9. During debriefing that afternoon, Subject 2 recalled his renumbering scheme from Stage 7, which ultimately removed some misunderstandings observable (but not remarked upon) during the conversation. Later that afternoon, in Stage 8, Subject 2 produced Diagram 7 with that renumbering (Figure 7), so that the patterns in nodes with the same function are symmetric to each other. Over the next few days, in Stage 9, Subject 2 revised and extended his protocol, and produced Diagram 8 (Figure 8). It classifies states and moves according to their safety, while maintaining vertical, horizontal, and through the center symmetry with respect to states and edges.

Subject 2 now felt he was an expert, and had been since Stage 5, although he had only competed once since then, as described in the next section. "…in a way I had reformulated the game in my own terms and then detached it to some extent from its original shape. My (final) game consisted of boxes … and arrows and I had mastered these …." Why then had he continued to draw? His additional work, he believed, increased his subtle "…knowledge of the structure underlying the game. Knowledge that might not directly influence performance." Even as he dissected his experience, Subject 2 remained puzzled by his own lack of rigor. "… I probably could have proceeded

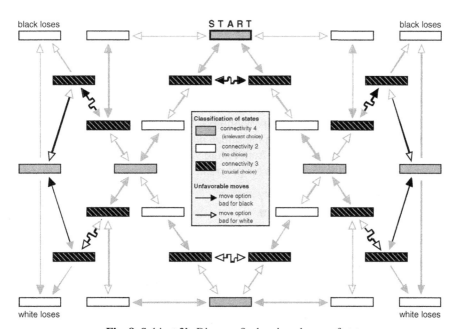

Fig. 8. Subject 2's Diagram 8, showing classes of states.

in a much more intelligent/elegant way and in less time. I am not ashamed of this but rather surprised. For before [writing this protocol] … I [thought I had] approached the task in a more controlled and directed way than this documentation has brought to light." Nonetheless, it was a remarkable intellectual journey.

5   Subject 3 — An Automated Approach

Subject 3 is a graduate student in psychology who ostensibly chose not to draw any diagrams at all. He was among the students at lunch that day with Subject 2, but professed disinterest. Eventually the fun drew him in, and he began to play against Subject 2. "After a while the game became dull, because neither [of us]… understood the point of it…. We started to discuss, that there must be some good strategy, and we must find it…. [Subject 2 began] to draw a tree-like decision graph [Diagram 2], where every branch represented a possible move. …The only thing I remember of his drawing was that he was thinking of a good labelling of the positions. We agreed that numbers are good indicators of the locations, but we didn't talk about [how to assign numbers to locations]. He wanted to count how many [distinct states there were]. But it was really time to go back to [class]."

Subject 3 is not a trained programmer. Nonetheless, alone later that evening he wondered "why people should bother with counting [moves] so much … if computers are much more suitable…. Therefore I decided to build a program that can play this game. My original plan was to play with the computer" and have it count the states for him. He began to write the algorithms but then realized that the program might also perform as an opponent. "At this point, I did not think of any strategy, only the rules [of the game]… I remembered the lunchtime games, and [Subject 2's] idea that a good labelling is needed. Without much thinking I came up with [Figure 9]. The reason I chose this labelling is because if you make a 'line' from the original shape by 'pushing' from the top and from the bottom the numbers will increase from 1 to 5."

Two hours later Subject 3 had a program. The program reproduced the pong hau k'i game board, with red and blue (rather than black and white) circles for the playing pieces. The code, however, did not employ Subject 3's numbering in Figure 9. The order in which the circles were drawn on the screen on each turn, and the order in which they were stored, using his numbering from Figure 9, was 2, 1, 3, 5, 4.

Subject 3 now began to play against his program, which made random legal moves.

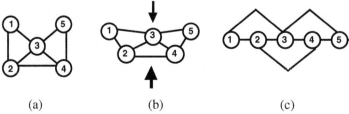

| (a) | (b) | (c) |

Fig. 9. Subject 3's compression of the board to create a numeric encoding, as given in his protocol. (a) The board itself, (b) the initial pressure , and (c) the collapse into a line.

"This was the first time, I wanted to understand the strategy." Subject 3 soon won against his program, played "10 –30" more contests, and remained convinced that he was correct. Still later the same evening, Subject 2 appeared in Subject 3's room. "He [Subject 2] … seemed very tired. He became delighted that" another person was working on pong hau k'i so intensely. "So we quickly started to play," but Subject 2 warned that he was undefeatable. "After 5 –8 steps we gave up, because I couldn't lose [either]. Tie game. By that time, we realised that both of us [had become experts at pong hau k'i]. He did his way theoretically, I [did] mine practically."

6 Discussion

6.1 The Right Answers

Pong hau k'i indeed includes only 30 possible piece configurations, and 60 possible states. Subject 1 found all 30 piece configurations (several more than once), but overlooked several edges and used a backward edge to pair two different states, asserting a repetition that was not there. Subject 2's first three diagrams included all the configurations too, but were missing edges; the others are complete. Subject 3 wrote code that was capable of drawing all possible board positions, although he only experienced the ones to which he led it to during competition.

Pong hau k'i is a draw game; one cannot win unless one's opponent errs; from most states, any move is safe. Only in the four states Subject 2 called "delicate" is it possible to make a fatal error, and only one of the two moves in each of those states will cause a loss. Even after the wrong move from a delicate state, one's opponent has two possible moves, only one of which wins. When questioned weeks later, Subject 1 wrote that after "a few games with one person" he realized that "it is quite difficult to lose the game, and I had an idea about situations I should avoid." Subject 3 made a similar discovery: "After my first [non-draw] I realised that the only way [to win] is to put the opponent into the side positions. [This describes both states in Figure 4.] … I tested this idea, and tried to avoid [having my pieces there] …." Thus, one would expect all three subjects to play expert pong hau k'i; drawing against each other, and winning if an opportunity arises.

The task was to construct a labeled diagram of the state space, a rule set, and a good heuristic. Subject 1 only constructed the space, but felt that he had completed the task. Subject 3 constructed a program that "knew" the space, without ever drawing it out. He learned rules and heuristics from experience that moved him through the space as he competed against his program, a kind of virtual representation. Subject 2 developed game graphs that proved the heuristics and the rule set that the other subjects arrived at. His final product is a powerful, rigorous, visual argument.

6.2 The Diagrams

Although all the subjects received the same instructions, each of them approached the task differently. Subject 1 was strongly influenced by Figure 2. He explored the game by expanding the tree depth first, generating moves with his list of edges. He noted repetition in the diagram but did not attempt to winnow it out — rather, he used it to curtail further search until all nodes had been explored. His directed edges were an innovation not shown during the lecture. Subject 2 began with a similar approach, but soon worked breadth first, that is, level by level. In addition, he became interested in the *shape* of the space even more than its contents. He used repetition, both of states and of piece configuration, as a way to tease out all possibilities. Finally, Subject 3, although he claimed not to have drawn at all, effectively had his program draw diagrams for him. While he competed against it, he certainly saw on the screen far many more pictures of the board itself than either of the other two students.

To describe the problem, however, all three found the physical board itself confining and soon abandoned it. Each of them began by numbering the board: Subject 1 as if he were reading text, left to right and top to bottom; Subject 2 left to right but from the bottom up, and Subject 3 with a thoughtful, visually-described process which he quickly abandoned. Subjects 2 and 3 considered more than one numbering. In every numbering, however, the central position was always labeled 3, presumably because it is also the median of the numbers from 1 to 5. Numbers quickly led to a one-dimensional, horizontal format: Subject 1 used a string of B's and W's, Subject 2 used circles in a row, and Subject 3 encoded a state as an array of numbers that denote color. The colors of the playing pieces were also clearly irrelevant. Subject 1 used letters instead, Subject 2 shifted to red and green, and Subject 3 shifted to red and blue.

Although the horizontal axis is a neutral representation [7], it fails to convey key information: the legal moves as defined by the edges. The edges on the game board serve as a move generator. Once they abandoned the two-dimensional board, each subject needed a way to produce all possible moves methodically from a given state. Subject 1 extracted moves from an edge list that really contains indices into a string of B's and W's. Subject 2 memorized them as the arcs in the centers of Figures 7 and 8. Subject 3 wrote a simple routine that served the way Subject 1's edge list did.

A diagram about play must include the moves, the agent making them, and the ultimate outcome. Subject 1 labeled the edges to designate the mover, rather than labeling the states, as in Figure 2. Subject 2 colored his edges to show how pieces might move. Subject 3's program, which could have portrayed movement, simply makes the next state "appear" on the screen, without any transitional sliding. Both Subject 1 and Subject 2 highlighted winning nodes in their diagram, imputing importance to the absence of a cycle. Subject 2 went further, aligning these highlighted nodes. Subject 3 had no tangible record of winning states, only his memory of them.

The subjects' diagrams served a variety of purposes. Subject 1 used his tree diagram to check whether or not he had considered all possible states, and his occasional sketches to check his reasoning. Subject 3 used his computer-drawn diagrams to provide him with playing experience; those states were connected temporally but visible only one at a time. Subject 2 used his first diagrams the way Subject 1 used his tree, and subsequently used one diagram to check another, and portions of one diagram to generate other portions. He suspects that he began to think about strategy around the same time; his "structural" concerns in Diagram 3 were replaced by his "semantic" considerations in Diagram 4.

Some states were particularly salient. Subject 1 sketched (i.e., translated from a string back to a diagram) only two: Figure 4(b) where the mover wins, and Figure 4(a) where white must win or lose. Subject 3 appears to have noticed these in passing. Subject 2, however, eventually focused on four states one play removed from Figure 4(b), and did not recognize the significance of those positions until he explained his diagram verbally to a friend — producing the diagram was not as powerful as interpreting it aloud.

Each numbering of the locations on the board drives exploration of the search space differently, but there are 120 possible numberings. It is likely more than coincidental that Subject 2's second numbering, for Diagram 7, matched Subject 3's. Indeed, Subject 3's original idea was to explore several such numberings as if they were preference strategies, playing one against the other. Remarkably, the other numbering he planned to explore was the numbering used by Subject 1.

Finally, none of these subjects used their diagrams to compete. Each of them believed he had internalized the most important facets of his diagram(s), and could play perfectly without reference to them. The diagrams were thus an exploratory device, and, once completed, not essential for expert performance.

6.3 Reproducing This Behavior in a Program

For a game as small as this, it is not difficult to write a program that searches exhaustively, noting potential cycles. In that way, a program could generate its own lookup table to provide a correct move from every state. From such a table, moreover, it is easy enough to extract correct (i.e., safe) moves and even delicate states — otherwise, any random move will do.

Nonetheless, pong hau k'i presents a variety of challenges for cognitive modeling. For example, a game-learning program called *Hoyle*, known for its prowess on far more difficult games, developed basic skill at pong hau k'i quickly. After observing 5 error-free contests played by others, Hoyle never lost a contest at pong hau k'i. An expert player should also win if the opposition errs. Given the opportunity, however, Hoyle never won a contest. Inspection of its knowledge base indicated that the program knew how a win would arise at the game, it just was not winning. Hoyle's standard approaches to diagrammatic reasoning were of no help here [8]. Although the program detected significant visual patterns formed by playing pieces and empty locations on the board [9], patterns that were associated with a win also occurred in contests that went on to a draw. Hoyle therefore saw no advantage to producing them.

The secret to computer expertise at pong hau k'i, for Hoyle at least, turned out to be the notion of delicate states and the appropriate way to exploit them. What Subjects 2 and 3 learned to do was to lure their opponents toward delicate states. An agent that expects flawless opposition will not do this, since pong hau k'i is, after all, a draw game. Given this understanding and a fallible opponent, Hoyle wins often and quickly. The key situations for Hoyle were described not by configurations of pieces but by paths in the search space, the same paths Subject 2 eventually saw and highlighted in his final diagrams.

7 Conclusions

AI researchers devised the game tree diagram to facilitate their analysis of two-person games. A game tree captures turn-taking and the impact of sequences of decisions, and serves as a repository for states and their relationships. Indeed, this diagram has supported the development of many sophisticated search mechanisms that now underlie champion game-playing programs. Nonetheless, a game tree does not facilitate either the correct and complete generation of moves, or the detection of duplicates in cyclic games, both of which are crucial to expert play.

The subjects found 60 states far too many game boards to draw. Instead, each of them encoded the two-dimensional board in a single dimension. Ironically, the linear representations they devised to simplify their task made move generation considerably more difficult. (It is, we suspect, far easier to imagine sliding along visible lines than jumping in a set of apparently disconnected leaps, as the centers of Figures 6 and 7, and Subject 1's need to sketch, demonstrate.) Thus they were forced to find some methodical way to generate moves: indexing into a sequence, envisioning with arrows, or writing a routine. Furthermore, Figure 9 demonstrates that the mapping from two dimensions to one is itself a diagram.

While a game tree represents cycles by endless repetition, a *game graph* links moves backwards to earlier states. Subject 1 quickly arrived at this device, using predominantly one-headed arrows to indicate repetition. This also allowed him to retain the game tree's convention that represents the mover by a node's level. In contrast, Subject 2 used two-headed arrows, but then found that in his new game graph he could no longer "read" the board to identify the mover. This gave rise to incorrect associations among states with the same visual properties but different movers. In response, Subject 2 confirmed correctness and completeness visually, relying on the inherent symmetries of the game.

One might hypothesize some process for the representational shifts that abound in these protocols, a "think-choose-try-evaluate-revise" cycle. Certainly, each subject first selected a representation and then marched determinedly through move generation. (Subjects 2 and 3 had enough qualms about their initial choices to explore others later; they seemed most disturbed by the way the symmetries on the board failed to match the properties of these few integers.) Subject 2 became dissatisfied and abandoned several approaches while he was engaged in them, which suggests that a metalevel critic ran in parallel as he drew. He also used the properties of the current diagram (e.g., the number of edges to a node, intersecting edges, asymmetry) to structure the next one. By Diagram 4 he had abandoned any pretense at a tree; he now sought to convey not only sequences of states in play, but also the symmetries of their values. His revisions included the state representation, the mover designation, and the larger structure.

The protocols reveal a wealth of cognitive activity in which diagrams play a central role. The game tree diagram alone cannot, and did not, guarantee completeness, correctness, or well-managed repetition. Subject 1's diagram served to satisfy his curiosity, not to organize his thoughts about the problem. Subject 3's diagrams were a kind of random travel through the space, generated by his code. That travel was enough to support the development of expertise. Only with Subject 2 might we argue that diagrams led to deeper understanding of the game, an understanding unnecessary to expertise, but satisfying in its level of knowledge organization. His Diagrams 6 through 8 imposed order and a value system on a set of repetitive chronological sequences.

Acknowledgements. Thanks to Boicho Kokinov for his invitation to teach at The Cognitive Science Summer School, and for attracting such thoughtful students. Thanks too to the anonymous referees for their suggestions, and to our three anonymous, enthusiastic, and cooperative subjects, who transformed a homework assignment into an intriguing demonstration of the role of diagrams in human cognition.

References

1. Zaslavsky, C.: Tic Tac Toe and Other Three-in-a-Row Games, from Ancient Egypt to the Modern Computer. Crowell, New York (1982)
2. Anzai, Y., Simon, H., The Theory of Learning by Doing. Psychological Review. 36 (1979) 124-140
3. Cross, N., Christaans, H., Dorst, K. (eds.): Analyzing Design Activity. Series Vol. 30. John Wiley and Sons, Chichester (1996)
4. Suwa, M., Tversky, B.T., Gero, J.S., Purcell, T.: Seeing into sketches: Regrouping Parts Encourages New Interpretations. In: J.S. Gero, B.T. Tversky, and T. Purcell (eds.): Visual and Spatial Reasoning in Design II. Key Centre of Design Computing and Cognition, University of Sydney, Sydney, Australia (2001) 207-219
5. Pearson, D.G., Alexander, C., Webster, R.: Working Memory and Expertise Differences in Design. In: J.S. Gero, B.T. Tversky, and T. Purcell (eds.): Visual and Spatial Reasoning in Design II. Key Centre of Design Computing and Cognition, University of Sydney, Sydney, Australia (2001) 237-251
6. Saariluoma, P., Maarttola, I.: Spatial Mental Content and Visual Design. In: J.S. Gero, B.T. Tversky, and T. Purcell (eds.): Visual and Spatial Reasoning in Design II. Key Centre of Design Computing and Cognition, University of Sydney, Sydney, Australia (2001) 253-268
7. Tversky, B.T.: Multiple Mental Models. In: J.S. Gero, B.T. Tversky, and T. Purcell (eds.): Visual and Spatial Reasoning in Design II. Key Centre of Design Computing and Cognition, University of Sydney, Sydney, Australia (2001) 3-13
8. Epstein, S.L.: Learning to Play Expertly: A Tutorial on Hoyle. In: Fürnkranz, J. Kubat, M. (eds.) Machines That Learn to Play Games. Nova Science (2001) 153-178
9. Epstein, S.L., Gelfand, J., Lock, E.T., Learning Game-Specific Spatially-Oriented Heuristics. Constraints. 3 (1998) 239-253

Using Animation in Diagrammatic Theorem Proving

Daniel Winterstein[1], Alan Bundy[1], Corin Gurr[1], and Mateja Jamnik[2]

[1] Division of Informatics, University of Edinburgh, 80 South Bridge, Edinburgh, EH1 1HN, U.K.
{danielw,bundy}@dai.ed.ac.uk, corin@cogsci.ed.ac.uk
[2] School of Computer Science, University of Birmingham, Birmingham, B15 2TT, U.K.
M.Jamnik@cs.bham.ac.uk

Abstract. Diagrams have many uses in mathematics, one of the most ambitious of which is as a form of proof. The domain we consider is real analysis, where quantification issues are subtle but crucial. Computers offer new possibilities in diagrammatic reasoning, one of which is animation. Here we develop animated rules as a solution to problems of quantification. We show a simple application of this to constraint diagrams, and also how it can deal with the more complex questions of quantification and generalisation in diagrams that use more specific representations. This allows us to tackle difficult theorems that previously could only be proved algebraically.

1 Introduction

In adapting diagrammatic reasoning to computers, there is the exciting possibility of developing diagrams in new ways. Diagrams in textbooks are necessarily static. However, if we consider the very real differences between text and hypertext, we see that diagrammatic reasoning on computers need not be just a straight conversion of diagrammatic reasoning on paper. One such new direction is the use of animation. In this paper we describe how animation can be applied to represent and reason about quantification. Other applications are possible: although unrelated to our present work, animated diagrams may also be useful in representing and reasoning about temporal relations. There is an obvious attraction in using time to represent itself.[1]

From the very beginning of mathematics, diagrams have been used to give proofs in subjects such as geometry and number theory. Nevertheless diagrammatic proof is only partially understood today. In particular, we do not have a general theory for handling quantification and the related topic of generalisation.

---

[1] This would probably not be suitable for domains which involve precise time calculations, as these would be hard to judge in an animation. For qualitative reasoning though, or as part of a mixed system, it seems an interesting line for future research.

M. Hegarty, B. Meyer, and N. Hari Narayanan (Eds.): Diagrams 2002, LNAI 2317, pp. 46–60, 2002.
© Springer Verlag Berlin Heidelberg 2002

We identify three distinct problems in this area: quantifier hierarchy, quantifier type and identifying generalisation conditions. In sentential reasoning, quantifier hierarchy is determined by reading from left-to-right. Quantifier type is determined by the semantics attached to the symbols $\forall, \exists$ (which are part of the common language of scientists and mathematicians), and generalisation is controlled by explicit conditions (e.g. $\forall A, B, C.\texttt{triangle}(ABC) \wedge \texttt{angle}(A, B) = 90^o \Rightarrow ...$). Unfortunately these solutions do not naturally carry over to diagrams.

The use of two dimensions with several spatial relations being significant removes the neat left-to-right ordering on objects that we have in sentential reasoning. For quantifier type, we could try to label objects with the $\forall, \exists$ symbols as in algebra, but it is not always clear which object such a label applies to, especially if the diagram involves composite objects. Fig. 1 gives an example of how this could be less than clear (does the $\forall$ symbol apply to the closest line, triangle or square?). Unlike algebra, where the conditions on a theorem are explicitly stated, diagrams often contain a lot of information that may or may not be relevant. Thus several generalisations are possible, some of which may be false. For example in Fig. 1, we might generalise to 'all triangles', 'all right angled triangles', or 'all similar triangles[2] to the one drawn'. Often there is a clear intuitive generalisation, but this must be formalised if we are to give rigorous proofs (c.f. [9] for a detailed discussion of this). So diagrams require a fresh approach to all three problems.

Fig. 1. Problems in Pythagoras' Theorem: It is not clear what the $\forall$ symbol applies to, and we have not specified which aspects of the triangle are important.

1.1   Related Work

Several approaches have tackled the problem of quantification in diagrams by introducing new notation. This can produce systems as powerful as predicate logic. However, it generally leads to more complex diagrams, and great care is required if these are to retain their intuitive feel. For example, in 1976 Schubert, starting from semantic nets, developed (by adding more and more notation) a diagrammatic representation that is as expressive as modal lambda calculus (see Fig. 2) [12]. Unfortunately, the resulting diagrams are extremely difficult to read and, to the best of our knowledge, were not generally used.

---

[2] Two triangles are similar if they differ only in scale or left-right orientation.

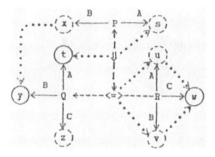

Fig. 2. A Schubert diagram for "Several children are on the playground. Most of them are playing in the sandbox."

A more recent example is [8], which extends Venn diagrams with extra notation and some inference rules to give a heterogenous system[3] as expressive as first-order predicate logic. It is shown to be sound and complete, but also seems too difficult to use. As Ambrose Bierce's inventor said: "I have demonstrated the correctness of my details, the defects are merely basic and fundamental" [2]. By extending diagram systems on purely logical criterion without considering ease-of-use issues, the final systems lost the very qualities that make diagrams attractive.

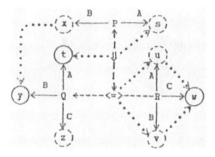

Fig. 3. We are unsure what this Schubert diagram means.

A more successful diagrammatic system with quantifiers is Gil *et al's con-straint diagrams*, which are an extension of *spider diagrams* [5]. The objects in spider diagrams are oval contours (representing sets with overlap=intersection), points (representing members of sets) and 'spiders' (linked points representing statements of the form $x \in A \cup B$). Constraint diagrams introduce arrows (representing relationships) and quantifiers. Only points and spiders are quantified over. Quantifier type is handled by drawing points in different ways. That is, they have different primitive objects for the different quantifiers: • = ∃point,

---

[3] A system whose representations mix diagrams and text [1].

$\star = \forall$points. Quantifier ordering problems are dealt with by labelling quantifiers with numbers.

The solutions in [5] for handling quantification have similarities to the ones we will present here. However the abstract indirect nature of constraint diagrams makes them 'closer' to algebra, and thus the problems are easier than for more specific diagrams (c.f. §2.2).

The generalisation problem arises only in a limited form for [5]. This is because, except for set membership, all conditions must be made explicit – as in algebra. Set membership conditions are handled by fixing the interpretation: a point x is interpreted as belonging to the smallest region containing it. 'Spiders' provide a means for over-riding this default interpretation. We shall look at how this idea of a default reading with extra syntax to give other readings can be extended to cover diagrams where a wide range of conditions can be represented.

Using different objects to represent quantifier type would not be suitable for our analysis diagrams, where we variously wish to quantify over points, sets, functions, lengths, etc., as it would involve introducing multiple primitives for each type of object. This would quickly get confusing.

2   Prerequisites: Some Definitions

2.1   Real Analysis

Our work is in the domain of real analysis (which gives a rigorous underpinning for calculus, and leads on to fields such as topology). Analysis is a form of geometry, but one whose dry algebraic formalism can make it hard to learn. This makes it an attractive area for applying diagrammatic reasoning. We have implemented a prototype interactive theorem prover using a diagrammatic logic [14]. Our aim is to produce a teaching system based on this, so issues of comprehension and understanding are paramount.

We follow Cauchy's $\epsilon - \delta$ analysis (also known as *standard analysis*), based on arbitrarily small error terms [11]. Definitions often involve arbitrarily small open balls. We write $B_r(x) = \{x' : |x - x'| < r\}$ for the ball of radius r with centre x (this notation is common but not universal). Our examples in this paper will be based on open sets, which are defined as follows:

Definition 21. If X is open... $\operatorname{open}(X) \Rightarrow \operatorname{set}(X) \wedge \forall x \in X, \exists \epsilon > 0. B_\epsilon(x) \subset X$

Definition 22. X is open if... $\operatorname{set}(X), \forall x \in X, \exists \epsilon > 0. B_\epsilon(x) \subset X \Rightarrow \operatorname{open}(X)$

2.2   Diagrams

Diagrammatic representations are by their nature quite specific, however the level of specificity varies. We introduce the term *direct* to informally describe the degree of this. A more direct diagram is one where the representation used is closely linked to its meaning. For example drawing a triangle to reason about triangles. By contrast, an indirect diagram is one where the relation between sign and meaning is arbitrary, and based on convention. Constraint diagrams are an example of this, where a dot can represent anything from a spatial point to a person. Textual representations are always indirect.

In general, the closer the link between signifier and signified,[4] the more specific the representations are (i.e. more direct diagrams tend to be more specific). Specific representations lead to the generalisation problem outlined in §1. Also, it seems that the more specific the representations are, the harder it is to properly perform universal quantification. This is because there is extra information that the user must ignore. For example, it is easier to reason with *'let X represent any man...'* than *'let the late Jon Barwise, who had fading brown hair and contributed so much to diagrammatic reasoning, represent any man...'*. Nevertheless, specific representations do seem to have strong advantages. In particular, more direct diagrams give representations for geometric objects which are both very natural, and seem to lend themselves well to diagrammatic reasoning. See [7] or [13] for discussions of this issue. Our domain is in geometry, hence we have adopted a system based on fairly direct diagrams. This gives us quite natural representations for many of the objects in the domain, but makes quantification a difficult issue.

Our diagrams consist of labelled graphical objects with relations between them. Relations may be represented either graphically or algebraically (this makes our representations heterogenous, although we will continue to refer to them as diagrams).

2.3 Proof System

Often diagrammatic reasoning is presented as a question of interpreting static diagrams. Here we consider dynamic diagrammatic reasoning, where the process of drawing is important, as opposed to just the finished diagram.

Our logic is defined using redraw rules, which are similar to rewrite rules but transform diagrams rather than formulae. This reflects our belief that diagrammatic reasoning is often linked to the drawing process, rather than just the finished diagram. These rules are expressed diagrammatically by an example transformation. A *simple redraw rule*, $D_0 \hookrightarrow D_1$, consists of an initial diagram (D_0, the antecedent or pre-condition) and a modified diagram (D_1, the consequent, or post-condition). Fig. 4 gives an example redraw rule.

Theorems are stated as rules rather than statements (e.g. $sin^2\theta + cos^2\theta \Rightarrow 1$, $1 \Rightarrow sin^2\theta + cos^2\theta$ rather than $\forall\theta.sin^2\theta + cos^2\theta = 1$), and are also expressed diagrammatically (i.e. as redraw rules). A proof consists of a demonstration that the antecedent of the theorem can always be redrawn to give the consequent diagram using an accepted set of rules (i.e. the axioms). Hence a proof is a chain of diagrams, starting with the theorem antecedent and ending with the theorem consequent. We refer to an incomplete or complete proof as a *reasoning chain*.

Informally, the procedure for applying a simple rule is:

---

[4] A *signifier* is the method (e.g. a word or picture) used to represent a concept (the *signified*); together they make up a *sign* [4].

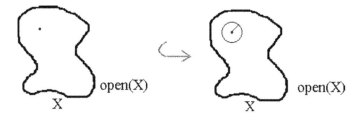

Fig. 4. Definition 21 as a redraw rule.

1. The antecedent diagram is matched with some part of the working diagram (i.e. the last diagram in the reasoning chain).
2. The working diagram is modified in an equivalent way to the modification between the antecedent and consequent diagrams. This modified diagram is added to the end of the reasoning chain.

The principal differences from rewrite rules are:

- There can be an infinite number of valid (but equivalent) redrawings for a given diagram, a given rule and a given matching (e.g. a rule may specify that a point should be drawn, but leave open the choice of which point to draw).
- Due to the problem of multiple possible generalisations, there is no clear choice for how the matching algorithm should work.

Fig. 4 shows how Definition 21 can be implemented as a redraw rule. The antecedent will match any point y in any open set Y; the consequent guarantees the existence of a ball $B_\epsilon(y) \subset Y$.

Consider implementing the converse rule (Definition 22). This definition can be read as 'X is open if, given any point x in X, we can find an $\epsilon > 0$ such that $B_\epsilon(x) \subset X$'. Note the verb 'we can find...' – this condition can be thought of as dynamic: it gives a type of behaviour which we must demonstrate to show that X is open (by contrast, the conditions in Definition 21 can be thought of as *adjectival*). Static diagrams are not well suited to representing behaviour. They are better suited to adjectives than verbs. Instead, we introduce *animated redraw rules*. An *animation* here is a chain of diagrams. Animated redraw rules have an animation as their pre-condition. Where simple redraw rules match the last diagram in the reasoning chain, animated rules must match a section of the reasoning chain.[5] Fig. 5 gives an example redraw rule with an animated antecedent.

---

[5] The full system includes two further types of rule for case-split introduction and case elimination.

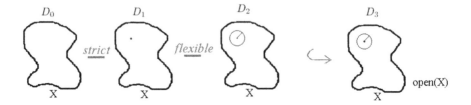

Fig. 5. Definition 22 as an animated redraw rule. The terms *strict* and *flexible* are explained in section 3.3.

3   Using Animation for Quantification

3.1   Quantifier Hierarchy

As with sentential reasoning, quantifier order can be important. Animation gives a reliable and intuitive ordering without introducing extra notation. This is because of causality: it is obvious that object A cannot depend on object B if B was drawn after A. Fig. 6 gives an example based on the joke "Every minute, somewhere in the world a woman gives birth. We must find this woman and stop her." This shows how animation in a very simple way eliminates the ordering ambiguity which allows two conflicting interpretations of the first sentence. Structurally this is equivalent to quantifier numbering in [5], although in presentational terms it is very different.

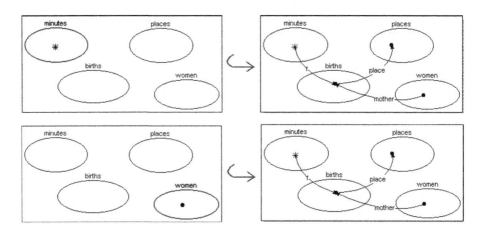

Fig. 6. 'For any minute, there is a woman who gives birth' vs. 'There is a woman who gives birth every minute'. The second diagram is identical for both cases, but the starting diagram shows which object comes first.

3.2   Generalisation

In our logic, a matching algorithm allows redraw rules to be applied to a wide range of diagrams. Thus the matching algorithm determines generalisation of the rule (and vice-versa: specifying generalisations would establish a matching criterion) [14]. As discussed in §1, there are often several possible generalisations, hence several matching algorithms are possible. It seems unlikely that there will be a canonical answer to the question of what aspects of a diagram should be read as generalisation conditions.

We could simply make all the relevant information explicit in the diagram, and assume everything else is unimportant. However this would make for cluttered, less legible, diagrams. A more sensible approach is to have a default interpretation that certain aspects of a diagram are assumed to be important. Ideally, this should be the same as the intuitive reading of the diagram. The conditions specified by this default interpretation can be strengthened or relaxed, but only through explicit conditions in the diagram. Spider diagrams show how this can be used for representing set membership, where spiders are used to override the default reading. It can be extended to cover other relations. Some of the default readings we use are:

- Lengths are considered unimportant (to be generalised), unless a statement of the form $\mathtt{length}(A) < \mathtt{length}(B)$ or $\mathtt{length(A)} = \mathtt{B}$ is added.
- Right angles are considered salient,[6] unless the angle is tagged with a $\angle$ symbol indicating an aribtrary angle.

3.3   Quantifier Behaviour

Statements in our logic are expressed as rules, hence the question of 'how do quantifiers behave?' becomes 'when should a rule antecedent match a reasoning chain?' That is, the question of what does a diagram/animation mean is recast as 'what diagrams/reasoning chains does it match?'

We have to be careful working in direct diagrams, as a quantified object is also a specific example.[7] For example, when reasoning about an abstract universally quantified point, we must nevertheless draw a particular point, and this point will have properties that do not hold universally. However, as long as such properties are not used in the reasoning that follows, they will not affect the generality of the proof. The reasoning that follows would work for any point, so it does not matter which point was actually drawn. The specific case that is drawn comes to represent a class of equivalent cases. What matters is that the reasoning is generic (with indirect representations, this is automatically enforced by using generic objects; with direct representations the generality of the reasoning must be checked).

Consider again Fig. 5, where there is a universally quantified point x in the middle of the rule antecedent. Suppose we wish to apply this rule to show that

---

[6] Assumed to be an intentional feature and therefore not to be generalised.

[7] To be precise, it is interpreted as a specific example.

the set $Y = B_1((0,0))$ is open. First we introduce an arbitrary point $y \in Y$ to match the point x in the rule antecedent. We still have further reasoning to do before the rule will match: we have to find an ϵ-ball about y that lies within the set Y. The reasoning that follows must be universally applicable, which means that it must not use the specific nature of the point y, only the fact that $y \in Y$. For example, suppose we concluded $B_{0.1}(y) \subset Y$ from $y = (0.7, 0.8)$. The reasoning is sound, but it does not apply to other values of y. Our chain of reasoning finds an ϵ for $y = (0.7, 0.8)$, but this reasoning could not be applied to any point. Hence the rule – which requires that such an ϵ exists for any point – is not applicable If the reasoning that follows draws on non-universal aspects of the point, then we say that the point y has been *compromised*; it is no longer universally quantified. This is simple to check: an object is compromised if later reasoning alters its generalisation (by adding extra conditions).

This leads us to the following method for reliably enforcing generic reasoning: suppose an animated redraw rule has the antecedent $D_0 - D_1 - \dots - D_n$, where the D_i are diagrams. When a universally quantified object is introduced into the proof, it must be done exactly as shown in the rule. The interpretation of the object introduced into the reasoning chain must be equivalent to the interpretation of the object introduced in the rule antecedent. If diagram D_i introduces a universally quantified object, we call the transition $D_{i-1} - D_i$ a *strict* transition, since it will only match a transition $P_j - P_{j+1}$ if $P_j - P_{j+1}$ shows equivalent modifications to $D_{i-1} - D_i$ *and no other modifications* (i.e. no extra constructions or conditions). Moreover, subsequent reasoning on P_{j+1} must not compromise the new universally quantified object drawn. This ensures that the reasoning that follows will be as general as the rule requires.

When the rule antecedent contains an existentially quantified object, all that must be shown is that some matching object can be constructed in the reasoning chain. How this is done does not matter (as long as it does not compromise a universally quantified object). Hence an existentially quantified object can be drawn in any manner using several redraw operations, since all we require for the rule antecedent to match is that some such object exists. If diagram D_j introduces an existentially quantified object we call the transition $D_{j-1} - D_j$ a *flexible* transition. A flexible transition allows arbitrary other constructions to be drawn in the reasoning chain when moving from one diagram in the rule to the next.

For example, to prove the theorem $\mathsf{open}(B_r(x))$ takes 11 steps in our logic. A sketch of this proof is given in Fig. 7. We start with the set $B_r(x)$ (diagram P_0 in Fig. 7). The first step is to introduce an arbitrary point in $B_r(x)$ to match the universally quantified point in diagram D_1, Fig. 5. It then takes three steps to construct a suitable ϵ-ball (P_5 in Fig. 7) and five more steps to show that it lies inside $B_r(x)$ (diagram P_{10}). All these steps are performed using simple redraw rules. The final step is to apply the animated rule shown in Fig. 5. Let $P_0 - \dots - P_{11}$ be the proof. Then D_0 matches P_0, D_1 matches P_1 and D_2 matches P_{10}, as shown in Fig. 8.

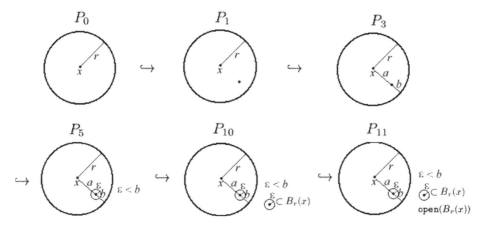

Fig. 7. Sketch proof for $\mathbf{open}(B_x(r))$. Space limitations prevent us giving the full proof or the rules used.

$$D_0 - strict - D_1 - flexible - D_2 \hookrightarrow D_3$$
$$\downarrow \qquad\qquad \downarrow \qquad\qquad \downarrow \quad \downarrow$$
$$P_0 \quad \hookrightarrow \quad P_1 \hookrightarrow \quad ... \quad \hookrightarrow P_{10} \hookrightarrow P_{11}$$

Fig. 8. Antecedent matching with strict and flexible transitions.

The method described above syntactically prescribes quantifier type in rule antecedents in terms of the matching criterion regarding transitions.

3.4   Quantification in the Rule Consequent

Rule consequents are always a single diagram containing new objects or new conditions.[8] New objects are assumed to be existentially quantified. More complex inferences can be expressed in this formalism by two rules linked with a syntactic tag. For example, the statement $p \Rightarrow \exists x.\forall y.q(x,y)$ would be converted into two rules: $p \Rightarrow \exists x.r(x)$ and $r(x), y \Rightarrow q(x,y)$, where r is the syntactic tag created to link the two.

3.5   Representing Quantifier Type

A side-benefit of animation is that it clears up labelling ambiguities. When dealing with emergent or composite objects (i.e. objects formed as a result of drawing other objects, such as some of the squares in Fig. 1), it is not necessarily clear what object a label applies to. However if objects are introduced one at a time

---

[8] If we allowed objects/conditions to be deleted or changed it would give a non-monotonic logic.

with their labels, then this ambiguity vanishes. In our logic we require a reasoning step to recognise a composite object, so composite objects are 'drawn' (and labelled) after their parts.

With labelling ambiguity removed, we could simply reinstate the algebraic symbols $\forall, \exists$ and represent quantifier type by labelling objects with them. Diagrams also allow other, potentially more interesting, possibilities. These include some form of drawing convention, such as colour-coding or using different shaped objects. Or – since quantifiers are introduced one at a time – quantifier type can be represented by having different transitions between frames in the animation.

Any of these representation methods would be sufficient to distinguish the two quantifier types, but they have different advantages. Using the established symbols gives the user something they may already be familiar with. Colour-coding is 'cleaner' since it does not introduce extra labelling, and this may aid comprehension.

Both of the above methods rely purely on convention for their meaning. However, since quantifier behaviour (i.e. their syntactic meaning) comes from the diagram transitions, we could also represent quantifiers by labelling the transitions. A more interesting option is to use special transitions that can attempt to convey the difference in meaning. These special transitions are animations of a different kind. They are independent of the reasoning rather than a part of it. For example, a universally quantified point could 'roam it's habitat', indicating that it is not a specific point. With an existentially quantified object in a rule antecedent, the transition could indicate 'miscellaneous drawing' to illustrate that, when applying the rule, other unspecified constructions will be necessary at this stage. Such an approach would not be of interest to professional users, but could be helpful in teaching applications. However the quantifier type is not visible in the final diagram. To a certain extent, the strengths and weaknesses of these representation methods complement each other, and so they can be combined. We currently use a combination of colour-coding and special transitions, although in the future the system will be customisable to a user's preferences. Colour-coding is identical at the syntactic level to the different primitive objects used in [5], plus it can be used uniformly across types of object that are drawn in quite different ways – although it does restrict the use of colour for representing other properties.

4   Open Issues

Consider the rule in Fig. 5. The natural way of using this rule requires at least one point in the set - so it cannot be applied to the empty set. This introduces a dilemma: our definition of an open set is slightly different from the standard one in that it excludes the empty set. The cause of this discrepancy is that our diagrammatic universal quantifier has existential import. That is, the statement $\forall x \in X$ implies $\exists x \in X$. This is also true in aristotelian logic [10] and natural language, but of course false in predicate calculus. Currently we handle the empty set as a special case. However, since correspondence with conventional logic is desirable in mathematical domains, this is not ideal.

5   Converting Animated Rules into Quantified Rules

So far we have described how sentential rules correspond to animated diagrammatic rules. Now we look into how animated diagrammatic rules correspond to the sentential ones. The concept of well-formed formulae (wff) can be translated to diagrams. We assume the following loose definition of a well formed diagram (wfd) here:

Definition 51. Suppose we have a diagram language L consisting of graphical objects, labelling constants and first-order predicates, and for each property that can be inferred from the diagrams, there is a corresponding predicate in L (i.e. any diagram in L can also be described purely algebraically in L). Then say:

- The empty diagram is a wfd.
- If A is a wfd, X an object within L, X has label l_X and l_X is not already used in A, then $A \cup X$ is a wfd.
- If A is a wfd, X some objects in A and $p(X)$ a predicate in L, then $A \cup p(X)$ is a wfd, where $p(X)$ could be drawn either graphically or algebraically using object labels.

Are any conjunctions of predicates allowed? Since we are using heterogenous diagrams, any combination can be stated, but there are sensible restrictions that could be made (e.g. disallowing $point(X) \wedge line(X)$).

5.1   Well Formed Formulae

Let us assume we have an interpretation function I that maps single diagrams to unquantified algebraic statements in a suitable domain. Let X denote a vector of variables, let p, q denote any conjunction of predicates in L (e.g. $\texttt{equal\_area}(X, Y)$, $\texttt{point}(X) \wedge \texttt{in}(X, Y)$, etc.)
Given diagrams D_0, D_1 the simple rule $D_0 \hookrightarrow D_1$, is well-formed if D_0, D_1 are wfd and $D_0 \subset D_1$. Simple rules correspond to statements of the form:

$$\lambda X. p(X) \ \Rightarrow \ q(X)$$

Here, $p(X) = I(D_0), q(X) = I(D_1) \backslash I(D_0)$ (with the natural mapping between labels and variables).
Given any animated antecedent $D_0 - ... - D_n$, we can add another diagram D_{n+1} to it in two ways:

1. With a flexible transition, introducing existentially quantified objects:
   If $A \Rightarrow B$ is a wff, $\texttt{var}(A)$ are the variables (free and bound) in A, and $X \cap \texttt{var}(A) = \emptyset$ then

$$A, \exists X. p(var(A), X) \ \Rightarrow \ B \text{ is a wff}$$

2. With a strict transition, introducing universally quantified objects:
   If $A \Rightarrow B$ is a wff, $\text{var}(A)$ are the variables (free and bound) in A, and $X \cap \text{var}(A) = \emptyset$ then

$$A, \exists X'.p(var(A), X'), \forall X.p(var(A), X) \Rightarrow B \text{ is a wff}$$

5.2   Example Animated Redraw Rule (Fig. 5) revisited

Following the correspondences given above and assuming the behaviour of interpretation function I, we get:

$$I(D_0) = \lambda X.\text{set}(X)$$
$$I(D_0 - D_1) = I(D_0), \exists x'.\text{point}(x'), x' \in X, \forall x.\text{point}(x), x \in X$$
$$I(D_0 - D_1 - D_2) = I(D_0 - D_1), \exists \epsilon.\text{real}(\epsilon)\epsilon > 0, B_\epsilon(x) \subset X$$
$$I(D_0 - D_1 - D_2 \hookrightarrow D_3) = I(D_0 - D_1 - D_2) \Rightarrow \text{open}(X)$$

Hence the redraw rule converts to the algebraic rule:

$$\text{set}(X), X \neq \emptyset, \forall x \in X, \exists \epsilon > 0.B_\epsilon(x) \subset X \Rightarrow \text{open}(X)$$

which is almost Definition 22. As explained in §4 we currently need a seperate redraw rule to cover the case $X = \emptyset$.

6   Conclusions & Future Work

Real analysis is a domain where great care is required to avoid mistakes, and where quantifier ordering is often important. As part of our project to produce a diagrammatic formalisation for this subject, we have developed rules with animated pre-conditions. This paper demonstrates how these rules work, showing how they allow us to perform generic quantified reasoning with specific representations. We hope that this has applications to other fields; any domain that considers dynamic behaviour seems promising. A second type of animation (special transitions for representing quanifier type) is also introduced to give more meaningful representations.

The treatment given in §5 is quite loose. We intend to develop a formal semantics for animated redraw rules, plus a formal definition of our interpretation function and matching criterion (whose behaviour we have assumed here). This should then allow us to show equivalence between redraw rules and algebraic rules.

There are drawbacks to our method of representing and reasoning about quantification. The issue of existential import raises serious questions. The use of colour to represent quantifier type severely limits the way in which colour can be used elsewhere in the diagram. Also, the extra dimension of time in the representations might prove to be harder for users because of 'overloading' understanding through extra demands on working memory. Perhaps the greatest drawback of animation is that it is not suited to being printed (e.g. in textbooks or papers), except as cumbersome comic strips where the simplicity of the representation is lost. Note that to a certain extent, this does not apply to its use on blackboards, where animation can be performed, albeit a little crudely.

However the advantages are a logic that is, we hope, more elegant and natural to use. Using animation to extend diagrams avoids extra labelling and should be more intuitive, since it draws on cause-and-effect for meaning rather than requiring conventions. Moreover rules with animated pre-conditions focus attention on the reasoning used in a proof. This could be beneficial from an educational point of view. Using extra notation it is possible to avoid the use of animation within our system. For example, Fig. 9 gives a non-animated version of the redraw rule in Fig. 5. We feel that by compressing all the information into one diagram, the non-animated version obscures the relations between the objects. We will test student responses to this difference as part of our system evaluation. We are also investigating how expressive our analysis diagrams are (i.e. how good a coverage of theorems we can achieve). Our preliminary work suggests that diagrammatic reasoning can be successful in teaching this domain [14].

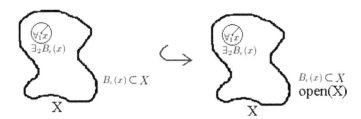

Fig. 9. Static redraw rule defining an open set.

References

1. J.Barwise & J.Etchemendy "Heterogeneous Logic" in Diagrammatic Reasoning: Cognitive and Computational Perspectives. AAAI Press/The MIT Press, 1995.
2. A.Bierce "Fantastic Fables: The Flying Machine" 1899. Project Gutenberg e-text 1995, ftp://ftp.ibiblio.org/pub/docs/books/gutenberg/etext95/fanfb10.txt
3. J.Howse, F.Molina & J.Taylor "On the Completeness and Expressiveness of Spider Diagram Systems" proceedings of the 1st International Conference on Theory and Application of Diagrams. pp26-41, Springer-Verlag, 2000.
4. B.Fraser "Structuralism" course notes, Cambridge University. Available online at http://www.classics.cam.ac.uk/Faculty/structuralism.html, 1999.
5. J.Gil, J.Howse & S.Kent "Towards a Formalization of Constraint Diagrams" Symposium on Visual Languages and Formal Methods. IEEE, 2001.

6. J.Gil & Y.Sorkin "The Constraint Diagrams Editor" Available online at http://www.cs.technion.ac.il/Labs/ssdl/research/cdeditor, 1999 – 2000.

7. C.Gurr, J.Lee & K.Stenning "Theories of Diagrammatic Reasoning: Distinguishing Component Problems" in Minds & Machines 8(4) pp533-557, Kluwer Academic, 1998.

8. E.Hammer "Reasoning with Sentences and Diagrams" in "Working Papers on Diagrams and Logic" editors G.Allwein & J.Barwise, Indiana University Logic Group, 1993.

9. P.J.Hayes & G.L.Laforte "Diagrammatic Reasoning:Analysis of an Example" in "Formalizing Reasoning with Visual and Diagrammatic representations: Papers from the 1998 AAAI Fall Symposium" editors G.Allwein, K.Marriott & B.Meyer, 1998.

10. G.Klima "Existence and Reference in Medieval Logic" in "New Essays in Free Logic" editors A.Hieke & E.Morscher, Kluwer Academic, 2001.

11. T.W.Körner "Analysis" lecture notes, Cambridge University. Available online at http://ftp.dpmms.cam.ac.uk/pub/twk/Anal.ps, 1998.

12. L.K.Schubert "Extending the Expressive Power of Semantic Networks" in Artificial Intelligence. 7(2) pp163-198, Elsevier, 1976.

13. A.Sloman "Interactions Between Philosophy and A.I.: The Role of Intuition and Non-Logical Reasoning in Intelligence" proceedings 2nd International Joint Conference on Artificial Intelligence, 1971.

14. D.Winterstein "Diagrammatic Reasoning in a Continuous Domain" CISA Ph.D. Research Proposal, Edinburgh University. Available online at http://www.dai.ed.ac.uk/homes/danielw/academic, 2001.

Generating Euler Diagrams

Jean Flower and John Howse

School of Computing and Mathematical Sciences, University of Brighton,
Lewes Road, Brighton BN2 4GJ,UK
{J.A.Flower,John.Howse}@bton.ac.uk fax +44 1273 642405,
http://www.it.bton.ac.uk/research/vmg/VisualModellingGroup.html

Abstract. This article describes an algorithm for the automated generation of any Euler diagram starting with an abstract description of the
diagram. An automated generation mechanism for Euler diagrams forms
the foundations of a generation algorithm for notations such as Harel's
higraphs, constraint diagrams and some of the UML notation. An algorithm to generate diagrams is an essential component of a diagram tool
for users to generate, edit and reason with diagrams.
The work makes use of properties of the dual graph of an abstract diagram to identify which abstract diagrams are "drawable" within given
wellformedness rules on concrete diagrams. A Java program has been
written to implement the algorithm and sample output is included.

1   Introduction and Background

Euler diagrams consist of contours, simple closed curves, which split the plane
into zones. A concrete Euler diagram is a drawing which represents information
about sets and their intersections. This information can be encapsulated by an
abstract diagram. An abstract (Euler) diagram consists of contours, which are
just abstract notions, and information about how those contours are used to
give a set of zones. An abstract diagram has zero or many concrete representations, and this paper is primarily concerned with the construction of concrete
representations from an abstract description. An important consequence is the
identification of a complete set of drawable abstract diagrams, with tests for
drawability made at the abstract level. This work supports the development of
a constraint diagram tool, and this application is discussed in subsection 1.1.
Definitions of the terms "concrete (Euler) diagram", "abstract (Euler) diagram"
are in subsection 1.3, and the algorithm for construction of concrete diagrams is
in section 2.

1.1   An Application to Motivate the Work

The construction of a concrete representation of an abstract diagram is useful
in many environments. In this section we describe one application for the work.
Constraint diagrams [6] convey a subset of first-order predicate calculus concerning sets and their elements. The simplest kind of constraint diagrams are

M. Hegarty, B. Meyer, and N. Hari Narayanan (Eds.): Diagrams 2002, LNAI 2317, pp. 61–75, 2002.
© Springer-Verlag Berlin Heidelberg 2002

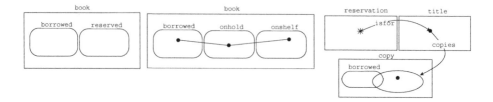

Fig. 1. An Euler diagram, a spider diagram and a constraint diagram

Euler diagrams [4], which make statements about the enclosure and intersection of sets in the system. The first of three diagrams in figure 1 is an Euler diagram which conveys the fact no book is both reserved and borrowed.

The second diagram in figure 1 is a *spider diagram* [5] showing three states for a book: 'borrowed', 'on the shelf', or 'on hold'. No book may be in two states, and there is a book in one of the states. A further extension of the notation adds universal elements, and relational navigation [6], giving the full notation for constraint diagrams. The third diagram in figure 1 is a constraint diagram which states that reservations are only held on titles for which there is a copy in stock. Furthermore, there is a copy that is in stock and not in the borrowed state.

Textual notations (eg. *OCL* [17]) and diagrammatic notations (eg. constraint diagrams) can be used for the same purpose. Many statements can be more clearly expressed in diagrammatic form. Diagrams can also be used fruitfully for reasoning ([7], [8]), with animation showing the logical steps involved in an argument.

Reasoning by transforming diagrams holds potential for the future, especially if are tools available to assist. Such a tool would be an editor (be able to assist the user with creation and editing of diagrams) but also needs an understanding of the diagram. This understanding determines whether two constraints are logically related and enables animated reasoning with diagrams. Flexible and combined use of textual and diagrammatic notations requires easy transformation from one notation to the other, in particular generating diagrams from textual notation.

1.2   Related Work

There is much work on graph-drawing algorithms, see e.g. [1].

Euler diagrams are similar to the hypergraphs discussed in [2]. A hypergraph is given by a set of contours, and a set of points. Each contour is specified as containing some points and excluding others. This is similar to the problem addressed here, except the description of an Euler diagram places conditions on the make-up of the *regions* of the diagram, and not just on a set of specific points. Our Euler diagrams could be thought of as examples of hypergraphs.

There is also work on string graphs in e.g. [14]. The authors cast the problem of the drawability of Euler diagrams as equivalent in complexity to the problem of recognising string graphs.

In the above references the authors use weaker definitions of a well-formed diagram.

1.3 Key Definitions

An *abstract Euler diagram* comprises a set of *contours* and a set of *zones* which are subsets of the set of contours. This corresponds to *type-syntax* in [9].

Definition 1. *An* abstract (Euler) diagram *is a pair:* $d = \langle \mathcal{C}(d), \mathcal{Z}(d) \rangle$ *where*
  (i) $\mathcal{C}(d)$ *is a finite set whose members are called* contours.
  (ii) $\mathcal{Z}(d) \subseteq \mathcal{PC}(d)$ *is the set of* zones *of d, so* $z \in \mathcal{Z}(d)$ *is* $z \subseteq \mathcal{C}(d)$.
  (iii) $\bigcup_{z \in \mathcal{Z}(d)} z = \mathcal{C}(d)$
  (iv) *The empty set* $\{\} \in \mathcal{Z}(d)$.
The set of abstract diagrams is denoted $\mathcal{D}$.

Example 1 (An abstract diagram). $\langle \{a, b, c\}, \{\{\}, \{a\}, \{a, b\}, \{b\}, \{c\}\} \rangle \in \mathcal{D}$.

The condition that the empty set is included as a zone could be omitted, and this work on converting abstract to concrete diagrams would hold for the subset of abstract diagrams which did include the empty set as a zone. In a concrete diagram, the zone corresponding to the empty set will be the zone outside all contours of the diagram.

A *concrete Euler diagram* is a set of labelled *contours* (simple closed curves) in the plane, each with a unique label. A *zone* is a connected component of the complement of the contour set. Each zone is contained in a set of contours. This corresponds to *token-syntax* in [9].

Definition 2. *A* concrete (Euler) diagram *is a triple* $\hat{d} = \langle \widehat{\mathcal{L}}(\hat{d}), \widehat{\mathcal{C}}(\hat{d}), \widehat{\mathcal{Z}}(\hat{d}) \rangle$ *whose components are defined as follows:*
  (i) $\widehat{\mathcal{C}}(\hat{d})$ *is a finite set of simple closed curves,* contours, *in the plane* $\mathbb{R}^2$. *Each contour has a unique label from the set* $\widehat{\mathcal{L}}(\hat{d})$.
    1. Contours meet transversely and without triple points.
    2. Each component $\hat{z} \in \mathbb{R}^2 - \bigcup_{\hat{c} \in \widehat{\mathcal{C}}(\hat{d})} \hat{c}$ *is uniquely identified by a set of contours*
$$X \subset \widehat{\mathcal{C}}(\hat{d}) \text{ with } \hat{z} = \bigcap_{\hat{c} \in X} interior\,(\hat{c}) \cap \bigcap_{\hat{c} \in \widehat{\mathcal{C}}(\hat{d}) - X} exterior\,(\hat{c}).$$
  (ii) *A* zone *is a connected component of* $\mathbb{R}^2 - \bigcup_{\hat{c} \in \widehat{\mathcal{C}}(\hat{d})} \hat{c}$. *The set of zones of* $\hat{d}$ *is denoted* $\widehat{\mathcal{Z}}(\hat{d})$.
  (iii) *A zone* $\hat{z}$ *is uniquely determined by the set of contour labels* $\widehat{\mathcal{L}}(\hat{z})$ *for the contours which contain the zone.*
The set of concrete diagrams is denoted $\widehat{\mathcal{D}}$.

Example 2 (A concrete diagram). Let $\hat{d}$ be the concrete diagram given in figure 2. $\widehat{\mathcal{C}}(\hat{d})$ has three elements (the three contours shown) $\widehat{\mathcal{L}}(\hat{d}) = \{a, b, c\}$ and $\widehat{\mathcal{Z}}(\hat{d})$ has five elements, uniquely determined by the label sets $\{\}$, $\{a\}$, $\{a, b\}$, $\{b\}$ and $\{c\}$.

a concrete Euler diagram   illegal Euler diagrams

Fig. 2. Well-formed and not well-formed concrete diagrams

The rules about transverse crossing, absence of triple points and connectedness of zones are the chosen well-formedness rules for this paper. Figure 2 shows a well-formed concrete diagram and some which are not well-formed. In future work, it is intended to extend this work to accommodate different definitions of "well-formed" concrete diagrams.

Definition 3. *The mapping* $ab : \widehat{\mathcal{D}} \to \mathcal{D}$ *("ab" for "abstractify") forgets positioning of the contours. It is defined by*

$$ab\left(\langle \widehat{\mathcal{L}}(\hat{d}), \widehat{\mathcal{C}}(\hat{d}), \widehat{\mathcal{Z}}(\hat{d}) \rangle\right) = \langle \widehat{\mathcal{L}}(\hat{d}), \left\{ \widehat{\mathcal{L}}(\hat{z}) : \hat{z} \in \widehat{\mathcal{Z}}(\hat{d}) \right\} \rangle$$

Example 3 (Abstractification). Let $\hat{d}$ be the concrete diagram given in figure 2. $\mathcal{C}(ab(\hat{d})) = \{a, b, c\}$ and $\mathcal{Z}(ab(\hat{d})) = \{\{\}, \{a\}, \{a, b\}, \{b\}, \{c\}\}$.

Definition 4. *A concrete diagram* $\hat{d}$ *represents or complies with an abstract diagram* d *if and only if* $d = ab(\hat{d})$. *An abstract diagram which has a compliant concrete representation is* drawable.

Example 4 (Drawability). The abstract diagram $\langle \{a, b\}, \{\{\}, \{a, b\}\} \rangle$ is undrawable. This can be investigated by futile attempts to produce a diagram (the temptation to draw two coincident concrete contours violates the condition that contours meet transversely). Later in the paper, corollary 1 is used to show that there is no concrete representation for the abstract diagram.

This paper is about the construction of concrete representations of abstract diagrams. We seek an inverse of the map $ab : \widehat{\mathcal{D}} \to \mathcal{D}$.

2   A Sketch of the Algorithm

We use the concept of a *plane dual graph* of a concrete diagram.

Definition 5. *A concrete labelled graph* $\hat{G}$ *is a triple* $\langle \widehat{\mathcal{L}}(\hat{G}), \widehat{\mathcal{V}}(\hat{G}), \widehat{\mathcal{E}}(\hat{G}) \rangle$ *where the components are defined as follows:*

(i) $\widehat{\mathcal{L}}(\hat{G})$ *is a set of labels for the graph*

(ii) $\widehat{\mathcal{V}}(\hat{G})$ *is a set of vertices. Each vertex* $\hat{v}$ *is labelled with* $\widehat{\mathcal{L}}(\hat{v}) \subseteq \widehat{\mathcal{L}}(\hat{G})$ *and each vertex has a position in the plane* $\mathbb{R}^2$.

(iii) $\widehat{\mathcal{E}}(\hat{G})$ *is a set of edges. Each edge* $\hat{e}$ *is a pair of vertices from* $\widehat{\mathcal{V}}(\hat{G})$. *The label sets on adjacent vertices must have singleton symmetric difference (one set is the other with a single additional element). The edges can be associated with the label which distinguishes the labels of the end-vertices.*

The set of concrete labelled graphs is denoted $\widehat{\mathcal{LG}}$.

Note that although the vertices of a concrete labelled graph have a position in the plane, the edges are simply pairs of vertices. The edges are not associated with curves in the plane ie. we have a *geometric graph* in [1].

Definition 6. *The map pdual : $\widehat{\mathcal{D}} \to \mathcal{P}\widehat{\mathcal{LG}}$ ("pdual" for "plane dual") is defined by*

$$\hat{G} \in pdual\langle \widehat{\mathcal{L}}(\hat{d}), \widehat{\mathcal{C}}(\hat{d}), \widehat{\mathcal{Z}}(\hat{d}) \rangle$$

if and only if $\widehat{\mathcal{L}}(\hat{G}) = \widehat{\mathcal{L}}(\hat{d})$ and there is a bijection $\widehat{\mathcal{V}}(\hat{G}) \to \widehat{\mathcal{Z}}(\hat{d}); v \mapsto \hat{z}$ if and only if v is inside the part of the plane specified by $\hat{z}$ and the labelling matches: $\widehat{\mathcal{L}}(\hat{v}) = \widehat{\mathcal{L}}(\hat{z})$. Finally, $e \in \widehat{\mathcal{E}}(\hat{G})$ if and only if the corresponding zones are topologically adjacent in the plane.

Any concrete labelled graph gives an *abstract labelled graph*, by forgetting vertex positions.

Definition 7. *An* abstract labelled graph *is a triple $\langle \mathcal{L}(G), \mathcal{V}(G), \mathcal{E}(G) \rangle$ where the components are defined as follows:*

   (i) $\mathcal{L}(G)$ *is a set of labels for the graph*
   (ii) $\mathcal{V}(G)$ *is a set of vertices. Each vertex $\hat{v}$ is labelled with $\mathcal{L}(v) \subseteq \mathcal{L}(G)$.*
   (iii) $\mathcal{E}(G)$ *is a set of edges. Each edge $\hat{e}$ is a pair of vertices in $\mathcal{V}(G)$, where the vertex labels must have a singleton symmetric difference (one vertex set exceeds the other by a single additional element). The label which distinguishes the end vertices can be used to label the edge.*
The set of abstract labelled graphs is denoted $\mathcal{LG}$.

Example 5 (An abstract labelled graph). The abstract labelled graph G has two labels: $\mathcal{L}(G) = \{a, b\}$, three vertices: $\mathcal{V}(G) = \{v_1, v_2, v_3\}$, the vertices labelled as follows: $\mathcal{L}(v_1) = \{\}$, $\mathcal{L}(v_2) = \{a\}$, $\mathcal{L}(v_3) = \{a, b\}$ and one edge: $\mathcal{E}(G) = \{\{v_1, v_2\}\}$. Note that a second edge $\{v_2, v_3\}$ could have been added to this example, but the pair $\{v_1, v_3\}$ would not be admitted as an edge because of the labels on the vertices v_1 and v_3.

Definition 8. *The map $f : \widehat{\mathcal{LG}} \to \mathcal{LG}$ ('f' for 'forgetful') is defined by forgetting positional information in the vertex set $\widehat{\mathcal{V}}(\hat{G})$.*

Proposition 1. *Given a concrete diagram $\hat{d} \in \widehat{\mathcal{D}}$, and two plane duals $\hat{G}_1, \hat{G}_2 \in pdual(\hat{d})$, their abstract labelled graphs are equal: $f(\hat{G}_1) = f(\hat{G}_2)$.*

Definition 9. *Although a concrete diagram $\hat{d}$ has many plane dual graphs given by the set pdual($\hat{d}$), we can refer to "the" abstract dual graph of a concrete diagram, abG($\hat{d}$) $\in \mathcal{LG}$.*

Definition 10. *The map diag : $\mathcal{LG} \to \mathcal{D}$ ("diag" for "diagrammise") is defined by $\langle \mathcal{L}(G), \mathcal{V}(G), \mathcal{E}(G) \rangle \mapsto \langle \mathcal{C}(d), \mathcal{Z}(d) \rangle$ where the abstract contour set is the label set $\mathcal{L}(G) = \mathcal{C}(d)$ and the zones are the vertices $\mathcal{V}(G) = \mathcal{Z}(d)$.*

This map can be thought of as forgetting the edge information in an abstract labelled graph.

Example 6 (The diag mapping). Let G be the abstract labelled graph given in example 5. The abstract diagram $diag(G) = d$ has contours $\mathcal{C}(d) = \{a, b, c\}$ and zones $\{\{\}, \{a\}, \{a, b\}\}$

What we have so far is illustrated in the following commutative diagram (shown on the left) which shows "forgetful" mappings as information is lost moving from a concrete environment to the abstract level.

$$
\begin{array}{ccc}
\widehat{\mathcal{D}} \xrightarrow{\;pdual\;} \widehat{\mathcal{LG}} & \qquad & \widehat{\mathcal{D}} \longleftarrow \widehat{\mathcal{LG}} \\
{\scriptstyle ab}\downarrow \qquad\;\; \downarrow{\scriptstyle f} & & \uparrow \qquad\quad \uparrow \\
\mathcal{D} \xleftarrow[\;diag\;]{} \mathcal{LG} & & \mathcal{D} \longrightarrow \mathcal{LG}
\end{array}
$$

The strategy of the algorithm is to attempt to find inverses of the functions *pdual*, f and *diag* (inverses shown on the right). A mapping from abstract diagrams to concrete diagrams will be found which factors through abstract dual graphs and plane dual graphs. Factoring the problem through dual graphs reduces one task to three steps, and allows the use of existing knowledge from graph theory and graph drawing. As the work progresses, we will see that some inverse functions cannot be defined on the whole domain, and in this way, some abstract diagrams become classified as undrawable. The following three sections of the paper cover three key steps:

(i) map from abstract diagrams to abstract labelled graphs: $\mathcal{D} \to \mathcal{LG}$

(ii) map from abstract labelled graphs to concrete labelled graphs: $\mathcal{LG} \to \widehat{\mathcal{LG}}$

(iii) map from concrete labelled graphs to concrete diagrams: $\widehat{\mathcal{LG}} \to \widehat{\mathcal{D}}$

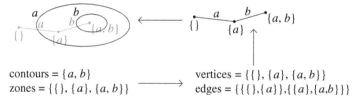

contours $= \{a, b\}$ vertices $= \{\{\}, \{a\}, \{a, b\}\}$
zones $= \{\{\}, \{a\}, \{a, b\}\}$ edges $= \{\{\{\}, \{a\}\}, \{\{a\}, \{a, b\}\}\}$

Fig. 3. The steps of the algorithm

2.1 Creation of an Abstract Labelled Graph from an Abstract Diagram

Definition 11. *The map superDual* $: \mathcal{D} \to \mathcal{LG}$ *is defined by* $\langle \mathcal{C}(d), \mathcal{Z}(d) \rangle \mapsto \langle \mathcal{C}(d), \mathcal{Z}(d), \mathcal{E}(G) \rangle$ *where the edges include all possible* $e = (v_1, v_2)$ *where* v_1 *and* v_2 *have singleton symmetric difference.*

The superdual of an abstract diagram uses the contours for labels and the zones for vertices. Deriving the superdual of an abstract diagram d and mapping back to an abstract diagram recovers d, having constructed edges between some vertices and then forgotten the edges again.

Proposition 2. *If $d \in \mathcal{D}$ then $diag(superDual(d)) = d$.*

Figure 3 illustrates vertex and edge-labelling. In small cases (all examples with three or fewer contours), the abstract labelled dual graph of a concrete diagram $abG(\hat{d})$ is exactly the super-dual of its abstract diagram $superDual(ab(\hat{d}))$. However, figure 4 shows that vertex labels can differ by a single contour label even when the zones are not adjacent in the concrete diagram. The vertices $\{s\}$ and $\{p, s\}$ in $superDual(ab(\hat{d}))$ are adjacent, but they are not adjacent in $abG(\hat{d})$.

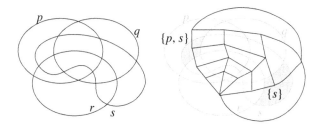

Fig. 4. The dual graph of the Venn diagram on four contours

Proposition 3. *Let $\hat{d}$ be a concrete diagram. Then $abG(\hat{d}) \in \mathcal{LG}$ is a subgraph of $superDual(ab(\hat{d}))$ which includes all vertices (sometimes called a* wide sub-graph*).*

The abstract dual of a concrete diagram satisfies *connectivity conditions*:

Definition 12 (The connectivity conditions). *An abstract labelled graph $\langle \mathcal{L}(G), \mathcal{V}(G), \mathcal{E}(G) \rangle$ satisfies the* connectivity conditions *if it is connected and, for all labels $l \in \mathcal{L}(G)$, the subgraphs $G^+(l)$ generated by vertices whose labels include l, and $G^-(l)$ generated by vertices whose labels exclude l are connected.*

Theorem 1 (The connectivity theorem). *Let $\hat{d}$ be a concrete diagram. Then $abG(\hat{d})$ passes the connectivity conditions.*

The proof of this result uses some results from topology concerning paths in the plane.

Corollary 1 (The connectivity test). *If d is an abstract diagram whose super-dual fails the connectivity conditions then d is undrawable.*

This corollary provides a practical condition for drawability at the abstract diagram level.

Example 7. The abstract diagram of example 4 $\langle \{a, b\}, \{\{\}, \{a, b\}\} \rangle$ has super-dual with two vertices, labelled $\{\}$ and $\{a, b\}$ and no edges. The super-dual is disconnected, and so the abstract diagram is undrawable.

2.2   Creation of a Plane Dual Graph from an Abstract Labelled Graph

If we are given an abstract labelled graph which passes the connectivity conditions, it is potentially the abstract graph of a concrete diagram. To begin the construction of such a concrete diagram (if it exists), assign points in the plane to vertices of the graph to give a concrete labelled graph. Recall that the vertices of a concrete labelled graph have positions in the plane, but the edges are simply pairs of vertices, rather than curves in the plane.

The dual of a concrete diagram can be drawn as a plane graph. Given an abstract labelled graph G, it is potentially the abstract graph of a concrete diagram only if it is planar. If G is non-planar, it may still be possible to remove edges from it to leave a reduced planar graph which still passes the connectivity conditions. For example, edge removal is necessary to produce a plane dual for the Venn diagram on four contours. Different choices of edge-removal lead to different concrete representations.

For non-planar abstract labelled graphs, further work is needed to ascertain a sound strategy for edge-removal (maintaining connectivity conditions) to give a planar graph where possible. Strategies exist in the literature for removing edges to gain planarity, but in this context, attention must be paid to the connectivity conditions.

Example 8 (Edge-removal and connectivity). The following abstract diagram has a superdual which passes the connectivity conditions. However, it is non-planar and if any edge is removed, the connectivity conditions fail. It has five contours $\{a, b, c, d, e\}$ and fifteen zones. $\{\{\}, \{b\}, \{c\}, \{d\}, \{e\}, \{a, b\}, \{a, c\}, \{a, d\},$ $\{a, e\}, \{a, b, c\}, \{a, b, d\}, \{a, b, e\}, \{a, c, d\}, \{a, c, e\}, \{a, d, e\}\}$ (inspired by an example in [14]). The superdual is homeomorphic to the complete graph K_5.

In the current implementation of the algorithm, an iterative planarising step is used which is not optimal but works in small cases. An improved planarising step should take account of the rich structure of the abstract graph given by the labelling on the vertices. Connectivity conditions reveal the graph as connected and bipartite (by cardinality of vertex labels). The fact that edges only join vertices with singleton symmetric difference provides more structure within the dual graph.

Whatever planarising step is used, the aesthetics of the result are unimportant at this stage. The placing of the vertices determines faces of the plane graph. The next step extracts the faces as as combinatorial constructions and makes a new embedding in the plane.

2.3   Creation of Concrete Contours from a Plane Labelled Graph

Given a concrete labelled graph with labelled vertices and edges, the faces of the graph are the starting point for constructing the contours for a concrete diagram. We would like to be able to construct concrete contours which give one zone for each graph vertex, and no other zones. The labels on the zones should

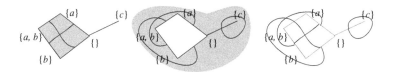

Fig. 5. Arcs across dual faces

match the labels on the graph vertices. The strategy is to draw the edges of the concrete labelled graph, and to draw arcs across faces to join the midpoints of edges. These arcs will combine to create the closed contours of a concrete diagram (see figure 5). The labels on the edges around each face determine how the arcs will intersect.

Definition 13. *A face-cycle of face f in concrete labelled graph G is a cycle of edges in the boundary of f which make up a cycle in G. The contour labels in a face-cycle can be cycled, or reversed, giving another representative of the same cycle.*

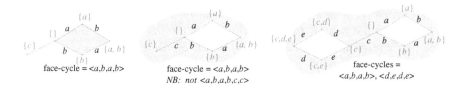

Fig. 6. Face-cycles of a concrete labelled graph

For internal (bounded) faces of a concrete labelled graph G, there is a single face-cycle made up of a subset of the boundary edges. However, an outside (unbounded) face of G may have multiple face-cycles as shown in figure 6.

Lemma 1. *Let f be a face of a concrete labelled graph. For any contour label c, there is an even number of occurrences of c in a face-cycle of f.*

The proof uses a count of symbols in vertex-labels.

Lemma 2. *Let G be a concrete labelled graph whose abstract labelled graph passes the connectivity conditions. Take a face f and a contour label c. There are zero or two occurrences of c in the face-cycle of f.*

The proof of this result is similar to a proof of the five-colouring theorem which uses *Kempe chains* [3]. Use connectivity and lemma 1.

When constructing the concrete contours, one zone will be constructed around each plane labelled vertex. It is important to ensure that the arcs across each face intersect so that no additional unwanted zone(s) appear in the face. Problematic cases with unwanted concrete zones are shown in figure 7. The unwanted zones could be either zones which aren't specified by the vertex set, or a second component of an existing zone, giving disconnected zones. The first dia-

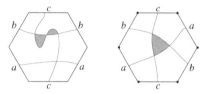

Fig. 7. Introducing unwanted zones

gram in figure 7 can be resolved by rendering the face convex and using linear arcs across the face.

In the second diagram, however the arcs are placed without triple points, an unwanted zone will appear because of the sequence of edges around the face. The potential introduction of unwanted zones in this way provides another set of conditions for construction of a concrete diagram. The following definition and theorem set up enough notation to determine whether or not a concrete labelled graph can be used to construct a compliant concrete diagram.

Definition 14. *Define the* crossing index *of a face-cycle. For each pair of contour labels which occur in the face-cycle, determine whether the pairs are nested. If the letters are not nested, the pair contributes 1 to the crossing index, otherwise the pair contributes 0 to the crossing index. Symbols a and b are nested in abba, but not nested in abab.*

Example 9 (Crossing index). The cycle $\langle a, b, c, a, b, c \rangle$ has crossing index equal to 3, because all pairs $\{a, b\}$, $\{b, c\}$, and $\{a, c\}$ give non-nested sub-cycles $\langle a, b, a, b \rangle$, $\langle b, c, b, c \rangle$ and $\langle a, c, a, c \rangle$. Another example, $\langle a, b, c, b, a, c \rangle$, has crossing index equal to 2.

If a face-cycle has n symbols, each occurring twice, then its crossing index x is bounded by $0 \leq x \leq \frac{n(n-1)}{2}$.

Theorem 2 (The face-conditions). *Let d be a concrete diagram, and P a plane dual graph. For each face-cycle of a plane dual graph, with crossing index x and length $2n$, the crossing index is $x = n - 1$.*

The proof uses Euler's formula for plane graphs and the handshaking lemma.

Corollary 2 (The face-cycle conditions). *If P is a concrete labelled graph with a face-cycle whose crossing index is x and number of edges is $2n$ with $x \neq n - 1$, then P cannot be used to construct concrete contours with one zone containing each vertex.*

Example 10 (Face-conditions). In the left-hand diagram of figure 8, there are three faces and two face-cycles. The four-sided faces have crossing index $x = 1$ and cycle length $n = 2$, and the six-sided face has crossing index $x = 2$ and cycle length $n = 3$. The concrete labelled graph passes the face-conditions.

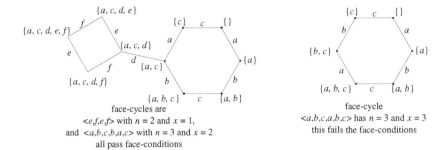

Fig. 8. Face-conditions

The abstract diagram with zones $\{\{\}, \{a\}, \{a, b\}, \{a, b, c\}, \{b, c\}, \{c\}\}$ is undrawable (see the second graph in figure 8). Its plane dual graph is a cycle of six edges. The faces both have face-cycles $\langle a, b, c, a, b, c \rangle$ with crossing indices $x = 3$, and the lengths given by $n = 3$. It is not the case that $x = n - 1$.

Proposition 4. *If a concrete labelled graph passes the face-conditions and connectivity conditions, and removal of an edge maintains the connectivity conditions, then removal of that edge also maintains the face-conditions.*

Removal of edges (e.g. to ease planarisation) cannot jeopardise the existence of a plane representation which passes the face-conditions.

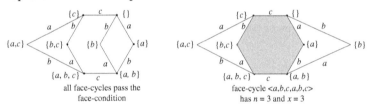

Fig. 9. Face-conditions and layout

One might hope that, given an abstract dual graph, all plane representations of it either pass or fail the face-conditions. However, this is not the case. Figure 9 shows a dual of the Venn diagram on three contours, after removal of the edges $\{\{a\}, \{a, c\}\}$ and $\{\{b\}, \{b, c\}\}$. The absence of these two edges does not cause failure of the seven connectivity conditions. The super-dual has a plane representation which passes the face-conditions (in fact *all* plane representations of the super-dual will pass the face conditions), so proposition 4 says that after removing two edges, there must also be a plane representation which passes the face conditions. This is shown on the left of the figure. But removal of these edges also introduces the existence of a plane representation which fails the face-conditions, shown on the right. This example shows that failure of the face-conditions in a single plane subgraph of $superDual(d)$ does not necessarily render the underlying abstract diagram d undrawable. To complete the argument for example 10 we have to say that not only does the presented plane graph fail the face-conditions, but *all* plane representations of the abstract dual will fail the face conditions. For the example shown, this is easy, but for larger graphs, this kind of argument poses a problem.

Future work on improving the planarising step in this algorithm should seek plane representations which pass the face-conditions, if they exist.

Given an abstract labelled graph which passes the connectivity conditions, and a plane graph representation which passes the face-conditions, to complete the construction of the contours we draw the plane graph with all faces convex and draw arcs linear across faces. Of course, it is impossible to draw a graph with *all* faces convex, but the faces can be placed as convex faces in a disc with one new non-convex outside face. This step, called *circularisation* of the dual, is discussed in the next section.

2.4   Circularisation

Given a concrete labelled graph which passes the connectivity and face conditions, we seek a plane representation with convex faces. The barycentric approach to graph drawing [1] gives all but one face convex. Circularisation reproduces all the faces of a graph inside a disc, introducing a new (non-convex) face outside the disc. Some vertices of the original graph may appear more than once, and some edges will be duplicated. Circularisation can be achieved by taking faces

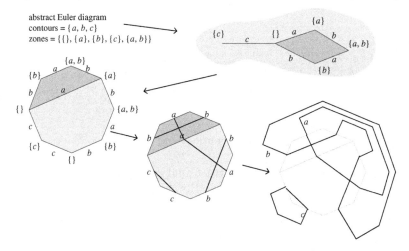

Fig. 10. The circularisation process and addition of arcs

in turn, inserting them into a disc. After the first face, further faces are chosen and inserted by identifying common edges which are already present in the disc. The edges which remain un-identified after all faces are inserted are the edges around the edge of the disc (polygon). These correspond to edges making up a spanning tree of the original graph.

An alternative view of the same process begins by choosing a spanning tree of the original graph. Split the edges into pairs, and fatten the spanning tree into a new polygonal face. Move the infinity point into this new face and squeeze the original faces so that all vertices lie around a circle (the boundary of the new face). Edges in the chosen spanning tree are duplicated around the outside of the circularised graph whereas edges not in the spanning tree appear once each, across the disc.

The process of *circularisation* is illustrated in figure 10 as the second step (top right to bottom left diagram): a dual graph with two faces is circularised to give two faces in a disc and one additional non-convex face outside the disc.

The rewards of circularisation come from the construction of the concrete Euler contours. As in figure 5, arcs will be constructed across faces of the concrete labelled graph, but now the arcs can be drawn as straight lines. A $2n$-sided face will have n arcs drawn across it. The arcs join the midpoints of edges with shared labels. They are guaranteed to only cross other arcs in the same face (because the faces are convex), and they are guaranteed to meet transversely (because they are linear). The face-conditions guarantee that no new zones will appear as the linear arcs are added.

The added arcs contribute to piecewise-linear contours of the required concrete diagram. The contours need to be completed by drawing arcs outside the disc (the last step shown in figure 10). Starting at a vertex labelled {}, read the labels from the edges around the circle, giving a word of contour labels. This word is made up of nested pairs of contour labels. In figure 10, the nested word is *babbabcc*. Use the nesting to determine which labels are pairs. The labels which pair identify edges which will be joined by arcs outside the disc. Draw arcs outside the disc joining the innermost pairs: *bb* and *cc*. Join the two edges labelled *a* and join the remaining two edges labelled *b*.

The algorithm to this point has been implemented in the Java programming language, with outcomes shown in figure 12. The results can be difficult to interpret by eye. The final proposition suggests one resolution of this problem. The resulting concrete diagram can appear somewhat convoluted. Measure

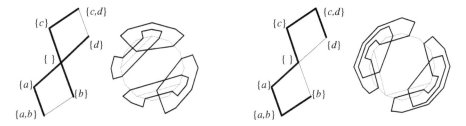

Fig. 11. Spanning trees and circularisation

the lengths of arcs drawn outside the disc, and take this to be a measure of convolution of the concrete diagram. Recall that the decisions made during circularisation give a spanning tree of the dual.

Proposition 5. *Convolution of the concrete diagram is minimised by choosing a spanning tree with the vertex labelled {} as its centre, with vertices as close as possible to the centre (for centre of a tree, see [3]).*

3   Conclusions

Figure 12 shows the outcome for all drawable diagrams with two or three contours. Some diagrams could be made less convoluted by application of the span-

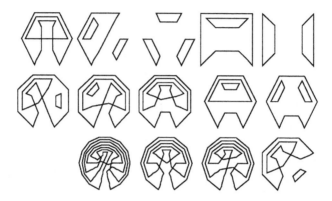

Fig. 12. Program output

ning tree with null centre (see prop 5). All other abstract diagrams with two or three contours were determined to be undrawable by testing for connectivity conditions and face-conditions.

The algorithm has been proved to be practical in small cases, as it has been implemented in the Java programming language. The program accepts a string description of a set of sets as zone descriptors. It first constructs the super-dual graph (def 11) before checking the connectivity conditions (thm 1). If the connectivity conditions pass, then edges are removed to assist with an iterative planarising step (sec 2.2). If the face-cycle conditions (cor 2) fail, then alternative planarisations are sought. If a planar representation of the dual is found which passes the face conditions, then the circularisation process (sec 2.4) is applied to construct contours and resulting concrete diagram is drawn.

Remaining questions include:

(i) If an abstract diagram has a non-planar super-dual (e.g. the Venn diagram on four contours), what is a good strategy for selecting edges (maintaining connectivity conditions) to get a planar dual graph?

(ii) What is the most effective planarising algorithm for a labelled dual graph?

(iii) Is it possible to adapt the algorithm to allow inclusion of triple points, or new zones, in the concrete diagrams? This would make more abstract diagrams drawable.

(iv) How can the resulting concrete diagram be manipulated to maintain the topological properties of contour intersection, and enhance the clarity of the concrete diagram?(prettification)

One intended application of this work is to enable a constraint diagram tool to construct appropriate diagrams which are equivalent to textual constraints in software modelling. The kinds of constraints which occur normally have few contours. If a diagram is considered as a set of "nested" components - diagrams within diagrams - then the number of contours will be further reduced. The algorithm as it is presented here is effective for such small examples, and after more work on the planarising step, would be practical for larger examples too.

Current work includes the study of existing Venn diagram algorithms to address edge-removal and smart planarising steps for Euler duals. We are also looking at "nesting" diagrams and the impact this has on layout algorithms.

Thanks to John Taylor, Gem Stapleton, and the conference referees for constructive comments on drafts of this article. This work was partially supported by UK EPSRC grant GR/R63516.

References

1. G. Di Battista, P. Eades, R. Tamassia, I. G. Tollis. *Graph Drawing: algorithms for the visualization of graphs.* Prentice Hall, 1999.
2. F. Bertault, P. Eades. *Drawing hypergraphs in the subset standard.* Springer Verlag, Graph Drawing proceedings, LNCS 1984, 2000.
3. N. Biggs, E. K. Lloyd, R. J. Wilson. *Graph Theory 1736-1936* OUP, 1976.
4. L. Euler. *Lettres a Une Princesse d'Allemagne,* vol 2. 1761. Letters No. 102–108.
5. J. Gil, J. Howse, S. Kent. Formalising Spider Diagrams. *Proc. IEEE Symposium on Visual Languages (VL99),* Tokyo, Sept 1999. IEEE Comp Soc Press, 130-137.
6. J. Gil, J. Howse, S. Kent, J. Gil. Towards a formalisation of constraint diagrams. Proc IEEE Symp on Human-Centric computing (HCC'01). Stresa, Sept 2001, p.72-79.
7. E. Hammer. *Logic and Visual Information.* CSLI Publications, Stanford, 1995.
8. J. Howse, F. Molina, J. Taylor. On the completeness and expressiveness of spider diagram systems. *Proc. Diagrams 2000,* Edinburgh, Sept 2000. LNAI 1889, Springer-Verlag, 26-41.
9. J. Howse, F. Molina, S.-J. Shin, J. Taylor. On diagram tokens and types. Accepted for Diagrams 2002.
10. S. Kent. Constraint diagrams: Visualising invariants in object oriented models. *proceedings of OOPSLA97,* ACM SIGPLAN Notices 32, 1997.
11. F. Molina. Reasoning with extended Venn-Peirce diagrammatic Systems. PhD Thesis, University of Brighton, 2001.
12. Object Management Group. UML Specification, Version 1.3: www.omg.org.
13. C. Peirce. *Collected Papers.* Harvard University Press, 1933.
14. M. Schaefer and D. Štefanikovič. *Decidability of string graphs.* Proc 33rd ACM Symposium on the Thy of Comp, p.241-246, 2001.
15. S.-J. Shin. *The Logical Status of Diagrams.* CUP, 1994.
16. J. Venn. On the diagrammatic and mechanical representation of propositions and reasonings. *Phil.Mag.,* 1880. 123.
17. J. Warmer and A. Kleppe. *The Object Constraint Language: Precise Modeling with UML.* Addison-Wesley, 1998.

Corresponding Regions in Euler Diagrams

John Howse, Gemma Stapleton, Jean Flower, and John Taylor

School of Computing & Mathematical Sciences
University of Brighton, Brighton, UK
{John.Howse,G.E.Stapleton,J.A.Flower,John.Taylor}@bton.ac.uk
http://www.it.bton.ac.uk/research/vmg/VisualModellingGroup.html

Abstract. Euler diagrams use topological properties to represent set-theoretical concepts and thus are 'intuitive' to some people. When reasoning with Euler diagrams, it is essential to have a notion of correspondence among the regions in different diagrams. At the semantic level, two regions correspond when they represent the same set. However, we wish to construct a purely syntactic definition of corresponding regions, so that reasoning can take place entirely at the diagrammatic level. This task is interesting in Euler diagrams because some regions of one diagram may be missing from another. We construct the correspondence relation from 'zones' or minimal regions, introducing the concept of 'zonal regions' for the case in which labels may differ between diagrams. We show that the relation is an equivalence relation and that it is a generalization of the counterpart relations introduced by Shin and Hammer.

1   Introduction

Euler diagrams [1] illustrate relations between sets. This notation uses topological properties of enclosure, exclusion and intersection to represent the set-theoretic notions of subset, disjoint sets, and intersection, respectively. The diagram d_2 in figure 1 is an Euler diagram with interpretation A is disjoint from B. Venn [13] adapted Euler's notation to produce a system of diagrams representing logical propositions. In a Venn diagram all intersections between contours must occur. The diagram d_1 in figure 1 is a Venn diagram. Some extensions of Euler diagrams allow shading, as in Venn diagrams, but since we are interested in a syntactic correspondence between regions shading is irrelevant. Thus we treat Venn diagrams as a special case of Euler diagrams, and ignore shading. Peirce [10] extended Venn's notation to include existential quantification and disjunctive information.

Shin [11] developed sound and complete reasoning rules for a system of Venn-Peirce diagrams. This work was seminal in that the rules were stated at the diagrammatic level and all reasoning took place at that level. This was the first complete formal diagrammatic reasoning system; until then diagrammatic reasoning was a mixture of informal reasoning at the diagrammatic level and formal (and informal) reasoning at the semantic level. Hammer [3] developed a sound and complete set of reasoning rules for a simple Euler system; it only

M. Hegarty, B. Meyer, and N. Hari Narayanan (Eds.): Diagrams 2002, LNAI 2317, pp. 76–90, 2002.
© Springer-Verlag Berlin Heidelberg 2002

considered inferences from a single diagram and contained only three reasoning rules.

In order to compare regions in different diagrams, Shin and Hammer developed counterpart relations [4,11]. This paper considers an alternative, but related, approach to these counterpart relations and generalizes it to comparing regions in Euler diagrams. The solution of this problem is very important in extending diagrammatic reasoning to systems which have practical applications. Euler diagrams form the basis of more expressive diagrammatic notations such as Higraphs [5] and constraint diagrams [2], which have been developed to express logical properties of systems. These notations are used in the software development process, particularly in the modelling of systems and frequently as part of, or in conjunction with, UML [9]. Indeed, some of the notations of UML are based on Euler diagrams. The development of software tools to aid the software development process is very important and it is essential that such tools work at the diagrammatic level and not at the underlying semantic level so that feedback is given to developers in the notations that they are using and not in some mathematical notation that the developers may find difficult to understand. Thus it is necessary to construct a purely syntactic definition of corresponding regions across diagrams.

The task of defining such a correspondence relation is interesting, and very much non-trivial, in Euler diagrams because some regions of one diagram may be missing from another. For example, in figure 1 the region within the contours A and B in d_1 is missing from d_2. Diagram d_1 asserts that $A \cap B$ may or may not be empty, whereas d_2 asserts that $A \cap B = \emptyset$. What are the corresponding regions in this case?

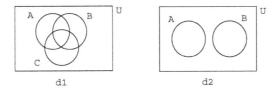

Fig. 1. A Venn diagram and an Euler diagram.

In §2 we give a concise informal description of Euler diagrams and a formal definition of its syntax. In §3 we define the correspondence relation between regions in the more straightforward case of Venn diagrams. In §4 we discuss the problems of defining corresponding regions in Euler diagrams and in the particularly difficult case of a system involving the disjunction of diagrams, before giving a general definition of the correspondence relation and showing that it is an equivalence relation. We then show, in §5, that it is a generalization of the counterpart relations developed by Shin and Hammer.

2   Syntax of Euler Diagrams

We now give a concise informal description of Euler diagrams. A *contour* is a simple closed plane curve. A *boundary rectangle* properly contains all other contours. Each contour has a unique label. A *district* (or *basic region*) is the bounded area of the plane enclosed by a contour or by the boundary rectangle. A *region* is defined, recursively, as follows: any district is a region; if r_1 and r_2 are regions, then the union, intersection and difference of r_1 and r_2 are regions provided these are non-empty. A *zone* (or *minimal region*) is a region having no other region contained within it. Contours and regions denote (possibly empty) sets. Every region is a union of zones. In figure 2 the zone within A, but outside B is missing from the diagram; the set denoted by such a "missing" zone is empty. An Euler diagram containing all possible zones is called a Venn diagram.

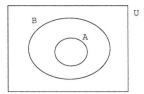

Fig. 2. An Euler diagram.

Given two diagrams we can connect them with a straight line to produce a *compound diagram* [6]. This connection operation is interpreted as the disjunction of the connected diagrams. A multi-diagram is a collection of compound diagrams and is interpreted as the conjunction of the compound diagrams. In this system a multi-diagram is in conjunctive normal form (cf. Shin's Venn II system [11]). In figure 3 diagrams d_1 and d_2 are to be taken in disjunction, thus $\{d_1, d_2\}$ is a compound diagram, as is $\{d_3\}$ (any unitary diagram is a compound diagram); the diagram $\{\{d_1, d_2\}, \{d_3\}\}$ is a multi-diagram.

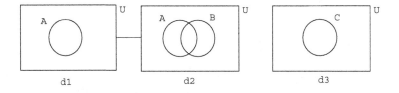

Fig. 3. Two compound diagrams.

A *unitary Euler diagram* is a tuple $d = \langle L, U, Z \rangle = \langle L(d), U(d), Z(d) \rangle$ whose components are defined as follows:

1. L is a finite set whose members are called *contours*. The element U, which is not a member of L, is called the *boundary rectangle*.
2. The set $Z \subseteq \mathbb{P}L$ is the set of *zones*. A zone $z \in Z$ is *incident* on a contour $c \in L$ if $c \in z$. Let $R = \mathbb{P}Z - \emptyset$ be the set of *regions*.

If $Z = \mathbb{P}L$, d is a *Venn diagram*. At this level of abstraction we identify a contour and its label. A zone is defined by the contours that contain it and is thus represented as a set of contours. The set of labels of a zone, z, is thus $L(z) = z$. A region is just a non-empty set of zones. The Euler diagram d in figure 2 has $L(d) = \{A, B\}$ and $Z(d) = \{\emptyset, \{B\}, \{A, B\}\}$.

A *compound diagram*, D, is a finite set of unitary diagrams taken in disjunction. A *multi-diagram*, Δ, is a finite set of compound diagrams taken in conjunction [6]. The set of labels of a compound diagram, D, is $L(D) = \bigcup\limits_{d \in D} L(d)$. The set of labels of a multi-diagram, Δ, is $L(\Delta) = \bigcup\limits_{D \in \Delta} L(D)$. In figure 3

$$L(\{\{d_1, d_2\}, \{d_3\}\}) = \{A, B, C\}.$$

3   Venn Diagrams

We will identify corresponding regions across Venn diagrams that do not necessarily have the same label sets. As an example, region $\{z_1, z_2, z_3, z_4\}$ in d_1 and region $\{z_5, z_6\}$ in d_2 in figure 4 are corresponding. We introduce the concept of a zonal region in order to identify this formally. Intuitively a zonal region is a region that becomes a zone when contours are removed. This is illustrated in figure 4. The contour with label C is removed and region $\{z_1, z_2\}$ becomes a zone, $\{z_5\}$, in the second diagram.

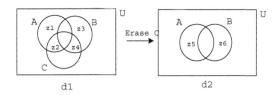

Fig. 4. Two Venn diagrams with different label sets.

3.1   Zonal Regions and Splits

In figure 4, consider how we might describe or identify the region $\{z_1, z_2\}$. Informally, it has description 'everything inside A but outside B'. Thus we associate $\{z_1, z_2\}$ with an ordered pair of sets, $\{A\}$ and $\{B\}$, which we shall write as

$\langle\{A\},\{B\}\rangle$. Similarly the region $\{z_1\}$ is associated with $\langle\{A\},\{B,C\}\rangle$, intuitively meaning 'everything inside A, but outside B and C'. In order to define zonal regions formally, and to allow us to compare regions across diagrams, we introduce the notion of a 'split'.

Definition 1. *A* **split** *is a pair of sets,* $\langle P,Q\rangle$, *such that* $P\cap Q=\emptyset$; *if* $P\cup Q\subseteq X$ *then* $\langle P,Q\rangle$ *is said to be a* **split on** X.

Addition is defined on splits with the following axioms:

1. $\langle P_1,Q_1\rangle=\langle P_2,Q_2\rangle\Leftrightarrow P_1=P_2\wedge Q_1=Q_2$
2. $\forall A\notin P\cup Q,\quad\langle P,Q\rangle=\langle P\cup\{A\},Q\rangle+\langle P,Q\cup\{A\}\rangle$
3. $\sum_{i=1}^{n}\langle P_i,Q_i\rangle=\sum_{j=1}^{m}\langle R_j,S_j\rangle$ if $\forall i\exists j\bullet\langle P_i,Q_i\rangle=\langle R_j,S_j\rangle$ and $\forall j\exists i\bullet\langle R_j,S_j\rangle=\langle P_i,Q_i\rangle$

Lemma 1. *Addition is commutative and associative. Each element is idempotent. If* $\langle P,Q\rangle$ *is a split and* S *is a finite set such that* $(P\cup Q)\cap S=\emptyset$ *then*

$$\langle P,Q\rangle=\sum_{W\subseteq S}\langle P\cup W,Q\cup(S-W)\rangle$$

This lemma follows from axioms 2 and 3. The last part of the lemma generalizes axiom 2 and is illustrated below.

$$\langle\{A\},\{B\}\rangle=\sum_{W\subseteq\{C,D\}}\langle\{A\}\cup W,\{B\}\cup(\{C,D\}-W)\rangle$$
$$=\langle\{A\},\{B,C,D\}\rangle+\langle\{A,D\},\{B,C\}\rangle+$$
$$\langle\{A,C\},\{B,D\}\rangle+\langle\{A,C,D\},\{B\}\rangle$$

Definition 2. *For unitary Venn diagram* d, *let* $\langle P,Q\rangle$ *be a split on* $L(d)$. *Then the* **zonal region** *associated with* $\langle P,Q\rangle$ *is*

$$\{z\in Z(d):P\subseteq L(z)\wedge Q\subseteq\overline{L(z)}\}$$

where $\overline{L(z)}=L(d)-L(z)$, [8].

In figure 5, zonal regions $\{z_1\}$, $\{z_2\}$, $\{z_3\}$ and $\{z_4\}$ are associated with $\langle\{A\},\{B,C,D\}\rangle$, $\langle\{A,D\},\{B,C\}\rangle$, $\langle\{A,C,D\},\{B\}\rangle$ and $\langle\{A,C,\},\{B,D\}\rangle$. The zonal region $\{z_1,z_2,z_3,z_4\}$ is associated with $\langle\{A\},\{B\}\rangle$. We have $\{A\}=L(z_1)\cap L(z_2)\cap L(z_3)\cap L(z_4)$ and $\{B\}=\overline{L(z_1)}\cap\overline{L(z_2)}\cap\overline{L(z_3)}\cap\overline{L(z_4)}$.

Lemma 2. *For any unitary Venn diagram* d, *if a zonal region* zr *is associated with* $\langle P,Q\rangle$ *then* $P=\bigcap_{z\in zr}L(z)$ *and* $Q=\bigcap_{z\in zr}\overline{L(z)}$.

Hence each zonal region is associated with a unique split. There is a parallel between axiom 2, $\langle P,Q\rangle=\langle P\cup\{A\},Q\rangle+\langle P,Q\cup\{A\}\rangle$, and lemma 3 below.

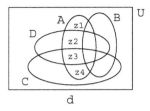

Fig. 5. Venn-4.

Lemma 3. *The split rule. Let zr be a zonal region of unitary Venn diagram d. If zr is associated with $\langle P, Q \rangle$ and $A \in L(d) - (P \cup Q)$ then $zr = zr_1 \cup zr_2$ where zr_1 and zr_2 are zonal regions associated with $\langle P \cup \{A\}, Q \rangle$ and $\langle P, Q \cup \{A\} \rangle$ respectively.*

For example, consider the diagram in figure 6. Zonal regions $zr_1 = \{z_2, z_4, z_5, z_6\}$, $zr_2 = \{z_2, z_4\}$ and $zr_3 = \{z_5, z_6\}$ are associated with $\langle \{B\}, \emptyset \rangle$, $\langle \{A, B\}, \emptyset \rangle$ and $\langle \{B\}, \{A\} \rangle$ respectively. From the split rule $zr_1 = zr_2 \cup zr_3$ and from axiom 2,

$$\langle \{B\}, \emptyset \rangle = \langle \{A, B\}, \emptyset \rangle + \langle \{B\}, \{A\} \rangle.$$

Informally, this is splitting zr_1 into the part contained in A and the part excluded from A. A more general version of the split rule now follows.

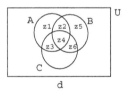

Fig. 6. Venn-3.

Corollary 1. *The derived split rule. If zr is a zonal region of unitary Venn diagram d associated with $\langle P, Q \rangle$ and $S \subseteq L(d) - (P \cup Q)$ then*

$$zr = \bigcup_{W \subseteq S} zr_W$$

where zr_W is the zonal region associated with $\langle P \cup W, Q \cup (S - W) \rangle$.

Taking $zr = \{z_1, z_2, z_3, z_4\}$ to be the zonal region associated with $\langle \{A\}, \emptyset \rangle$ in figure 6, using the derived spit rule with $S = \{B, C\}$ gives

$$zr = \{z_1\} \cup \{z_2\} \cup \{z_3\} \cup \{z_4\}$$

since $\{z_1\}$, $\{z_2\}$, $\{z_3\}$ and $\{z_4\}$ are associated with $\langle\{A\}\cup\emptyset, \emptyset\cup\{B,C\}\rangle$, $\langle\{A\}\cup \{B\},\emptyset \cup \{C\}\rangle$, $\langle\{A\} \cup \{C\},\emptyset \cup \{B\}\rangle$ and $\langle\{A\} \cup \{B,C\},\emptyset \cup \emptyset\rangle$ respectively. In general, if we set $S = L(d) - (P \cup Q)$ in the lemma above and take $zr = \{z_1, z_2, ..., z_n\}$ we get $zr = \bigcup_{i=1}^{n}\{z_i\}$.

Note that the split associated with a zone involves all the labels in the diagram: $\{z\}$ is associated with $\langle P,Q\rangle$ where $P = L(z)$ and $Q = \overline{L(z)} = L(d) - L(z)$. Since any region is a set of zones, we can use this to define a function, ρ, from regions to splits.

Definition 3. *Let d be a unitary Venn diagram.*

(i) *If $z \in Z(d)$ then $\rho(\{z\}) = \left\langle L(z), \overline{L(z)} \right\rangle$*

(ii) *If $r = \{z_1, z_2, ..., z_n\} \in R(d)$ then $\rho(r) = \sum_{i=1}^{n}\rho(\{z_i\})$*

For example, in figure 6, $\rho(\{z_1\}) = \langle\{A\}, \{B,C\}\rangle$. Under ρ the region $\{z_3, z_5, z_6\}$ maps to

$$\langle\{A,C\},\{B\}\rangle + \langle\{B\},\{A,C\}\rangle + \langle\{B,C\},\{A\}\rangle = \langle\{A,C\},\{B\}\rangle + \langle\{B\},\{A\}\rangle$$

Lemma 4. *Let zr be a zonal region of unitary Venn diagram d associated with $\langle P,Q\rangle$. Then $\rho(zr) = \langle P,Q\rangle$.*

The zonal region $\{z_1, z_2, z_3, z_4\}$ in figure 6 is associated with $\langle\{A\},\emptyset\rangle$ and

$$\begin{aligned}
\rho(\{z_1, z_2, z_3, z_4\}) &= \langle\{A\}, \{B,C\}\rangle + \langle\{A,B\}, \{C\}\rangle + \\
&\quad \langle\{A,C\}, \{B\}\rangle + \langle\{A,B,C\}, \emptyset\rangle \\
&= \langle\{A\}, \{C\}\rangle + \langle\{A,C\}, \emptyset\rangle \\
&= \langle\{A\}, \emptyset\rangle
\end{aligned}$$

Lemma 4 does not follow over to Euler diagrams, as we shall see in section 4. If we know certain relationships between zonal regions, we can make deductions about their images under ρ, and vice versa.

Lemma 5. *Let zr_1 and zr_2 be zonal regions of unitary Venn diagram d. If $\rho(zr_1) = \langle P_1, Q_1\rangle$ and $\rho(zr_2) = \langle P_2, Q_2\rangle$ then*

$$zr_1 \subseteq zr_2 \Leftrightarrow P_2 \subseteq P_1 \wedge Q_2 \subseteq Q_1$$

If $(P_1 \cup P_2) \cap (Q_1 \cup Q_2) = \emptyset$ then $zr_1 \cap zr_2 = zr_3$ where

$$\rho(zr_3) = \langle P_1 \cup P_2, Q_1 \cup Q_2\rangle$$

From lemma 5 we can deduce that the zonal region associated with $\langle\{A\},\emptyset\rangle$ in diagram d, figure 6, is not a subset of the zonal region associated with $\langle\{B\},\emptyset\rangle$. The zonal region associated with $\langle\{A\},\emptyset\rangle$ is $\{z_1, z_2, z_3, z_4\}$. The zonal region associated with $\langle\{B\},\emptyset\rangle$ is $\{z_2, z_4, z_5, z_6\}$ and $\{z_1, z_2, z_3, z_4\} \not\subseteq \{z_2, z_4, z_5, z_6\}$. Lemma 5 also tells us the zonal regions associated with $\langle\{A\},\emptyset\rangle$ and $\langle\{B\},\emptyset\rangle$ intersect to give a zonal region associated with $\langle\{A, B\},\emptyset\rangle$, that is $\{z_1, z_2, z_3, z_4\} \cap \{z_2, z_4, z_5, z_6\} = \{z_2, z_4\}$. We now define correspondence between zonal regions. Corresponding zonal regions have the same semantic interpretation.

Definition 4. *Let zr_1 and zr_2 be zonal regions of Venn diagrams d_1 and d_2 respectively. Regions zr_1 and zr_2 are **corresponding** zonal regions [8], denoted $zr_1 \equiv_c zr_2$, if and only if $\rho(zr_1) = \rho(zr_2)$.*

In figure 4 zonal region $\{z_1, z_2\}$ in d_1 corresponds to zonal region $\{z_5\}$ in d_2 since $\rho(\{z_1, z_2\}) = \rho(\{z_5\}) = \langle\{A\},\{B\}\rangle$.

Theorem 1. *The relation $\equiv_c$ is an equivalence relation on zonal regions.*

3.2   Corresponding Regions in Venn Diagrams

The definition of correspondence is now extended to regions.

Definition 5. *Let r_1 and r_2 be a regions of Venn diagrams d_1 and d_2 respectively. Regions r_1 and r_2 are **corresponding** regions [8], denoted $r_1 \equiv_c r_2$, if and only if $\rho(r_1) = \rho(r_2)$.*

At the semantic level, corresponding regions represent the same set [6]. In figure

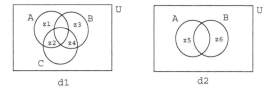

Fig. 7. Two Venn diagrams.

7, region $r_1 = \{z_1, z_2, z_3, z_4\}$ in diagram d_1 has

$$\rho(r_1) = \langle\{A\},\{B, C\}\rangle + \langle\{A, C\},\{B\}\rangle + \langle\{B\},\{A, C\}\rangle + \langle\{B, C\},\{A\}\rangle$$

Region $r_2 = \{z_5, z_6\}$ in diagram d_2 has

$$\rho(r_2) = \langle\{A\},\{B\}\rangle + \langle\{B\},\{A\}\rangle$$

Using axiom 2, $\langle P, Q\rangle = \langle P \cup \{A\}, Q\rangle + \langle P, Q \cup \{A\}\rangle$, we obtain

$$\langle\{A\},\{B\}\rangle + \langle\{B\},\{A\}\rangle =$$
$$\langle\{A\},\{B, C\}\rangle + \langle\{A, C\},\{B\}\rangle + \langle\{B\},\{A, C\}\rangle + \langle\{B, C\},\{A\}\rangle$$

Thus $r_1 \equiv_c r_2$.

Theorem 2. *The relation $\equiv_c$ is an equivalence relation on regions.*

Proofs for some of the results in this section can be found in [12]. Ideally, we want to be able to reason with diagrams that are Euler diagrams. The focus of this paper now turns to diagrams of this nature. Of the definitions related to Venn diagrams, 2 and 3 carry over to Euler diagrams. Also lemma 3 and corollary 1 apply to Euler diagrams.

4   Euler Diagrams

In this section we investigate problems related to zonal regions and their associated splits in Euler diagrams. It is no longer necessarily true that, for a zonal region zr associated with $\langle P, Q \rangle$, $\rho(zr) = \langle P, Q \rangle$ because the associated $\langle P, Q \rangle$ is no longer unique. In figure 8, the zonal region $\{z_1\}$ is associated with both $\langle \{A\}, \emptyset \rangle$ and $\langle \{A\}, \{B\} \rangle$ but $\rho(\{z_1\}) = \langle \{A\}, \{B\} \rangle \neq \langle \{A\}, \emptyset \rangle$. Thus lemma 4 fails. However, we can think of $\langle \{A\}, \emptyset \rangle$ and $\langle \{A\}, \{B\} \rangle$ as being 'equivalent in the context of d' because $\langle \{A\}, \emptyset \rangle = \langle \{A, B\}, \emptyset \rangle + \langle \{A\}, \{B\} \rangle$ and the zone corresponding to $\langle \{A, B\}, \emptyset \rangle$ is 'missing' from the diagram.

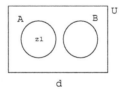

Fig. 8. An Euler diagram with a missing zonal region.

In some diagrams there may be a split on $L(d)$ with no zonal region associated with it. There is no zonal region associated with $\langle \{A, B\}, \emptyset \rangle$, in diagram d, in figure 8. Informally, in our 'algebra of splits' we can think of $\langle \{A, B\}, \emptyset \rangle$ as representing zero. If we allow this, we see that

$$\rho(\{z_1\}) = \langle \{A\}, \{B\} \rangle$$
$$= \langle \{A\}, \{B\} \rangle + \langle \{A, B\}, \emptyset \rangle$$
$$= \langle \{A\}, \emptyset \rangle$$

We have here the idea of equality *in the context of* a diagram.

Definition 6. *The **context** of unitary diagram d, denoted $\chi(d)$, is*

$$\chi(d) = \{ \langle P, Q \rangle : P \in \mathbb{P}L(d) - Z(d) \wedge Q = L(d) - P \}$$

*If $\langle P, Q \rangle \in \chi(d)$ then $\langle P, Q \rangle$ is **zero in the context of** d, denoted $\langle P, Q \rangle =_d 0$.*

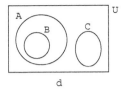

Fig. 9. A unitary Euler diagram.

The diagram in figure 9 has $Z(d) = \{\emptyset, \{A\}, \{A, B\}, \{C\}\}$, so

$$\chi(d) = \{\langle\{B\}, \{A, C\}\rangle, \langle\{A, C\}, \{B\}\rangle, \langle\{B, C\}, \{A\}\rangle, \langle\{A, B, C\}, \emptyset\rangle\}$$

corresponding to the four zones that are present in the Venn diagram with labels $\{A, B, C\}$ but are missing in d.

Lemma 6. *If d is a unitary Venn diagram $\chi(d) = \emptyset$.*

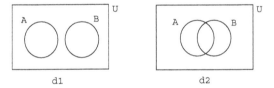

Fig. 10. Two Euler diagrams.

When considering more than one diagram, we need to take care when deciding what is the context. Considering diagrams d_1 and d_2, figure 10, in *conjunction* we may deduce that $\langle\{A, B\}, \emptyset\rangle$ is zero in context, since

$$\{z \in Z(d_1) : \{A, B\} \subseteq L(z) \land \emptyset \subseteq \overline{L(z)}\} = \emptyset$$

At the semantic level, the sets represented by the contours labelled A and B are disjoint. Thus we would want the zonal region associated with $\langle\{A\}, \emptyset\rangle$ in d_2 to correspond to that associated with $\langle\{A\}, \{B\}\rangle$, also in d_2. However if we were to take the diagrams in *disjunction*, we cannot deduce that the sets represented by the contours labelled A and B are disjoint. Thus we would not want $\langle\{A, B\}, \emptyset\rangle$ to be zero. In the disjunctive case it is incorrect for $\rho(\{z_1\}) = \langle\{A\}, \emptyset\rangle$.

In order to define the context of compound and multi-diagrams we first define a function ζ_δ, called *zonify*, from splits on $L(\delta)$ to sets of splits on $L(\delta)$, where δ is a unitary, compound or multi-diagram,

$$\zeta_\delta(\langle P, Q\rangle) = \{\langle P_i, Q_i\rangle : P \subseteq P_i \land Q_i = L(\delta) - P_i\}$$

The zonify function delivers the set of splits corresponding to the zones that are elements of the zonal region associated with $\langle P, Q \rangle$ in the Venn diagram with labels $L(\delta)$. Taking $\Delta = \{\{d_1, d_2\}, \{d_3\}\}$ in figure 11,

$$\zeta_\Delta(\langle\{A\}, \{B\}\rangle) = \{\langle\{A\}, \{B, C\}\rangle, \langle\{A, C\}, \{B\}\rangle\}$$

Consider the compound diagram $D = \{d_1, d_2\}$ in figure 12. The shaded zones

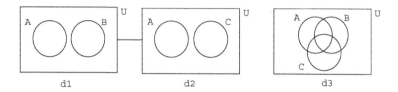

Fig. 11. A multi-diagram.

in the Venn diagram, d, with $L(d) = L(D)$, represent those sets we can deduce empty at the semantic level. Each of these shaded zones is associated with a split that partitions $L(D)$.

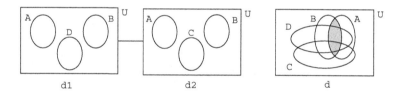

Fig. 12. A compound diagram and a Venn diagram.

Definition 7. *Let D be a compound diagram. The **context** of D is defined to be*

$$\chi(D) = \bigcap_{d \in D} \left(\bigcup_{\langle P, Q \rangle \in \chi(d)} \zeta_D(\langle P, Q \rangle) \right)$$

*If $\langle P, Q \rangle \in \chi(D)$ then $\langle P, Q \rangle$ is **zero in the context of** D, denoted $\langle P, Q \rangle =_D 0$.*

The context of $\{d_1, d_2\}$ in figure 13 is $\chi(\{d_1, d_2\}) = \emptyset$, since $\chi(d_2) = \emptyset$. The contexts of diagrams d_3 and d_4 are

$$\chi(d_3) = \{\langle\{A\}, \{B\}\rangle\}$$
$$\chi(d_4) = \{\langle\{A, C\}, \{B\}\rangle, \langle\{A\}, \{B, C\}\rangle\}$$

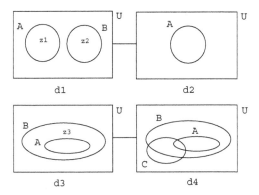

Fig. 13. A multi-diagram containing two compound diagrams.

We use the zonify function to find $\chi(\{d_3, d_4\})$.

$$\chi(\{d_3, d_4\}) = \left(\bigcup_{\langle P, Q \rangle \in \chi(d_3)} \zeta_D(\langle P, Q \rangle) \right) \cap \left(\bigcup_{\langle P, Q \rangle \in \chi(d_4)} \zeta_D(\langle P, Q \rangle) \right)$$

$$= \{\langle \{A, C\}, \{B\} \rangle, \langle \{A\}, \{B, C\} \rangle\} \cap$$
$$(\{\langle \{A, C\}, \{B\} \rangle\} \cup \{\langle \{A\}, \{B, C\} \rangle\})$$

$$= \{\langle \{A, C\}, \{B\} \rangle, \langle \{A\}, \{B, C\} \rangle\}$$

Definition 8. *Let Δ be a multi-diagram. The **context** of Δ is defined to be*

$$\chi(\Delta) = \bigcup_{D \in \Delta} \left(\bigcup_{\langle P, Q \rangle \in \chi(D)} \zeta_\Delta(\langle P, Q \rangle) \right)$$

*If $\langle P, Q \rangle \in \chi(\Delta)$ then $\langle P, Q \rangle$ is **zero in the context of** Δ, denoted $\langle P, Q \rangle =_\Delta 0$.*

The context of $\Delta = \{\{d_1, d_2\}, \{d_3, d_4\}\}$ in figure 13 is

$$\chi(\Delta) = \{\langle \{A, C\}, \{B\} \rangle, \langle \{A\}, \{B, C\} \rangle\}$$

Therefore $\langle \{A, C\}, \{B\} \rangle =_\Delta 0$ and $\langle \{A\}, \{B, C\} \rangle =_\Delta 0$.

Definition 9. *Let Δ be a multi-diagram, $\sum_{i=1}^{n} \langle P_i, Q_i \rangle$ and $\sum_{j=1}^{m} \langle P'_j, Q'_j \rangle$ be sums of splits. $\sum_{i=1}^{n} \langle P_i, Q_i \rangle$ and $\sum_{j=1}^{m} \langle P'_j, Q'_j \rangle$ are said to be **equal in the context of** Δ, denoted $\sum_{i=1}^{n} \langle P_i, Q_i \rangle =_\Delta \sum_{j=1}^{m} \langle P'_j, Q'_j \rangle$, if and only if there exists $\sum_{i=1}^{k} \langle R_i, S_i \rangle$ and $\sum_{j=1}^{l} \langle R'_j, S'_j \rangle$ such that*

$$\sum_{i=1}^{n} \langle P_i, Q_i \rangle = \sum_{i=1}^{k} \langle R_i, S_i \rangle, \ \sum_{j=1}^{m} \langle P'_j, Q'_j \rangle = \sum_{j=1}^{l} \langle R'_j, S'_j \rangle \ and$$

$$\forall i \, (\exists j \, \bullet \, \langle R_i, S_i \rangle = \langle R'_j, S'_j \rangle) \vee \langle R_i, S_i \rangle =_\Delta 0 \ and$$

$$\forall j \, (\exists i \, \bullet \, \langle R'_j, S'_j \rangle = \langle R_i, S_i \rangle) \vee \langle R'_j, S'_j \rangle =_\Delta 0.$$

In figure 13, taking $\Delta = \{\{d_1, d_2\}, \{d_3, d_4\}\}$ we have,

$$\langle\{A\}, \{B\}\rangle + \langle\{B\}, \{A\}\rangle \quad = \quad \langle\{A, C\}, \{B\}\rangle + \langle\{A\}, \{B, C\}\rangle + \langle\{B\}, \{A\}\rangle$$
$$=_\Delta \langle\{B\}, \{A\}\rangle$$

Definition 10. *Let Δ be a multi-diagram and d_1, d_2 be unitary diagrams such that $d_1 \in D_1, d_2 \in D_2$ where $\{D_1, D_2\} \subseteq \Delta$. Let r_1 and r_2 be regions of d_1 and d_2 respectively. Region r_1 is said to* **correspond in the context of** Δ *to region r_2, denoted $r_1 \equiv_\Delta r_2$, if and only if $\rho(r_1) =_\Delta \rho(r_2)$.*

Corresponding regions have the same semantic interpretation. Consider regions $r_1 = \{z_1\}$, $r_2 = \{z_1, z_2\}$, $r_3 = \{z_2\}$ and $r_4 = \{z_3\}$ in figure 13.

$$\begin{aligned}
\rho(r_1) &= \langle\{A\}, \{B\}\rangle =_\Delta 0 \\
\rho(r_2) &= \langle\{A\}, \{B\}\rangle + \langle\{B\}, \{A\}\rangle \\
&=_\Delta \langle\{B\}, \{A\}\rangle \\
\rho(r_3) &= \langle\{B\}, \{A\}\rangle \\
\rho(r_4) &= \langle\{B\}, \{A\}\rangle
\end{aligned}$$

Thus $r_1 \not\equiv_\Delta r_2$ and $r_2 \equiv_\Delta r_4$. Interestingly, we also have $r_2 \equiv_\Delta r_3$ (r_2 and r_3 are different regions in the same diagram).

Theorem 3. *The relation $\equiv_\Delta$ is an equivalence relation on regions of unitary diagrams contained in Δ.*

5   The Counterpart Relations of Shin and Hammer

The basic idea of the counterpart relation on Venn diagrams is to identify corresponding basic regions (i.e., the region enclosed by a closed curve) and then to recursively define the relation on unions, intersections and complements of regions. Shin only defines the counterpart on basic regions and leaves the rest implicit. Hammer defines the relation as follows for Venn diagrams:

The counterpart relation is an equivalence relation defined as follows. Two basic regions are counterparts if and only if they are both regions enclosed by rectangles or else both regions enclosed by curves having the same label. If r and r' are regions of diagram D, s and s' are regions of diagram D', r is the counterpart of s, and r' is the counterpart of s', then $r \cup r'$ is the counterpart of $s \cup s'$ and $\bar{r}$ is the counterpart of $\bar{s}$.

This definition works very well for Venn diagrams where all minimal regions must occur. In figure 14, the two regions enclosed by the rectangles are counterparts, and so are the two crescent-shaped regions within the circles labelled A but outside the circles labelled B; the region within all three curves in the

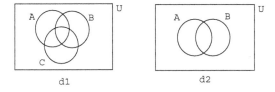

Fig. 14. Two Venn diagrams.

left-hand diagram has no counterpart in the other one. The counterpart relation is obviously equivalent to the correspondence relation defined in §3.

Hammer defines a counterpart relation on Euler diagrams, but only for diagrams with the same label set:

Suppose m and m' are minimal regions of two Euler diagrams D and D', respectively. Then m and m' are counterparts if and only if there are curves $B_1, \ldots, B_m$ and $B_{m+1}, \ldots, B_n$ of D and curves $B'_1, \ldots, B'_m$ and $B'_{m+1}, \ldots, B'_n$ of D' such that (1) for each i, $1 \leq i \leq n$, B_i and B'_i are tagged by the same label; (2) m is the minimal region within $B_1, \ldots, B_m$ but outside $B_{m+1}, \ldots, B_n$; and (3) m' is the minimal region within $B'_1, \ldots, B'_m$ but outside $B'_{m+1}, \ldots, B'_n$.

This definition is sufficient for Hammer's purposes, but it only covers a special case of Euler diagrams. Consider the two Euler diagrams in figure 15. Minimal region 1 is the counterpart of minimal region a, 2 is the counterpart of b and 3 is the counterpart of d. Minimal region c has no counterpart in the left-hand diagram.

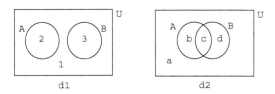

Fig. 15. Two Euler diagrams.

The correspondence relation defined in §4 agrees with this interpretation when the context is the disjunction of the two diagrams. It also agrees in the case in which the context is the conjunction of the two diagrams but adds in further correspondences such as region 2 corresponds with region $b \cup c$.

6   Conclusions and Further Work

We have constructed a purely syntactic definition of corresponding regions in Euler diagrams and shown it to be an equivalence relation and a generalization of the counterpart relations introduced by Shin and Hammer. At the semantic level, two corresponding regions represent the same set. The system of Euler diagrams we considered in this paper is in conjunctive normal form. However, we wish to reason in the more general case where we consider any combination of disjuncts and conjuncts of diagrams such as in constraint trees [7]. The correspondence relation defined in this paper can be adapted for such systems.

The general aim of this work is to provide the necessary mathematical underpinning for the development of software tools to aid reasoning with diagrams. In particular, we aim to develop the tools that will enable diagrammatic reasoning to become part of the software development process.

Acknowledgements. Author Stapleton would like to thank the UK EPSRC for support under grant number 01800274. Authors Howse, Flower and Taylor were partially supported by UK EPSRC grant GR/R63516.

References

1. L. Euler. Lettres a une princesse d'allemagne, 1761.
2. J. Gil, J. Howse, and S. Kent. Towards a formalization of constraint diagrams. In *Proc Symp on Human-Centric Computing*. IEEE Press, Sept 2001.
3. E. Hammer. *Logic and Visual Information*. CSLI Publications, 1995.
4. E. Hammer and S-J Shin. Euler's visual logic. In *History and Philosophy of Logic*, pages 1–29, 1998.
5. D. Harel. On visual formalisms. In J. Glasgow, N. H. Narayan, and B. Chandrasekaran, editors, *Diagrammatic Reasoning*, pages 235–271. MIT Press, 1998.
6. J. Howse, F. Molina, and J. Taylor. On the completeness and expressiveness of spider diagram systems. In *Proceedings of Diagrams 2000*, pages 26–41. Springer-Verlag, 2000.
7. S. Kent and J. Howse. Constraint trees. In A. Clark and J. Warner, editors, *Advances in object modelling with ocl*. Spinger Verlag, to appear, 2002.
8. F. Molina. *Reasoning with extended Venn-Peirce diagrammatic systems.* PhD thesis, University of Brighton, 2001.
9. OMG. UML specification, version 1.3. Available from www.omg.org.
10. C. Peirce. *Collected Papers*, volume Vol. 4. Harvard Univ. Press, 1933.
11. S.-J. Shin. *The Logical Status of Diagrams*. Cambridge University Press, 1994.
12. G. Stapleton. Comparing regions in spider diagrams. Available at www.it.brighton.ac.uk/research/vmg/papers.html.
13. J. Venn. On the diagrammatic and mechanical representation of propositions and reasonings. *Phil.Mag*, 1880.

CDEG: Computerized Diagrammatic Euclidean Geometry

Nathaniel Miller

University of Northern Colorado
Department of Mathematical Sciences
nat@alumni.princeton.edu

Abstract. **CDEG** (Computerized Diagrammatic Euclidean Geometry) is a computer proof system in which diagrammatic proofs of theorems of Euclidean Geometry can be given formally. The computer system manipulates geometric diagrams using an internal representation that is based on the idea that all the significant information in a geometric diagram is captured by its underlying topology. The proof system that **CDEG** implements is that of the author's diagrammatic formal system for geometry, **FG**. **CDEG** and **FG** are strong enough to be able to duplicate most, if not all, of the proofs in the first several books of Euclid's *Elements*. This paper explains **CDEG** and gives a brief example of how it works.

1 Formal Geometry

To begin, consider Euclid's first proposition, which says that an equilateral triangle can be constructed on any given base. While Euclid wrote his proof in Greek with a single diagram, the proof that he gave is essentially diagrammatic, and is shown in Figure 1. Diagrammatic proofs like this are common in informal treatments of geometry, but mathematicians have long assumed that such informal diagrammatic proofs cannot be made formal. However, the proof given in Figure 1 is in fact a formal proof in the author's formal system **FG** (Formal Geometry), which has been implemented in the computer system **CDEG** (Computerized Diagrammatic Euclidean Geometry). These systems are based on a precisely defined syntax and semantics of Euclidean diagrams. A diagram is defined to be a particular type of geometric object satisfying various conditions; this is the syntax of our system. Then, it is possible to give a formal definition of which arrangements of lines, points, and circles in the plane are represented by a given diagram; this is the semantics. Finally, we can give precise rules for manipulating the diagrams—rules of construction, transformation, and inference. These are the rules that are implemented by **CDEG**.

The details of how the syntax and semantics of these diagrams are formally defined are technical and will not be discussed in any detail here. The interested reader can find the details of the formal systems, as well as a detailed discussion of the history of the use of diagrams in geometry, in [2]. A crucial idea, however, is that all of the meaningful information given by a diagram is contained in its

M. Hegarty, B. Meyer, and N. Hari Narayanan (Eds.): Diagrams 2002, LNAI 2317, pp. 91–93, 2002.
© Springer-Verlag Berlin Heidelberg 2002

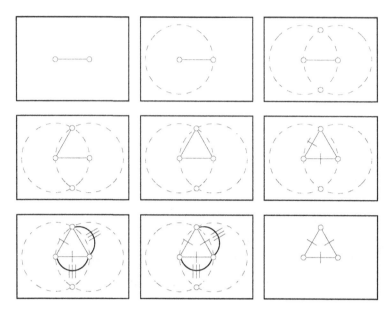

Fig. 1. Euclid's first proposition.

topology, in the general arrangement of its points and lines in the plane. Another way of saying this is that if one diagram can be transformed into another by stretching, then the two diagrams are essentially the same. This is typical of diagrammatic reasoning in general.

2   CDEG

The following picture shows a diagram output by **CDEG** which corresponds to the second diagram in Figure 1:

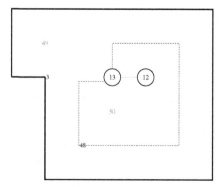

Notice that **CDEG** has labeled each segment, region, and dot with an identifying number. The diagram contains a frame, which contains two regions numbered 49

and 50; it also contains two dots numbered 12 and 13, which are connected by a straight line segment and enclosed by a segment numbered 48 which represents a circle centered about dot 12 going through the point represented by dot 13. The shape drawn by the computer is not at all circular, but all we care about here is the topology of the diagram. In fact, **CDEG** proper just outputs the topological structure of the diagram, which is then actually realized and drawn by a standard graph-drawing package. If we tell **CDEG** to add another circle centered at dot 13 and going through dot 12, we get the following diagram:

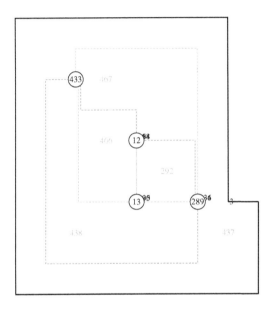

This diagram corresponds to the third diagram in Figure 1. In practice, the diagrams are much easier to read than the small black and white versions shown here, because **CDEG** colors each line and/or circle with a different color.

　　Why bother to computerize these diagrams and this formal system? There are at least two good reasons. First of all, to do so addresses any possible lingering doubts about the total formality of the system: the diagrams that the program manipulates are, by the nature of the computer, completely formal objects. And secondly, unlike with sentential formal systems, applying a construction rule to a diagram is a non-trivial operation that requires looking at many possible outcomes: the sort of operation that computers are good at and humans are not. For more details, see [2].

References

1. Euclid, *Elements*, T. L. Heath, ed., second edition, New York: Dover, 1956.
2. Miller, Nathaniel, *A Diagrammatic Formal System for Euclidean Geometry*, Ph.D. Thesis, Cornell University, 2001. Available at
   `http://hopper.unco.edu/faculty/personal/miller/diagrams/`.

Compositional Semantics for Diagrams Using Constrained Objects

Bharat Jayaraman and Pallavi Tambay

Department of Computer Science and Engineering
University at Buffalo (SUNY)
Buffalo, NY 14260-2000
{bharat,tambay}@cse.buffalo.edu

1 Motivation and Approach

We present a novel approach to the compositional semantics for a large family of diagrams. Examples include engineering drawings such as circuit diagrams, trusses, etc., process control diagrams such as traditional program flowcharts, data-flow diagrams depicting traffic flow or the dependency relationships between operators, etc. Providing a compositional semantics involves defining the meaning of the whole diagram in terms of the meanings of its parts. A common characteristic of these diagrams is that we need to depict both structure and behavior of some artifact, procedure, function, etc. We view these diagrams as graphs that are made up of nodes and edges. The meaning of each diagrammatic element (e.g., node) as well as the connections between elements (e.g., edge) are given in terms of a *constrained object* [3]. A constrained object is an object whose attributes may be subject to one or more constraints, i.e., relations among the attributes. While the concept of an object and its attributes captures the structural aspects of a diagrammatic element, the concept of constraint captures its behavioral properties. Constrained objects may be thought of as declarative counterparts of the traditional objects found in object-oriented languages.

To illustrate the notion of constrained objects as applied to diagrams, consider the components of an electrical circuit. The relevant attributes of a resistor are its current, voltage and resistance. The values of these attributes are governed by Ohm's law. When one or more resistors are connected together at a node, their currents must obey Kirchoff's law. This law is expressed as a constraint in the **node** class shown on the next page. Thus, the compositional semantics of a complete circuit diagram is obtained through a process of constraint satisfaction, i.e., solving the equations arising from the various resistors and nodes to determine the values of attributes that are unspecified.

$$V = I * R$$

```
class resistor {
  attributes    real V, I, R;
  constraints   V = I*R;
}
```

M. Hegarty, B. Meyer, and N. Hari Narayanan (Eds.): Diagrams 2002, LNAI 2317, pp. 94–96, 2002.
© Springer-Verlag Berlin Heidelberg 2002

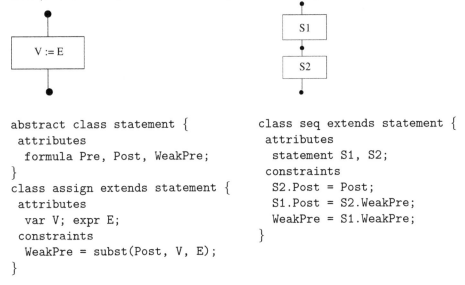

```
class node {
  attributes    resistor [] Rs;
  constraints    sum R in Rs: R.I = 0;
}
```

For another illustration of constrained objects, we show two familiar flowchart elements and their semantics in terms of constrained objects. A program flowchart can be understood in a few different ways (e.g., axiomatic, denotational, and operational semantics), and hence the constrained-object representation would be different in each case. Below we illustrate how the traditional axiomatic semantics of each flowchart element can be represented as constrained objects. Essentially the attributes of the constrained object capture not only structural information (i.e., flowchart node details) but also semantic information (i.e., weakest pre-conditions and post-conditions).

```
abstract class statement {
 attributes
   formula Pre, Post, WeakPre;
}
class assign extends statement {
 attributes
   var V; expr E;
 constraints
   WeakPre = subst(Post, V, E);
}
```

```
class seq extends statement {
 attributes
   statement S1, S2;
 constraints
   S2.Post = Post;
   S1.Post = S2.WeakPre;
   WeakPre = S1.WeakPre;
}
```

2   Current Status and Related Work

We have implemented the paradigm of constrained objects by translating the above forms of class definitions into constraint logic programs (CLP(R)) [3], and are also developing domain-specific drawing interfaces for different application domains. Constrained objects can be used to express the semantics of a variety of diagrams including functional (dataflow) diagrams, entity-relationship diagrams as well as data-structure diagrams. The foregoing class of diagrams express topologic but not geometric relationships between elements. We are currently examining the use of constrained objects to express 2-dimensional and 3-dimensional geometric structures. Such examples arise in modeling many physical and engineering systems. The flowchart example of the previous section illustrates that constrained objects are more suitable for reasoning about transition systems than for modeling them per se. This may be contrasted with approaches to the semantics of Petri nets, statecharts and other state transition systems [2,5,6]. Constrained objects, however, can be used to model the dynamic behavior of certain systems when it is possible to characterize the behavior of the system at a given time step in terms of its behavior at a previous time step.

There has been recent interest in the subject of semantics of drawings and understanding diagrams. Marriott and Meyer [7] discuss linear logic and situation theory for diagrammatic reasoning. While the former is resource-oriented, the latter is suitable for natural language. They also incorporate grammar-based interpretations into both logics. In comparison, our work stresses the role of objects to help facilitate a compositional semantics. Object structure is similar to grammatical structure. Our constraint language also comes under the umbrella of a logical language, although we have not considered resource-oriented or natural language semantics in our work. Our notion of constraints is sufficiently general to accomodate a large variety of semantic interpretations. However, in order to preserve computability and also computational tractability, the logical language of constraints must be limited in practice.

The recognition of engineering drawings and parsing diagrams has been a topic of recent interest. For example, reference [4] addresses the problem of constructing the 3-dimensional representation of an object from an engineering drawing of the object. To enable automatic recognition, paper line drawings are initially scanned. The reference [1] describes the analysis (parsing) of diagrams using constraint grammars, where diagrams are provided in terms of graphics primitives such as lines, polygons, etc. These works are representative of approaches that are aimed at a different set of issues (lower-level recognition and parsing) than what we consider in this paper (higher-level semantics).

References

1. R. Futrelle and N. Nikolakis. Efficient Analysis of Complex Diagrams using Constraint-based Parsing. In *Intl. Conf. on Document Analysis and Recognition*, 1995.
2. D. Harel and A Naamad. The STATEMATE Semantics of Statecharts. *ACM Trans. on Software Engineering and Methodology*, 5(4):293–333, October 1996.
3. B. Jayaraman and P. Tambay. Modeling Engineering Structures with Constrained Objects. In *Symp. on Practical Aspects of Declarative Languages*, pages 28–46, 2002.
4. T. Kanungo, R. Haralick, and D. Dori. Reconstruction of CAD Objects from Engineering Drawings: A Survey. In *First IAPR Workshop on Graphics Recognition*, pages 217–228, 1995.
5. L. Priese and H. Wimmel. On Some Compositional Petri Net Semantics. Technical Report 20–95, Institut fur Informatik, Universitat Koblenz, 1995.
6. G. Luettgen, M. Beeck, and R. Cleaveland. A Compositional Approach to Statecharts Semantics. ICASE Report No. 2000-12, NASA/CR-2000-210086,2000, 2000.
7. K. Marriot and B. Meyer. Non-standard Logics for Diagram Interpretation. In *Intl. Conf. on the Theory and Application of Diagrams*, 2000.

Retrieving 2-D Line Drawings by Example

Patrick W. Yaner and Ashok K. Goel

College of Computing, Georgia Institute of Technology
{yaner,goel}@cc.gatech.edu

Abstract. We present a two-phase retrieval process for finding a target query within a set of 2-D line drawings, where the query is itself a line drawing. We also describe a logic-based matching method that uses unification and resolution to compare the semantic networks representing the spatial structure of two 2-D line drawings.

Retrieval of diagrams from an external memory (e.g., a computer-based library of diagrams) is a key problem in reasoning about diagrams. The task in our work assumes a computer-based library of 2-D line drawings, takes as input a representation of a 2-D line drawing as a query, and its desired output is a set of drawings retrieved from the library that are similar to the query (meaning that the target can be found within these sources based on a qualitative high-level comparison). The goal of the work described in this paper is to develop a retrieval process and a matching procedure for enabling querying by example.

Retrieval Process and Architecture. We propose a computational architecture containing six basic components to support a retrieval process consisting of two phases. As illustrated in figure 1, the architecture is composed of the following components:

1. An external memory of diagrams indexed by feature vectors
2. A feature extractor for generating feature vectors
3. A process that generates a semantic network describing the contents (spatial structure in this case) of a diagram.
4. A process that matches a target's description (semantic network) to that of a given source from memory
5. A working memory with potential sources to match with the target query
6. A user interface for drawing the query and visualizing the responses

This architecture supports a retrieval process consisting of two phases: reminding and selection. The first phase takes as input a target example and returns as output pointers to stored diagrams whose feature vectors match that of the target. The stored diagrams—which in general are created elsewhere, i.e. external to this system—are indexed by feature vectors describing their spatial elements; a feature vector for the target is constructed dynamically. Pointers to those images with sufficiently similar feature vectors (according to some appropriate criteria) are then brought into working memory.

M. Hegarty, B. Meyer, and N. Hari Narayanan (Eds.): Diagrams 2002, LNAI 2317, pp. 97–99, 2002.

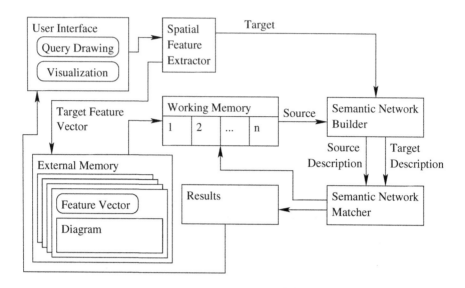

Fig. 1. Proposed System Architecture

Diagrams in our system are represented as sets of graphical objects (lines and other shapes, generally). A vector of features (by which we mean attribute-value pairs) of a diagram can be extracted and used to gauge the likelihood that a given diagram will match another one. For example, a simple feature vector might simply be a multiset of the object types contained in the diagram. A multiset is a set allowing for possible multiplicity of its members. So, the set of 3 A's and 4 B's might look like $\{3 \cdot A, 4 \cdot B\}$. Given a list of shapes—say, 2 rectangles, 3 circles, and a triangle—we can represent the feature vector as a mapping from object type to its multiplicity.

Under this system, a diagram could be retrieved if the multiset of shape types it contains is a superset of that of the target diagram The retrieval mechanism scans the external memory, calculating whether or not the multiset of objects in the target is a subset of the multiset of objects in each source diagram, returning those for which this is the case. That is, if T is the feature vector for the target, and $S_1, S_2, \ldots, S_k$ are the feature vectors of the diagrams currently in memory, then the retrieval method returns those diagrams for which $T \subset S_i$.

Selection and Matching. The second phase of the retrieval process involves selecting among the source diagrams in working memory those that most closely match the target. To determine this, first the system must generate a relational description of the contents of both the target and the sources. Here, we are dealing with spatial structure as the represented content, and have chosen to represent this using first-order logic (though it may be possible with a system such as this to represent semantic information about the diagram as well). A convenient way to think about descriptions in first-order logic is using semantic networks.

Semantic networks describing the spatial structures of the target example and the diagrams in the working memory are constructed for each image separately. The nodes of the network correspond to the objects and object types, and the links correspond to the relations between them. The semantic networks of the diagrams are matched with that of the target example. The matching process works by building a mapping from the target to the source, essentially finding the description of the target *within* that of the source, and making the appropriate correspondences. Because we are constructing a *mapping* from target to source, one could conceivably use this system as the first stages of a visual analogical reasoner of some sort.

The purpose of the matching process is to find the semantic network corresponding to the target diagram within that of the source. The system does this by constructing a mapping from the nodes in the target to the nodes in the source that preserves the structure of the semantic network. The process works by associating edges first. Once an edge is mapped, the nodes on either end of it are mapped. The process continues like this in a depth-first fashion, backtracking when necessary, until all of the nodes and edges in the target are mapped onto the source.

If this process succeeds, we have a match, and if not, we discard the source. In this way, we can use this method as a predicate, a test on an image from working memory. This test, then, is used to find all such matches in working memory, accumulating matching images and returning them all to the user upon completion.

Initial Experiments. We implemented the system in ANSI Common LISP. A vocabulary of 4 shape types was chosen, purely as proof of concept: lines, ellipses, triangles, and rectangles. The program reads in vector graphics files containing these shapes, and stores them in memory. The only properties dealt with in this first version of the program are those that define the shape, e.g. the height and width and location of a rectangle.

The relational descriptions drew from a vocabulary of four relations: `above`, `below`, `left-of`, and `right-of`. Note that in the methods given above, the relational description is generated when the matching is computed. Even for source diagrams, it is not generated when the diagram is placed in memory. If the relational description becomes more complex, it would be relatively straightforward to generate the relational description as the sources are added to memory, and store them with the diagram. It's not clear if there is an inherent advantage to either scheme; it almost certainly depends on the specifics of the application and the representations used.

Memory was populated with eight line drawings, and the program was run with nine separate target diagrams. In general, most of the targets recalled two or three diagrams, but all of them recalled at least one.

A System That Supports Using Student-Drawn Diagrams to Assess Comprehension of Mathematical Formulas

Steven Tanimoto, William Winn, and David Akers

University of Washington
Seattle, WA 98195, U.S.A.

Abstract. Graphical communication by students can provide important clues to teachers about misconceptions in fields like mathematics and computing. We have built a facility called INFACT-SKETCH that permits conducting experiments in the elicitation and analysis of student-drawn diagrams. A goal of our project is to achieve the integration of a construction-oriented image processing and programming system with a combination of automatic and manual assessment tools in order to have a highly effective learning environment for information technology concepts. Some of the diagrams we elicit from students represent image processing algorithms, and others represent predictions of what particular mathematical formulas will do to images. In order to serve its educational assessment and research purposes, INFACT-SKETCH provides an unusual combination of features: tight integration with a web-based textual communication system called INFACT-FORUM, administrative control of which drawing tools students will be permitted to use (freehand, rectangles, ovals, text labels, lines, etc.), complete event capture for timed playback of the drawing process for any sketch in the system, graphical quoting option in message replies, and structured annotations for educational assessment. We describe the rationale, intended use, design, and our experience so far with INFACT-SKETCH.

1   Sketching in Educational Communication

Understanding what students are thinking when they are learning new concepts and techniques is a major challenge of teaching. One approach to this challenge is to observe and make inferences from what students say and do, whereas another is to give them tests and questionnaires to fill out. We are taking the first approach. Because many topics that students learn have natural graphical representations, we decided to incorporate students' sketches into the mix of text and user-interface events that serve as inputs to assessment processes.

Our students are freshmen at the University of Washington who are enrolled in a seminar entitled, "Image Processing: Pixels, Numbers, and Programs" [2]. They learn about digital image representation, transformations using mathematical formulas, and Lisp programming in the context of image manipulation.

M. Hegarty, B. Meyer, and N. Hari Narayanan (Eds.): Diagrams 2002, LNAI 2317, pp. 100–102, 2002.
© Springer-Verlag Berlin Heidelberg 2002

As pointed out by Furnas et al [1], image processing is well-suited to the use of graphical representations in the course of problem solving.

We have developed a specialized software facility called INFACT-SKETCH that helps us to study the use of sketches in educational assessment.

2   The INFACT System

Our sketching facility is a Java applet that serves as an integrated part of a larger software system known as INFACT. This stands for "Interactive, Networked, Facet-based Assessment Capture Tool." It provides functionality for both students and teachers. For example, part of it called INFACT-FORUM provides a textual communication facility for students [3]. Teachers can make use of INFACT-MARKUP, which makes it easy to create structured annotations of student text and populate a database of "facet diagnoses" that represent student progress. Other parts of INFACT include facilities for viewing data, building facet catalogs, and importing and exporting data. In this paper, we describe INFACT's newest component: INFACT-SKETCH.

3   Drawing Affordances for Students

INFACT-SKETCH provides the student with a straightforward drawing tool that is easy to learn and use. It runs as an applet and comes up quickly if a student, working within INFACT-FORUM, elects to attach a sketch to her/his textual posting in the course of small-group online communication. Figure 1 shows the student's interface and a sample sketch.

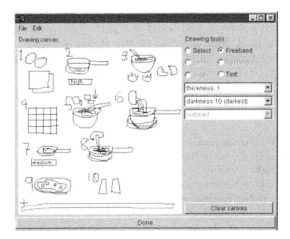

Fig. 1. The INFACT-SKETCH program as seen by the student. The sample sketch is one student's depiction of the process of making scrambled ham and eggs.

A freehand drawing tool (like a pencil) is provided. This tool does not incorporate any automatic beautification or intelligence, and responds as students tend to expect. We consider this to be the most important tool in the applet. A text tool makes it easy for students to place labels on their sketches. Text labels are typed at the keyboard. They can be positioned and repositioned easily. Geometric primitives such as rectangles, ovals, and line segments are also provided. The instructor, however, can enable or disable these tools. The rectangles and ovals can be either outlined or filled. The width and darkness of outlines and freehand strokes can be varied using choice boxes.

Students can select arbitrary subsets of the drawing objects they have placed on the canvas. Selections can be moved or erased. Any drawing action can be undone, say, in order to take back an action. We permit unlimited levels of undo. A sketch as part of a message, may be a reply to another message containing a sketch. The new sketch can include and build upon the first sketch much as a textual reply may contain a copy of the first message's text. We refer to this feature as sketch quoting.

4   Affordances for Instructors

When a sketch is being viewed by a user having administrative privileges, the user sees not only the sketch itself but also a playback interface that permits the drawing process to be inspected. All events, including undo events, can be played back at the original rate. It is also possible to rewind and fast-forward the process. Thus a teacher can inspect the sketch at any stage of its creation. An instructor can reply to any student's posting, and can include an annotated copy of a student's sketch in the reply.

Acknowledgements. We would like to thank R. Adams, M. Allahyar, C. Atman, N. Benson, A. Carlson, E. Hunt and T. VanDeGrift for their cooperation on the project of which INFACT-SKETCH is a part. We would also like to thank the students in Autumn 2001 offering of the Pixels seminar for being willing participants, and we thank the National Science Foundation for support under grant EIA-0121345.

References

1. Furnas, G., Qu, Y., Shrivastava, S., and Peters, G.: The Use of Intermediate Graphical Constructions in Problem Solving with Dynamic, Pixel-level Diagrams **Proc. DIAGRAMS 2000**. (2000).
2. Tanimoto, S. Pixels, Numbers, and Programs: A freshman introduction to image processing. Proc. IEEE Workshop on Combined Research and Curriculum Development in Computer Vision, December (2001).
3. Tanimoto, S., Carlson, A., Husted, J., Hunt, E., Larsson, J., Madigan, D., and Minstrell, J. Text forum features for small group discussions with facet-based pedagogy. Proc. CSCL 2002, Boulder, CO., (2002).

An Environment for Conducting and Analysing Graphical Communication Experiments[*]

Patrick G.T. Healey[1], Nik Swoboda[2], and James King[1]

[1] Queen Mary, University of London, London E1 4NS, UK
{ph, jking}@dcs.qmul.ac.uk
[2] DEIMOS Space S.L., Sector Oficios 34, Tres Cantos, Madrid 28760, Spain
nik.swoboda@deimos-space.com

Abstract. Drawing is a basic but often overlooked mode of human communication. This paper presents a shared whiteboard environment, written in Java, that was designed to be used to collect and analyse data gathered in interactive graphical communication experiments. Users of the software are presented with a 'virtual' whiteboard that is connected to another user's whiteboard to create a shared graphical communication space. In addition to logging all drawing activity between the pair and providing tools for the analysis of this data, the software can manipulate the layout and the degree of interactivity of the drawing being exchanged. The program can also be used to setup and manage multiple simultaneous shared whiteboard connections and subject groupings.

1 Introduction

This paper describes a tool for performing and analysing experiments on graphical interaction between people. This software exploits the potential that electronic whiteboards offer for the direct capture and control of drawing activity. Currently, studies of graphical interaction have relied on direct observation, video analysis, or analysis of photographs and screen shots. These data collection techniques are labour intensive and susceptible to error. This tool was designed with three basic aims: to provide a fine-grained and reliable method of capturing drawing activities, to support experimental investigation through the manipulation of aspects of the graphical interaction, and to aid data processing by providing specialised tools for analysing recorded graphical interactions. This tool has shown its usefulness in experiments conducted in Japan and the UK [1,2].

2 The Software Components

The program consists of three basic components. Firstly, a simple shared whiteboard that allows pairs of subjects to communicate through drawing. The whiteboard runs across a network allowing users to be either copresent or in different

---

[*] We gratefully acknowledge the support of the ERSC/EPSRC PACCIT initiative through the grant MAGIC: Multimodality and Graphics in Interactive Communication (L328253003). The support of ATR Media Information Science Laboratories has also been critical to the development of work in this paper.

M. Hegarty, B. Meyer, and N. Hari Narayanan (Eds.): Diagrams 2002, LNAI 2317, pp. 103–105, 2002.
© Springer-Verlag Berlin Heidelberg 2002

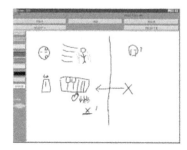

Fig. 1. Example Whiteboard Screen

locations. Users draw in freehand on the whiteboard selecting from a palette of colours in the editor bar. They see all their, and their partner's, drawing displayed in the whiteboard window. Secondly, a master program is used to configure, connect and monitor the individual whiteboard processes. These processes run separately on each subject's computer and record time stamped logs of both users drawing, mouse or pen movements, and button presses. Lastly, a suite of data analysis and presentation tools that is used to study the logs of data generated is provided.

2.1   The Whiteboard

The whiteboard window consists of three areas: the task bar, the editor bar, and the editing space. The task bar displays the buttons or other GUI components used by subjects in completing the given task e.g., buttons for response selection. The editor bar displays buttons to allow users to change colour or to enter erase mode. The majority of the window consists of the editing space where the users can draw. A screen shot of the whiteboard window can be seen in Fig. 1.

Central to the whiteboard is the concept of a *subject event*. Subject events consist of particular subject interactions with the task bar or the editor. An example of an event is a line drawn by one of the subjects which has a starting and ending time stamp, a list of points, and a colour. Subject events are sent between the pair of whiteboards and between the editor and the code that controls the experimental task being performed. Subject events are also written to a log file for subsequent data analysis. The logs contain a great deal of information including: each line or erase drawn as a time stamped list of points, the time of each button press in the task bar, and the time elapsed to complete the task. The logs also contain special task dependent information such as the result of the task and all experimental configuration parameters.

2.2   Master Program

Monitoring and control of the whiteboard processes is provided by the master program. The master reads a configuration file describing the task to perform.

This configuration file is a human readable text file containing parameters such as the number of experimental trials and subject pairings. Once a configuration file is loaded, the master displays the current options and checks the status of subject's whiteboards. The master program is used to start, monitor, and repeat experimental trials. This program has been used in experimental settings to control and manage the connections of fourteen pairs of subjects simultaneously.

2.3   The Data Analysis and Presentation Tools

The comprehensive nature of the data logged by the whiteboard means that accurate, down to pixel level, recreations of the drawings at any point in the experiment can be produced in the form of gif, jpeg, PDF or postscript images. It can also provide separate images of the activities of each person individually and each pair.

The data can also be used to generate a variety of statistics for each phase of an experiment for individuals and pairs. Individual statistics include: the total number of lines drawn and erases made, the total line and erase lengths, the total time spent drawing and erasing, the number of different colours used, and screen usage. Per pair statistics include: the number of colours used by a subject not used by their partner, and the amount of overlap in the drawing of the pair.

The data can also be used with a VCR-like playback tool. This tool allows the re-play, fast-forward, rewind, and stop of the drawing process at any point. It includes a time-line of the drawing activity of a pair which provides a visualisation of patterns of, e.g., turn-taking and overlap in a graphical exchange.

3   Controlling Subject Interaction

This tool is designed to support the experimental investigation of graphical interaction. Thus, a significant part of its functionality has been developed to support the manipulation of the level of communicative interaction possible between experimental subjects. Currently, this is implemented in two main ways. The first is the manipulation of the spatial organisation of the interaction, e.g, the division of the screen into separate drawing regions in which drawing can be selectively prevented. The second is the topological manipulation of the drawing in the form or transposition of regions of drawing from one part of a subject's screen onto another part of their partner's drawing space and the rotation of a subject's drawing before displaying it on their partner's drawing space.

References

1. Healey, P., Swoboda, N., Umata, I., Katagiri, Y.: Representational form and communicative use. In Moore, J., Stenning, K., eds.: Proceedings of the 23rd Annual Conference of the Cognitive Science Society. (2001) 411–416
2. Healey, P., Swoboda, N., Umata, I., Katagiri, Y.: Graphical representation in graphical dialgoue. To appear in a special issue of the *International Journal of Human Computer Studies* on Interactive Graphical Communication (forthcoming)

Grammar-Based Layout for a Visual Programming Language Generation System

Ke-Bing Zhang[1], Kang Zhang[2], and Mehmet A. Orgun[1]

[1] Department of Computing, ICS, Macquarie University, Sydney, NSW 2109, Australia
{kebing, mehmet}@ics.mq.edu.au
[2] Department of Computer Science, University of Texas at Dallas
Richardson, TX 75083-0688, USA
kzhang@utdallas.edu

Abstract. This paper presents a global layout approach used in a general-purpose visual language generation system system. Our approach is grammar-based graph drawing, in which layout rules are embedded in the productions of reserved graph grammars. Thus, the grammar formalism serves both the visual language grammar and the layout grammar. An example visual language is demonstrated.

1 Introduction

For a visual language user, the two most important aspects of a visual program are its physical layout (what the user sees and manipulates), and its meaning (what the user expresses with it). Any implementation of the language has to maintain the correspondence between these two aspects [1]. A good layout of a visual program could give users a clear view of the syntactic and the semantic relationships amongst the program components.

Brandenburg presented a grammar-based graph drawing algorithm [2], based on a sequential graph rewriting system or graph-grammar to replace some initial graph by productions. The algorithm is efficient in the construction of nice drawings of graphs. McCreary *et al.* proposed a grammar-based layout approach [4], that uses a new technique for hierarchical and directed graphs in clan-based decompositions. The key idea of that approach is to recognize intrinsic sub-graphs (clans) in a graph by producing a parse tree that carries the nested relationships among the clans. The parse tree of the graph can be given a variety of geometric interpretations.

This paper presents a layout approach that is based on Brandenburg's algorithm [2], adapted the techniques of Lai and Eades [3] and those of McCreary *et al.* [4] in the reserved graph grammar (RGG) formalism. The approach embeds layout rules in the productions of RGGs [6], which are specified graphically, rather than textually, through relative geometrical positions. Our approach is used for both local and global layout in a visual language generation system, called VisPro [7].

2 The Layout Approach

We embed layout rules in a RGG to define the layout graph grammar for handling local and global layout. In the layout mechanism, the right graph of a production rep-

M. Hegarty, B. Meyer, and N. Hari Narayanan (Eds.): Diagrams 2002, LNAI 2317, pp. 106–108, 2002.
© Springer-Verlag Berlin Heidelberg 2002

resents a pattern whose graph elements need to be spatially rearranged; the left graph represents a preferred layout for the elements.

The layout rules form a set of syntax-directed geometrical constraints for visual objects in the host diagram. In the global layout, the sub-graphs of a host graph are computed (for local layout) in isolation and then combined as a whole graph according to the reserved graph grammar during parsing and are then drawn. Therefore the parse tree of the graph grammar also directs the layout process, and the graph grammar formalism serves both the visual language grammar and the layout grammar.

Specifying layout rules in VisPro is highly intuitive and easy. When a user specifies a graph grammar for a visual programming language, he/she may specify relative geometrical position of each visual object (node) in the right graph of a production. The specification is similar to specifying a graph grammar for a visual language. A parser can be produced according to the graph grammar and the layout rules. Our approach is general enough to solve complicated layout problems in visual language generation systems.

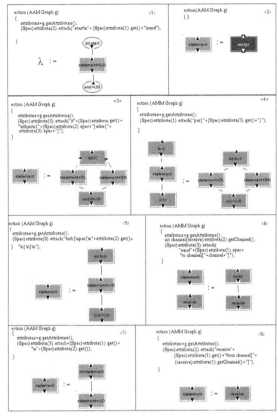

Fig. 1. RGG with layout rules for *Flowstart*

3   An Example

We now use the visual language *FlowChart* to demonstrate how our layout approach works. *FlowChart* offers a visual Flow Chart tool, with which a user can specify production rules graphically, and enter components such as a number or a string into a node of a diagram. After compilation, the *FlowChart* produces the result according to the user's inputs. The reserved graph grammar with layout rules is shown in Fig. 1 where <i> represents production i. Fig. 2(a) shows a *FlowChart* diagram designed and drawn by a user. After compilation, the diagram is laid out and automatically redrawn as shown in Fig. 2(b).

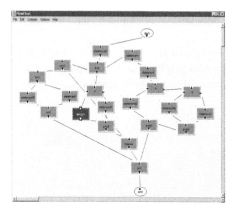
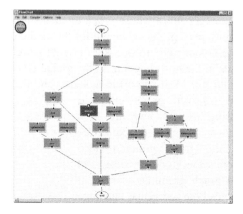

(a) before compilation   (b) after compilation

Fig. 2. A diagram of *FlowChart*

4   Conclusion and Future Work

Embedding layout rules in graph grammar makes layout process general enough for visual programming language generation. It is flexible and highly intuitive. So far, layout rules in the VisPro are specified manually. We plan to adapt the approach of Ryall *et al.* [5] into VisPro so that a user can specify layout of sub-graphs and overall visual organization of a diagram by selecting different constraints according to the user's aesthetic criteria, such as symmetry, hub shape, even spacing, and clustering.

References

1. Andries, M., Engels, G. and Rekers J.: How to represent a visual program? In *Proc. of Workshop on Theory of Visual Languages* (TVL96), Gubbio, Italy (1996).
2. Brandenburg, F. J.: Layout Graph Grammars: the Placement Approach. *Lecture Notes in Computer Science*, Vol. 532. Graph Grammars and Their Application to Computer Science. Springer-Verlag (1991) 144-156.
3. Lai, W. and Eades, P.: Structural Modeling of Flowcharts, In: Eades, P. and Zhang, K. (eds.): *Software Visualisation*, Series on Software Engineering and Knowledge Engineering, vol. 7, World Scientific Co. (1996) 232-243.
4. McCreary C. L., Chapman R., and Shieh F-S.: Using Graph Parsing for Automatic Graph Drawing, Vol. 28. *IEEE Transactions on Systems, Man and Cybernetics*, No. 5, Sept., (1998) 545-561.
5. Ryall, K., Marks, J. and Shieber, S.:An Interactive Constraint- Based System for Drawing Graphs. In *Proc. User Interface Software and Technology* (UIST 97), Banff, Alberta (1997) 97-104.
6. Zhang, D-Q. and Zhang, K.: Reserved Graph Grammar: A Specification Tool for Diagrammatic VPLs. In *Proc 1997 IEEE Symposium on Visual Languages*, Capri, Italy. IEEE CS Press, Los Alamitos, USA (1997) 284-291.
7. Zhang, D-Q. and Zhang, K.: VisPro: A Visual Language Generation Toolset, In *Proc.1998 IEEE Symposium on Visual Languages,* Halifax, Canada. IEEE CS Press, Los Alamitos, USA (1998) 195-201.

Heterogeneous Data Querying in a Diagrammatic Information System

Michael Anderson[1] and Brian Andersen[2]

[1]Department of Computer and Information Sciences
Fordham University
441 Fordham Road
Bronx, New York 10458
mianderson@fordham.edu
[2]Department of Computer Science
University of Hartford
200 Bloomfield Avenue
West Hartford, CT 06117
banderse@cs.hartford.edu

Abstract. We show how Diagrammatic SQL, a query language that can be used to query information represented as diagrams, can be combined with SQL to permit queries on combinations of diagrammatic and symbolic data producing both diagrammatic and symbolic responses.

1   Introduction

When dealing with different types of knowledge in some system, it may often be the case that a variety of knowledge representation and inferencing techniques may be required to manage each knowledge type effectively. A problem that can arise when multiple knowledge representation schemes are used is how these differing representations and inferencing mechanisms be forged into a cohesive whole that permits reasoning across the knowledge types represented.

We show how our diagrammatic extension to Structured Query Language (DSQL) can be combined with traditional Structured Query Language (SQL) to permit queries on both diagrammatic and symbolic data in a diagrammatic information system that uses a relational database as its storage component.

2   Diagrammatic Information Systems

We define a *Diagrammatic Information System* (DIS) as a system that allows users to pose queries concerning diagrams, seeking responses that require the system to infer information from combinations of both diagrammatic and non-diagrammatic data. We are developing a diagram-type and domain independent DIS core capable of accepting

M. Hegarty, B. Meyer, and N. Hari Narayanan (Eds.): Diagrams 2002, LNAI 2317, pp. 109–111, 2002.
© Springer-Verlag Berlin Heidelberg 2002

domain-dependent diagrammatic and non-diagrammatic knowledge, producing instantiations of DIS's. We have developed *Diagrammatic SQL* (DSQL), an extension of Structured Query Language (SQL), that supports querying of diagrammatic information. Just as SQL permits users to query relations in a relational database, DSQL permits a user to query collections of diagrams. DSQL is based on the *inter-diagrammatic reasoning* (IDR) architecture [1], [2].

Besides providing a basis for a diagrammatic query language, a relational database that stores image data can be used by the system as a storage management component. Further, as relational databases already manage other types of data, use of one as a storage management component with a diagrammatic extension to its SQL gives the system a means to query both diagrammatic and non-diagrammatic data simultaneously. This provides a linkage between heterogeneous data allowing whole new classes of queries. Our current instantiation of a diagrammatic information system diagrammatically infers appropriate responses to an interesting range of heterogeneous queries posed against cartograms of the United States and symbolic/numeric data associated with them.

2.1   Example Heterogeneous Data Queries

Two examples of such queries are "What is the total population of states that have desert?" and "Show a cartogram of those states that have populations under one million." The first query uses diagrammatically represented data (namely diagrams of individual states and individual vegetation types in the United States) and inter-diagrammatic reasoning techniques (via DSQL) to determine those states that intersect desert. When these states are so determined, their names are collected into a set and passed to SQL which accesses the symbolic data representing state population in the database. As state names are keys in this relational database, the population fields corresponding to these keys are gathered, summed, and returned by SQL as the overall response to the query. This example query and its response follow:

```
SELECT SUM (Population) FROM States WHERE Name IN
(DSELECT State FROM States WHERE Vegetation IS Desert)
67315255
```

The mechanism that is used to provide the communication link between DSQL and SQL is the IN operator. The IN operator is based upon set membership, returning true when the value on its left hand side is a member of the set on its right hand side. In the above example, DSELECT (the DSQL selection operator) diagrammatically determines those states that intersect desert, creates a diagram of these and, as a side-effect, returns the set of names of these states. SQL then uses the IN operator to collect and sum the populations of all states whose name is in that set.

Fig. 1. States with grassland and populations under one million.

The second query reverses the roles of DSQL and SQL while still using the IN operator to communicate results between them. In this query, SQL is used to determine the names of those states that have a population under one million. DSQL, using diagrammatically represented data and inter-diagrammatic reasoning techniques, determines those states that intersect grassland. The names of these states are then constrained to only those that are members of the set of names returned by the SQL portion of the query and a diagram is constructed as the overall response to the query. This query and both its symbolically and diagrammatically represented responses follow below and in Fig. 1:

```
DSELECT   State FROM States
WHERE Vegetation IS Grassland AND Name IN
(SELECT Name FROM States WHERE Population < 1000000)
Montana, North Dakota, South Dakota, Wyoming
```

These queries are but two examples of whole new classes of queries that have been made possible by the integration of results from DSQL and SQL queries. We intend to extend this work to deal with more extensive databases concerned with a variety of different domains. Further, we are investigating how we might use other SQL mechanisms to provide a tighter integration with DSQL.

This material is based upon work supported by the National Science Foundation under grant number IIS-9820368.

References

1. Anderson, M:. Toward Diagram Processing: A Diagrammatic Information System. In Proceedings of the 16th National Conference on Artificial Intelligence, Orlando, Florida. July, 1999.
2. Anderson, M. and McCartney, R.: Inter-diagrammatic Reasoning. In Proceedings of the 14th International Joint Conference on Artificial Intelligence, Montreal, Canada. August, 1995.

Visualization vs. Specification in Diagrammatic Notations: A Case Study with the UML

Zinovy Diskin*

Frame Inform Systems, Ltd., Riga, Latvia
zdiskin@acm.org

In the world of diagrammatic notations, the Unified Modeling Language (UML) should be of special interest for cognitive studies. On one hand, UML integrates a host of different diagrammatic languages in real engineering use, and thus studying it should be itself extremely interesting. On the other hand, UML is adopted as a standard in software industry, and has already become a standard *de facto* in object-oriented analysis and design. This property provides a real practical value for UML studies. However, despite its dramatically increasing popularity, UML's drawbacks are well known and widely criticized. Currently, users and vendors associate their hopes on a better UML with the next version of the standard, UML 2.0, but it is commonly recognized that the problems standing in front of UML 2.0 are extremely hard [3]. So, careful cognitive analysis of UML appears to be an interesting, beneficial and urgent issue in the diagrammatic world.

However, before the instrumentary of cognitive science could be applied to UML, some key preliminary step must be done. On Peirce's semiotic scale ranging signs from icons to indexes to symbols, UML's notational constructs reside in the indexical-symbolic area, and some of them are close to the extreme symbolic end. Indeed, the meaning of many UML's constructs lives in the world of fairly abstract concepts. Hence, any reasonable study of the UML notation must begin with specifying precise semantics of the modeling constructs, otherwise their cognitive analysis will be built on sand. Unfortunately, the UML semantic volume and manuals provide rather vague description of constructs' semantics, and actually the latter has to be discovered and made precise, in ideal, formalized. This is a highly non-trivial issue, as any other formalization of substantial intuition underlying a complex domain, but our case has one more additional peculiarity.

Let's suppose that we have somehow fulfilled that hard work and formalized semantic meaning of a set of UML's constructs in, say, first order logic (FOL). Then we can recover a formal semantic meaning $M(D)$ of any UML diagram D using these constructs, and specify it by a set of FOL-formulas (a theory), F. However, since F is just a set of strings, it would hardly help us in cognitive analysis of the diagram D. What we really need is a precise *diagrammatic* specification of $M(D)$, say, by a diagram D_0, so that D_0 could serve as a sort of template against which the UML-diagram D could be examined and analyzed with the entire power of cognitive science.

---

* Supported by Grants 93.315 and 96.0316 from the Latvia Council of Science

M. Hegarty, B. Meyer, and N. Hari Narayanan (Eds.): Diagrams 2002, LNAI 2317, pp. 112–115, 2002
© Springer-Verlag Berlin Heidelberg 2002

Fortunately, a language for specifying complex constructions in a graphic yet formal way was developed in mathematical category theory, and then was adapted for software engineering needs (see [2] for a brief presentation). The underlying logic might be called *Arrow Diagram Logic* (ArrDL), and its specification format is called *sketch*. Briefly, a sketch is a directed multigraph, some fragments of which are marked with predicate labels taken form a predefined signature. The entire program outlined above could be called *sketching UML diagrams* and it should result in presenting UML-diagrams as visual interfaces to sketches specifying their intended semantic meaning; Fig. 1 presents a general schema of the approach.

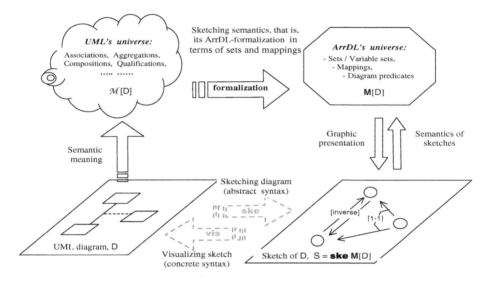

Fig. 1. Sketching UML-diagrams: A general schema.

Having a UML-diagram D on one hand, and another diagram – sketch S – on the other hand, we can start a reasonable cognitive analysis of notational mechanisms used in D. Sketching a major UML's construct of association, including N-ary and qualified associations, was performed in [1]; some preliminary semantically-based analysis of UML notation was also outlined there.

A typical sample of the results is presented in Table 1. In the left column are UML's qualification diagrams: D_0 is a pure qualification and D_1 and D_3 contain possible additional constraints, their intended meaning is explained in the textboxes between the columns. The precise semantics of the diagrams is specified by sketches in the right column. Nodes denote sets and arrows are mappings between them. Symbols [1-1], [inv] and [comp] are predicate labels hung on their diagrams (indicated by thin auxiliary lines) and denoting diagrams' properties, [comp] actually denotes a diagram operation.[1] Figurative arrows'

---

[1] When one draws sketches on a computer display, explicating the diagram on which a label is hung is really easy: clicking the label highlights the diagram.

tails and heads are also predicate labels hung on trivial diagrams consisting of just single arrows. Semantics of the labels and other details can be found in the full paper [1].

Table 1. Constraints for qualification

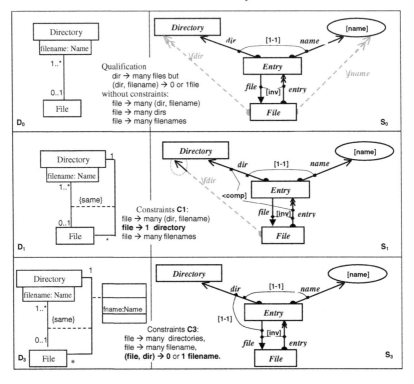

Compare textual descriptions of the universe in the middle "column" with its UML (on the left) and sketch (on the right) specifications. It is difficult to avoid a feeling that textual descriptions are much more clear than the corresponding UML diagrams. It looks like that these diagrams have resulted from an attempt to put quite clear and simple semantic pictures into an unsuitable syntactical framework. In contrast, sketches make textual descriptions precise (even formal) and clear.

The situation with UML-diagrams becomes even worse when we consider constraints C2, symmetric to C1, and C4, symmetric to C3. Here the term *symmetric* means that the sketch specifications of these constraints, S_2 and S_4, are direct right-hand-side mirror images of sketches S_1 and S_3 resp. (hopefully, the reader can easily imagine them). In contrast to this nice graphic symmetry of sketches, the UML-diagrams specifying the constraints, D_2 and D_4 resp., are rather complicated and entirely non-symmetric to diagrams D_1 and D_3 – details can be found in [1].

The problems with UML's representations of the constraints in question are caused by the fact that these constraints are *diagram* predicates and should be correspondingly specified (see [2] for some details of ArrDL's syntax). In contrast, their UML diagrammatic representations appear as a sort of arbitrary *ad hoc* visualizations; such visualizations may be apt but often they are awkward or/and obscure semantics.

The full paper [1] presents other interesting examples of notational problems caused by that UML misses the fundamental role of the notion of diagram predicate. An important one is the infamous problem of multiplicities for N-ary associations widely discussed in the literature for a long time. It is shown in [1] that the multiplicity problem is actually a pseudo-problem induced by improper notation rather than the subject matter as such.

References

1. Z. Diskin. Visualization vs. specification in diagrammatic notations: A case study with the UML. Full version of the material in the present poster, `ftp://ftp.fis.lv/pub/diskin/Diagrams02.ps`, 2001
2. Z. Diskin, B. Kadish, F. Piessens, and M. Johnson. Universal arrow foundations for visual modeling. In M. Anderson *et al*, *Diagrams'2000: 1st Int. Conf. on the Theory and Applications of Diagrams, Springer LNAI*, volume 1889, 2000, pp.345–360.
3. C. Kobryn. Will UML 2.0 be agile or awkward? *Communications of the ACM*, 45(1):107–110, 2002.

The Inferential-Expressive Trade-Off: A Case Study of Tabular Representations

Atsushi Shimojima

School of Knowledge Science
Japan Advanced Institute of Science and Technology
1-1 Asahi-dai, Tatsunokuchi, Nomi-gun, Ishikawa, 923-1292, Japan

Abstract. Many graphical systems (e.g., Euler diagrams, maps, pictorial images, and even tables) support efficient inferences or rich presentation of information apparently at the expense of expressive flexibility. This association of inferential efficiency, expressive richness, and expressive inflexibility in a graphical system has been pointed out by various researchers (e.g., Sloman [1], Stenning and Oberlander [2]). This paper investigates the semantic mechanism of the association by closely examining a particular system of tabular representations, which, despite its simplicity, clearly exhibits all those opposing functional traits. Using a semantic framework of channel theory (Barwise and Seligman [3]), we will show that the common mechanism is a parallelism between abstraction relations in represented properties and in representing properties.

Intuitions tell us that both positive and negative traits coexist in graphics. Graphics are rich in content; they facilitate our inferences on the depicted objects or situations; but they often have severe limitations in what can be expressed. Recent studies have started revealing the semantic mechanisms behind these functional traits of graphics individually: their expressive richness (e.g., Kosslyn [4]), their potentials for efficient inferences (e.g., Barwise and Etchemendy [5]), and their expressive inflexibility (e.g., Stenning and Oberlander [2]). Now, is there any relationship among these coexisting traits? Or are they all independent from each other?

Many researchers have suggested their connection. Sloman [1] mentions a trade-off between efficient "problem-solving power" and expressive generality, suggesting that "analogical" representation systems (such as graphical systems) sacrifice the latter for the former (pp. 217). Barwise and Etchemendy [5] mention a trade-off between expressive richness and expressive flexibility. Comparing pictorial images and first-order sentences, they point out that graphics are good at expressing conjunctive information but not at expressing disjunctive information (p. 22). Barwise and Hammer [6] state that the expressive generality of a system is often incompatible with its capacity for being homomorphic to its target domain, while this capacity is a root of a system's expressive richness (pp. 47–48).

It is, however, Stenning and Oberlander [2] who have offered the clearest insight into the connection between positive and negative functional traits of graphics. According to them, "graphical representation such as diagrams limit abstraction and thereby aid processibility" (p. 98). Here, "limiting abstraction" means limited capacities of expressing weak, non-specific information, while "processibility" is capacities of supporting

M. Hegarty, B. Meyer, and N. Hari Narayanan (Eds.): Diagrams 2002, LNAI 2317, pp. 116–130, 2002.
© Springer-Verlag Berlin Heidelberg 2002

	F_1	F_2	F_3	F_4	F_5	F_6
A	○		○			
B	○					
C	○					
D	○	○	○	○	○	
E		○		○	○	
G			○			○

Fig. 1. A well-formed representation of the system $\mathcal{R}_t$ of feature tables.

efficient inferences on the user's part. Stenning and Inder [7] even propose a general ordering of representation systems according to their expressive flexibility, and compare the inferential potentials of systems in different places in this ordering.

Thus, previous studies strongly suggest an association among (a) the potential of a graphical system to allow efficient inferences, (b) its capacity for rich presentation of information, and (c) its tendency to prohibit flexible presentations of information. What is then the semantic mechanism behind this association? Is there any common property of graphical representation systems from which these positive and negative traits all derive? Let us call this question *the trade-off problem*.

In this paper, we try to answer this question by investigating a particular system of graphical representations in detail. It is a system of tabular representations of a common type, which, despite its simple syntax and semantics, exhibit the apparent association in question. Relying on previous studies of inferential and expressive potentials of graphics, we will start with showing the exact senses in which this tabular system supports efficient inferences, enables rich expression of information, and prohibits flexible expression of information (section 1). After introducing basic semantic concepts, we will show that the tabular system has a property that can be called "trackings of capturing relations over homomorphic exhaustive sets" (section 2). This property is a near-sufficient condition for all the relevant functional traits, and hence can be considered the semantic mechanism behind the trade-off phenomenon. We will close the paper by discussing the extent and generalizability of our analysis (section 3). Discussions will be kept informal throughout this paper, for compact exposition of core ideas.

1   Reproducing the Trade-Off

The tabular system we investigate consists of simple "feature" tables such as the one in Figure 1. Here, the labels for rows, "A," "B," ... and "G," are the names of ink-jet printer models and the labels for columns, "F_1," "F_2," ... and "F_6," are the names of various functions that ink-jet printers may have. (For the ease of reference, we use these simple names for printer models and their potential functions, instead of real names and descriptions, such as "Epson PM750," "Canon M70," "Print on A3-size paper" and "Print in 720×360 dpi.")

The semantics of this system is natural one: if a circle appears in the intersection of the raw labeled by "X" and the column labeled by "Y," then it indicates that the

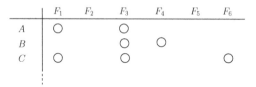

Fig. 2. The result of expressing the set of information $\{(1), (2), (3)\}$ in an $\mathcal{R}_t$-table.

printer model X has the function Y; if that position is blank, it indicates that the printer model X lacks the function Y. Thus, according to the table in Fig 1, the model A has the function F_3 but the model B does not; the model G has the function F_6 but the model E does not, and so on. The syntax of the system requires that all the six printer names appear as labels for rows in alphabetical order, and that all the six function names appear as labels for columns in numerical order. Each cell of a table must either be blank or have a circle in it. We call this system of tabular representations $\mathcal{R}_t$. [1]

Now, the trade-off problem is a question on the mechanism underlying the apparent association of several functional traits of representation systems. So far, those traits have been intuitively described as "inferential efficiency," "expressive inflexibility," and "expressive richness." The first step in the analysis of the phenomenon is then to unpack these intuitive descriptions. For this paper's plan, this amounts to specifying inferential efficiency, expressive inflexibility, and expressive richness as they are found in a system of tabular representations. We start with inferential efficiency.

1.1   Inferential Efficiency

In what respect does $\mathcal{R}_t$ support efficient inferences? Consider the question whether the information (4) is a consequence of the set of information $\{(1), (2), (3)\}$.

(1) The model A has the functions F_1 and F_3 and no other functions.
(2) The model B has the functions F_3 and F_4 and no other functions.
(3) The model C has the functions F_1, F_3 and F_6 and no other functions.
(4) None of the models A, B, and C has the function F_2.

Well, yes, it is, of course. One could answer this question by directly thinking about an arbitrary situation in which (1), (2), and (3) all hold, and asking oneself whether (4) necessarily holds in that situation. Alternatively, one could express all the information (1), (2), and (3) in an $\mathcal{R}_t$-table. Noticing that a blank column under the label "F_2" in the resulting table (Figure 2), one can read off the information (4) and validly conclude that (4) is a consequence of $\{(1), (2), (3)\}$.

This alternative way is an instance of the procedure characterized as "free ride" (Shimojima [8]), which various researchers have emphasized as a main root of inferential efficiency provided by graphical systems (Sloman [1], Lindsay [9], Barwise and

---

[1] We discuss this particular system for the sake of concreteness, but our discussions would apply to any feature tables and, less straightforwardly, to tabular representations in general.

Etchemendy [5]). In a free ride, one does not have to think directly about the consequence relation governing the printer situation; instead, one can just express one's premises in a graphic representation and observe the result to draw read off a consequence of the premises. Setting aside the issue of the semantic mechanism of free rides, we just note for now that there are numerous combinations of premises and conclusions whose derivability can be checked in similar ways with $\mathcal{R}_t$ tables. From the present table, for example, one notes a sequence of circles under "F_3" and can read off the consequence (5).[2]

(5) All of the models A, B, and C has the function F_3.

1.2   Expressive Inflexibility

Thus, we have identified the potentials for free rides and graphical consistency proofs as two main roots of the inferential efficacy of the system $\mathcal{R}_t$. In what respect is the system expressively inflexible then? Consider the following information:

(6) Exactly two models have the function F_1.

Can an $\mathcal{R}_t$-table express this information alone, without expressing any other information? First, in order to express this information, an $\mathcal{R}_t$-table must have two circles in the column labeled "F_1, but of course, these circles must be placed in some *particular* cells in that column. Thus, the $\mathcal{R}_t$-table ends up specifying *which two* of the models A, B, C, and D have the function F_1. Secondly, the $\mathcal{R}_t$-table must have either a circle or a blank in each cell of the other columns. Thus, the $\mathcal{R}_t$-table must also specify, for each of the printer models A, B, C, and D and each function other than F_1, whether that model has that function. For these reasons, no $\mathcal{R}_t$-table can express (6) alone.

   This inflexibility of expression in the system $\mathcal{R}_t$ is an instance of the property characterized as "content-specificity" or "over-specificity" by Stenning and Oberlander [2] and Shimojima [8]. Over-specificity is a system's incapability of expressing certain sets of information without expressing extra information, and hence it amounts to certain limitations on the expression of weak or abstract information. Philosophers have long considered this property as a distinguishing character of pictorial images (Berkeley [11], Hume [12], Dennett [13], Pylyshyn [14]), most notably of the system of geometry diagrams ("we can't draw a right triangle *per se*"). The property is now considered a feature of a wider range of graphical systems (such as Euler diagrams) and as the above example shows, it is also a feature of the tabular system $\mathcal{R}_t$. Note that there are many other instances of information that $\mathcal{R}_t$ is over-specific about, including:

(7) The model B has exactly two functions.

---

[2] Free rides are in turn one of a more general group of inferential procedures called "physical on-site inferences" (Shimojima [10]). The group contains three other inferential procedures, and precisely speaking, the inferential potential of $\mathcal{R}_t$ also consists in its capacities for these procedures. Due to space limitation, we have to omit discussions of these procedures here, but our analysis of $\mathcal{R}_t$ in terms of its potential for free rides largely applies to those procedures.

1.3   Expressive Richness

Intuitively, the $\mathcal{R}_t$-table in Figure 1 expresses the information that none of the models A, B, and C has the function F_2. Clearly, it expresses this information because, in that table, the entire area where the rows labeled "A," "B," and "C" intersect with the column labeled "F_2" is blank. Thus, in one way or another, this condition of the table, (4^*), indicates the information (4).

(4^*) The area where the rows labeled "A," "B," and "C" intersect with the column labeled "F_2" is all blank.

(4) None of the models A, B, and C has the function F_2.

This semantic relation, however, is distinct from the semantic rules specified in the beginning of this section. Those basic rules simply say that the appearance of a circle in a particular cell indicates the possession of a particular feature by a particular printer type, and that the blankness of a particular cell indicates the non-possession. They are, so to speak, concerned with the meanings of cell-wise states in a table. In contrast, the semantic relation from (4^*) to (4) is concerned with the meaning of the blankness of a *sequence* of cells. It is area-wise, so to speak.

This second semantic relation is an instance of the phenomenon called "derivative meaning" (Shimojima [15]). Although it is different from the basic semantic rules, it clearly depends on them. Just imagine there were not the basic rule concerning the meaning of the blankness of individual cells in am $\mathcal{R}_t$-table. Then the blankness of a sequence of cells would mean nothing, and the semantic relation from (4^*) to (4) would not hold. Note that the system $\mathcal{R}_t$ has many other semantic relations derived from its basic semantic relations. They include the one from (8^*) to (8) and from (9^*) to (9) below:

(8^*) The entire column labeled "F_1" is filled with circles.

(8) All of the models has the function F_1.

(9^*) More circles appear in the row labeled "D" than in the row labeled "E."

(9) The model D has more functions than the model E has.

Now, a system's capacity for such meaning derivation greatly contributes to the overall richness of expressive capacity. For example, Kosslyn [4] alludes to this phenomenon in describing the rich semantic capacities of scatter plots and other visualization forms of statistical data. The additional semantic relations such as the ones just specified is certainly a root of the semantic richness of the particular system $\mathcal{R}_t$.

Thus, we have succeeded in reproducing, in a simple tabular system, the opposition of functional traits that is apparently alluded to in the statement of the trade-off problem: (a) inferential efficiency arising from its potentials for free rides, (b) expressive inflexibility arising from its over-specific character, and (c) expressive richness arising from its potential for derivative meaning. Our underlying assumption is, of course, that this particular opposition found in $\mathcal{R}_t$ is an instance of the opposition responsible for the trade-off problem in general. However, let us defer discussions of the validity of this assumption until later, and let us concentrate on the issue of how we can identify the common semantic mechanism for these particular traits in $\mathcal{R}_t$.

2 Analysis

In our view, that common semantic mechanism consists in what we call "homomorphism of exhaustive sets," and all the three functional traits derive from it through various types of "tracking of exhaustive relations." In this section, we first define these two concepts, and then show how those semantic properties give rise to free rides, over-specificity, and derivative meaning in the system $\mathcal{R}_t$.

2.1 Homomorphism of Exhaustive Sets

Let us begin with introducing several basic terms. *Source tokens* of the system $\mathcal{R}_t$ are *particular* well-formed $\mathcal{R}_t$-tables drawn in different places, such as a piece of paper, a computer display, and a white board. Thus, a particular body of black ink printed as Figure 1 earlier in this paper is an example of a source token of $\mathcal{R}_t$. An $\mathcal{R}_t$-table printed on a different place would be counted as a different source token even if it has exactly the same structure as Figure 1. In contrast, *source types* are conditions or structures of source tokens, and therefore the same source type can hold of different source tokens. For example, the condition of an $\mathcal{R}_t$-table described as (9*) is a source type of the system $\mathcal{R}_t$. It holds of the particular $\mathcal{R}_t$-table in Figure 1, but it can hold of indefinite numbers of other $\mathcal{R}_t$-tables as far as they have more circles in the row labeled "D" than in the row labeled "E."

We distinguish *target tokens* and *target types* similarly. Target tokens are particular situations in different places and times concerning the six printer models and the six potential functions. In contrast, target types are conditions or structures of target tokens, and the same target type can hold of different target tokens. An example is the condition described as (9). It may hold of the current printer situation in Japan, but also can hold of the current printer situation in the US, as long as the printer model D has more functions than E has in both situations.

Given a source type σ and a target type θ of the system $\mathcal{R}_t$, we say σ *indicates* θ in $\mathcal{R}_t$, written as "$\sigma \Rightarrow_{\mathcal{R}_t} \theta$," if the semantic relation holds from σ to θ on the basis of the semantic rules for $\mathcal{R}_t$ or as "derivative" of them in the sense described in the last section. Γ of source types and a *set* Δ of target types, we say Γ is projected to Δ in $\mathcal{R}_t$, written as "$\Gamma \Rightarrow_{\mathcal{R}_t} \Delta$," if every member of Γ has some member of Δ that it indicates and if every member of Δ has some member of Γ that indicates it. We say a source token s *satisfies* a set Γ of source types if every condition in Γ holds of s. Similarly for target tokens and sets of target types. See Barwise and Seligman [3], especially chapter 20, for a more systematic presentation of these semantic concepts.

With this preparation, let us illustrate the notion of "exhaustive set." Recall that $\mathcal{R}_t$ is a special-purpose representation system, designed to display the functions of printer models A through G. The variety of functions to be displayed are F_1 through F_6. Thus, the semantically relevant structure of a table in this system can be specified by a set of 6×6 source types, specifying whether each cell has a circle or a blank space. When a set contains 6×6 of such source types corresponding to the 6×6 cells of a well-formed $\mathcal{R}_t$-table, we call the set the *cell-wise structure* of the $\mathcal{R}_t$-table. For example, the cell-wise structure of the table in Figure 1 is the set of source types listed in Figure 3, where "$\bigcirc(X, Y)$" denotes the condition that the intersection of the row labeled "X" and the

$\{ \bigcirc(A, F_1), \ \square(A, F_2), \ \bigcirc(A, F_3), \ \square(A, F_4), \ \square(A, F_5), \ \square(A, F_6),$
$\bigcirc(B, F_1), \ \square(B, F_2), \ \square(B, F_3), \ \square(B, F_4), \ \square(B, F_5), \ \square(B, F_6),$
$\bigcirc(C, F_1), \ \square(C, F_2), \ \square(C, F_3), \ \square(C, F_4), \ \square(C, F_5), \ \square(C, F_6),$
$\bigcirc(D, F_1), \ \bigcirc(D, F_2), \ \square(D, F_3), \ \bigcirc(D, F_4), \ \bigcirc(D, F_5), \ \square(D, F_6),$
$\square(E, F_1), \ \bigcirc(E, F_2), \ \square(E, F_3), \ \bigcirc(E, F_4), \ \bigcirc(E, F_5), \ \square(E, F_6),$
$\square(G, F_1), \ \square(G, F_2), \ \bigcirc(G, F_3), \ \square(G, F_4), \ \square(G, F_5), \ \bigcirc(G, F_6) \}$

Fig. 3. A cell-wise structure, Γ_{38}, in the system $\mathcal{R}_t$.

column labeled "Y" has a circle in it, and "$\square(X, Y)$" denotes the condition that that intersection has a blank space. We call this particular cell-wise structure "Γ_{38}." Since each well-formed table in $\mathcal{R}_t$ has $6 \times 6 = 36$ cells, there are a total of 2^{36} different cell-wise structures in $\mathcal{R}_t$.

By definition, every cell-wise structure is satisfied by some possible $\mathcal{R}_t$-table. And clearly, every $\mathcal{R}_t$-table has exactly one cell-wise structure. That is, if $\mathcal{G}$ is the set of all cell-wise structures, the following facts hold:

(10) Every source token of $\mathcal{R}_t$ satisfies at least one member of $\mathcal{G}$.

(11) Every source token of $\mathcal{R}_t$ satisfies at most one member of $\mathcal{G}$.

(12) Every member of $\mathcal{G}$ is satisfied by some source token of $\mathcal{R}_t$.

We will indicate these facts by calling $\mathcal{G}$ an *exhaustive set* in the source domain of the system $\mathcal{R}_t$.

Note that the condition (10) implies that the set of tokens satisfying at least one member of $\mathcal{G}$ exhaust the entire set of source tokens; the condition (11) implies that the members of $\mathcal{G}$ are mutually incompatible, while (12) implies that each member is singularly consistent. Thus, if we write "$\mathrm{tok}(\Gamma_i)$" to denote the set of tokens satisfying a member Γ_i of $\mathcal{G}$, the collection $\{\mathrm{tok}(\Gamma_i) \mid \Gamma_i \in \mathcal{G}\}$ partitions the entire set of source tokens of the system $\mathcal{R}_t$.

Interestingly, the system $\mathcal{R}_t$ has a "corresponding" exhaustive set in its target. Consider the set of target types listed in Figure 4, where "$F(X, Y)$" denotes the condition that model X features the function Y, and "$L(X, Y)$" denotes the type that model X lacks the function Y. Note that this set is a complete pair-wise specification of a binary "featuring" relation from the six printer models A through G to the six functions F_1 through F_6. Generally, when 6×6 of such target types specify a possible featuring relation among these models and functions, we call the set of those types a *pair-wise structure*. Assuming all combinations of the printer models and the functions are possible, there are a total of 2^{36} pair-wise structures in $\mathcal{R}_t$. The particular pair-wise structure listed in Figure 4 will be called "Δ_{38}" in the following discussions.

By definition, every possible printer situation within the coverage of $\mathcal{R}_t$ satisfies exactly one pair-wise structure. Conversely, every pair-wise structure is satisfied by some possible printer situation. Thus, calling the set of all pair-wise structures "$\mathcal{D}$," we can say:

(13) Every target token of $\mathcal{R}_t$ satisfies at least one member of $\mathcal{D}$.

$\{ F(A, F_1),\ L(A, F_2),\ F(A, F_3), L(A, F_4),\ L(A, F_5),\ L(A, F_6),$
$\ \ F(B, F_1),\ L(B, F_2),\ L(B, F_3), L(B, F_4),\ L(B, F_5),\ L(B, F_6),$
$\ \ F(C, F_1),\ L(C, F_2),\ L(C, F_3), L(C, F_4),\ L(C, F_5),\ L(C, F_6),$
$\ \ F(D, F_1),\ F(D, F_2),\ F(D, F_3), F(D, F_4),\ F(D, F_5),\ L(D, F_6),$
$\ \ L(E, F_1),\ F(E, F_2),\ L(E, F_3), F(E, F_4),\ F(E, F_5),\ L(E, F_6),$
$\ \ L(G, F_1),\ L(G, F_2),\ F(G, F_3), L(G, F_4),\ L(G, F_5),\ F(G, F_6)\ \}$

Fig. 4. A pair-wise structure, Δ_{38}, in the system $\mathcal{R}_t$.

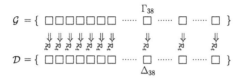

Fig. 5. The homomorphism h from the exhaustive set $\mathcal{G}$ in the source to the exhaustive set $\mathcal{D}$ in the target.

(14) Every target token of $\mathcal{R}_t$ satisfies at most one member of $\mathcal{D}$.

(15) Every member of $\mathcal{D}$ is satisfied by some target token of $\mathcal{R}_t$.

In other words, $\mathcal{D}$ is an exhaustive set in the target domain of the system $\mathcal{R}_t$.

It is this exhaustive set $\mathcal{D}$ that we earlier said to "correspond" to the exhaustive set $\mathcal{G}$. According to the semantic rules in the system $\mathcal{R}_t$, $\bigcirc(X, Y)$ indicates $F(X, Y)$, and $\square(X, Y)$ indicates $L(X, Y)$. Thus, for example, the set Γ_{38} is projected to Δ_{38} but to no other members of $\mathcal{D}$. Generally, every set in $\mathcal{G}$ has exactly one set in $\mathcal{D}$ that it is projected to. Thus, the projection in the system $\mathcal{R}_t$ provides a one-one mapping from $\mathcal{G}$ to $\mathcal{D}$. To indicate this fact compactly, we say that the exhaustive set $\mathcal{G}$ is *homomorphic* to the exhaustive set $\mathcal{D}$ in the system $\mathcal{R}_t$. We will also use "h" to denote the particular one-one mapping from $\mathcal{G}$ to $\mathcal{D}$.

Figure 5 pictures the homomorphism h from $\mathcal{G}$ to $\mathcal{D}$. Each point in the upper square represents an cell-wise structure in $\mathcal{G}$ such as Γ_{38}. Each arrow represents a semantic projection in $\mathcal{R}_t$ that relates the members of a cell-wise structure in $\mathcal{G}$ to the members of a pair-wise structure in $\mathcal{D}$ such as Δ_{38}. (The relation is a one-one correspondence in this particular case.) Then the entire collection of arrows represents the homomorphism h that maps each cell-wise structure in $\mathcal{G}$ to a unique pair-wise structure in $\mathcal{D}$. (The mapping is a one-one correspondence in this particular case.)

Now, due to this homomorphism, a large collection $\mathcal{G}$ of sets of source types semantically corresponds to a large collection of sets of target types. Moreover, by conditions (11), (12), (14), and (15) above, both collections consist of mutually incompatible, but singularly consistent sets of types. As we will see shortly, the extensive semantic correspondence between these special collections is a double-edged sword for the efficacy of the system $\mathcal{R}_t$: combined with "capturing" capacities discussed below, it creates constraint matching sufficient for on-site inferences and content inducement on the one hand, while creating constraint matching for over-specificity on the other hand.

$$\begin{aligned}
\{ &\bigcirc(A, F_1), \ \square(A, F_2), \ \bigcirc(A, F_3), \ \square(A, F_4), \ \square(A, F_5), \ \square(A, F_6), \\
&\square(B, F_1), \ \square(B, F_2), \ \bigcirc(B, F_3), \ \bigcirc(B, F_4), \ \square(B, F_5), \ \square(B, F_6), \\
&\bigcirc(C, F_1), \ \square(C, F_2), \ \bigcirc(C, F_3), \ \square(C, F_4), \ \square(C, F_5), \ \bigcirc(C, F_6), \\
&\square(D, F_1), \ \bigcirc(D, F_2), \ \bigcirc(D, F_3), \ \bigcirc(D, F_4), \ \bigcirc(D, F_5), \ \square(D, F_6), \\
&\square(E, F_1), \ \bigcirc(E, F_2), \ \bigcirc(E, F_3), \ \bigcirc(E, F_4), \ \bigcirc(E, F_5), \ \square(E, F_6), \\
&\square(G, F_1), \ \square(G, F_2), \ \bigcirc(G, F_3), \ \square(G, F_4), \ \square(G, F_5), \ \bigcirc(G, F_6) \}
\end{aligned}$$

Fig. 6. Another cell-wise structure, Γ_{269}, in the system $\mathcal{R}_t$.

2.2   Tracking of a Capturing Relation

We say a type *captures* a collection of sets of types if the set of tokens supporting the type is equal to the set of tokens satisfying at least one member of the collection. For example, compare the cell-wise structure Γ_{38} cited above and the following type:

(1*)   The row labeled "A" has circles in the columns labeled "F_1" and "F_3" and nowhere else.

Clearly, if an $\mathcal{R}_t$-table has the cell-wise structure Γ_{38}, then in that table, the row labeled "A" has circles in the columns labeled "F_1" and "F_3" and nowhere else. That is, every source token satisfying Γ_{38} supports the source type (1*). The same holds for the cell-wise structure Γ_{269} listed in Figure 6, for if a source token satisfies Γ_{269}, it necessarily supports (1*).

Suppose we enumerate all such cell-wise structures and call the resulting collection of cell-wise structures "$\mathcal{G}_{1^*}$." If the enumeration is complete, $\mathcal{G}_{1^*}$ should exhaust all possible "ways" in which (1*) holds. So, (1*) and $\mathcal{G}_{1^*}$ are in the following relationship:

(16)   Every source token supporting (1*) satisfies at least one member of $\mathcal{G}_{1^*}$.

(17)   Every source token satisfying some member of $\mathcal{G}_{1^*}$ satisfies (1*).

Exactly in this sense, we say that the type (1*) "captures" the collection $\mathcal{G}_{1^*}$. Intuitively, a type that captures a collection is a "short way" of saying the large disjunction of the conjunctions of 6×6 types contained in individual members of that collection.

The capturing relation holds between many source types of the system $\mathcal{R}_t$ and many subsets of the exhaustive set $\mathcal{G}$. For instance, the following source types capture distinct, but partly overlapping, subsets of $\mathcal{G}$:

(2*)   The row labeled "B" has circles in the columns labeled "F_3" and "F_4" and nowhere else.

(3*)   The row labeled "C" has circles in the columns labeled "F_1," "F_3" and "F_6" and nowhere else.

(4*)   The area where the rows labeled "A," "B," and "C" intersect with the column labeled "F_2" is all blank.

(6*)   The column labeled "F_1" has exactly two circles.

Let us use '$\mathcal{G}_i$' to denote the subset of $\mathcal{G}$ captured by a source type σ_i. The above six types therefore capture the subsets $\mathcal{G}_{2^*}, \mathcal{G}_{3^*}, \mathcal{G}_{4^*},$ and $\mathcal{G}_{6^*}$, respectively. The collections $\mathcal{G}_{1^*}, \mathcal{G}_{2^*},$ and $\mathcal{G}_{3^*}$ have 2^{30} members each, while $\mathcal{G}_{4^*}$ has 2^{33} members and $\mathcal{G}_{6^*}$ has 5×2^{30} members.

Now that we understand the notion of "capturing," let us illustrate what it is to "track" a capturing relation. We have seen that there is a one-one semantic mapping h from $\mathcal{G}$ to $\mathcal{D}$. Thus each subset $\mathcal{G}_i$ of $\mathcal{G}$ and its image $h(\mathcal{G}_i)$ under h are in a one-one correspondence. For example, $h(\mathcal{G}_{1^*})$ is the collection of pair-wise structures to which the cell-wise structures in $\mathcal{G}_{1^*}$ are projected by $\mathcal{R}_t$.

Interestingly, there is another way of describing this collection $h(\mathcal{G}_{1^*})$. It is the collection of all the "ways" in which the printer model A features the functions F_1 and F_3 and no other functions. In other words, $h(\mathcal{G}_{1^*})$ is the collection *captured* by the following target type:

(1) The model A has the functions F_1 and F_3 and no other functions.

To see this point clearly, let us see what sort of things are in $h(\mathcal{G}_{1^*})$. By definition, this collection contains all and only those sets of target types to which a member of $\mathcal{G}_{1^*}$ is projected. Since $\mathcal{G}_{1^*}$ is captured by the source type (1^*), $\mathcal{G}_{1^*}$ in turn consists of all and only those cell-wise structures that share the following source types:

(18) $\bigcirc(A, F_1), \square(A, F_2), \bigcirc(A, F_3), \square(A, F_4), \square(A, F_5), \square(A, F_6)$

Since each pair-wise structure in $h(\mathcal{G}_{1^*})$ is just the element-by-element translations of a cell-wise structure in $\mathcal{G}_{1^*}$, it must contain the corresponding target types below:

(19) $F(A, F_1), L(A, F_2), F(A, F_3), L(A, F_4), L(A, F_5), L(A, F_6)$

Moreover, every pair-wise structure containing (19) is clearly the element-by-element translation of a cell-wise structure containing (18). It follows that every cell-wise structure containing (19) is the element-by-element translation of a member of $\mathcal{G}_{1^*}$, and hence is a member of $h(\mathcal{G}_{1^*})$. In sum, $h(\mathcal{G}_{1^*})$ consists of all and only those pair-wise structures containing (19).

We have thus confirmed that (1) captures $h(\mathcal{G}_{1^*})$. But in the source domain of the system $\mathcal{R}_t$, (1^*) captures $\mathcal{G}_{1^*}$. This is what we call a "tracking of a capturing." Figure 7 shows the situation schematically. Intuitively, when a source type captures a collection whose correspondent collection under h is captured by a target type, it is a case of tracked capturing.

In the system $\mathcal{R}_t$, there are many cases of such tracking. For example, the source type (4^*) captures the collection $\mathcal{G}_{4^*}$ whose correspondent collection $h(\mathcal{G}_{4^*})$ is captured by the following target type:

(4) None of the models A, B, and C has the function F_2.

Similarly, the source types (2^*), (3^*), and (6^*) capture the collections whose correspondents are captured by the following target types respectively.

(2) The model B has the functions F_3 and F_4 and no other functions.
(3) The model C has the functions F_1, F_3 and F_6 and no other functions.
(6) Exactly two models have the function F_1.

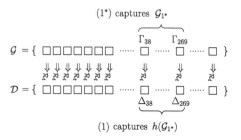

(1^*) captures $\mathcal{G}_{1^*}$

(1) captures $h(\mathcal{G}_{1^*})$

Fig. 7. The capturing of $\mathcal{G}_{1^*}$ by (1^*) tracks the capturing of $h(\mathcal{G}_{1^*})$ by (1).

2.3   Deriving the Phenomena

We have witnessed how two exhaustive sets are semantically homomorphic in the system $\mathcal{R}_t$ and how a capturing relation on one exhaustive set is tracked by a capturing relation on the other. Let us now see how these two properties of $\mathcal{R}_t$ give rise to free rides, derived meaning, as well as over-specific characters.

Derivative Meaning. Recall, from section 1.3, that the source type (4^*) indicates the target type (4) as derived meaning. This indication relation can be explained from the fact that the capturing of the collection $h(\mathcal{G}_{4^*})$ by (4) is tracked by the capturing of the collection $\mathcal{G}_{4^*}$ by (4^*). Suppose an arbitrary source token s (such as the table in Figure 1) supports the source type (4). Since (4^*) captures $\mathcal{G}_{4^*}$, s satisfies some member Γ_k of $\mathcal{G}_{4^*}$. But that member is semantically projected to some member Δ_k of $h(\mathcal{G}_{4^*})$. Since (4) captures $h(\mathcal{G}_{4^*})$, Δ_k entails (4). Thus, the token s ends up expressing (4). The indication from (4^*) to (4) is thus generated.

Similar analyses apply to other pairs of types such as (6^*) and (6). Generally, when a target type captures a collection of mutually incompatible, but singularly consistent sets of source types, and a source type tracks this capturing by capturing the semantically corresponding collection of source types, the source type has the target types as an inducible content. The homomorphism h between the exhaustive sets $\mathcal{G}$ and $\mathcal{D}$ generates quite extensive semantic correspondences of this kind, and therefore prepares much room of parallel capturings. Thus, the enrichment of the indication relation in $\mathcal{R}_t$ derived from this mechanism is quite significant.

Over-Specificity. Recall, from section 1.2, that any $\mathcal{R}_t$-table cannot express the information (6) alone, without expressing any other information. This over-specific character of $\mathcal{R}_t$ is accountable in the following way. Note (6) captures the collection $h(\mathcal{G}_{6^*})$, and this capturing is tracked by the capturing of $\mathcal{G}_{6^*}$ by (6^*). Obviously, $\mathcal{G}_{6^*}$ has a large number of elements. Correspondingly, there is a large number of elements of $h(\mathcal{G}_{6^*})$. Now, for a table s in this system to express (6), it must support the source type (6^*). Since (6^*) captures $\mathcal{G}_{6^*}$, s must satisfy one of the members of $\mathcal{G}_{6^*}$. But $\mathcal{G}_{6^*}$ contains more than one members, say, $\Sigma_1, \ldots, \Sigma_n$, which are semantically projected to distinct

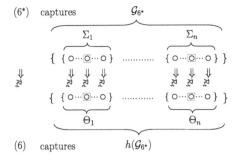

Fig. 8. Over-specificity in presenting (6), which is generated from the parallelism of the capturing of $\mathcal{G}_{6^*}$ by (6) and the capturing of $h(\mathcal{G}_{6^*})$ by (6*).

members $\Theta_1, \ldots, \Theta_n$ of $h(\mathcal{G}_{6^*})$. So, s ends up expressing one of these distinct members $\Theta_1, \ldots, \Theta_n$ of $h(\mathcal{G}_{6^*})$.

Recall $h(\mathcal{G}_{6^*})$ is a subset of the exhaustive set $\mathcal{D}$, and so its members are mutually incompatible, but singularly consistent sets of target types. It follows that for each member of $h(\mathcal{G}_{6^*})$, there is a target token that satisfies it while satisfying no other members. This in turn means that for each member of $h(\mathcal{G}_{6^*})$, there is a target token that does not satisfy it while satisfying some other member, and therefore supporting (6). Thus, each member of $h(\mathcal{G}_{6^*})$ contains some target type that does not follow from (6).

Figure 8 summarizes this situation. Each circle represents an individual source type or a target type, and each arrow represents an indication relation from a source type to an target type. Each member of $h(\mathcal{G}_{6^*})$ has a member (squared) that does not follow from (6), and each member of $\mathcal{G}_{6^*}$ has a corresponding member (squared). This mechanism forces the table s to express some unwarranted information, and generally, no $\mathcal{R}_t$-table can express (6) without thereby expressing some unwarranted information.

Thus, the inflexibility of expression in question derives from the tracking of capturing defined on homomorphic exhaustive sets. Generally, whenever a target type captures a collection with more than one members and this capturing is tracked by a capturing in the source domain, we have a case of over-specificity. This mechanism accounts for the system's inflexibility in expressing $\theta_1, \theta_2, \theta_3, \theta_4, \theta_5, \theta_7$, and numerous other target types whose capturings are thus tracked.

Free Rides. Recall, from section 1.1, that the system $\mathcal{R}_t$ allows a free ride from the set of information $\{(1), (2), (3)\}$ to its consequence (4). Recall that the target types (1),(2),(3), and (4) respectively capture the collections $h(\mathcal{G}_{1^*}), h(\mathcal{G}_{2^*}), h(\mathcal{G}_{3^*})$, and $h(\mathcal{G}_{4^*})$. These capturings in the target domain are tracked by other capturings in the source domain, namely, by the capturings of $\mathcal{G}_{1^*}, \mathcal{G}_{2^*}, \mathcal{G}_{3^*}$, and $\mathcal{G}_{4^*}$ by the source types (1*),(2*),(3*), and (4*). It is clear from the discussions on derivative meaning that this tracking makes the source types (1*),(2*),(3*), and (4*) respectively have the target types (1),(2),(3), and (4) as their derivative meaning. So, the set $\{(1^*), (2^*), (3^*)\}$ is projected to the set $\{(1), (2), (3)\}$ in $\mathcal{R}_t$, while (4*) indicates (4) in $\mathcal{R}_t$.

$$\{(1^*),(2^*),(3^*)\} \vdash \quad (4^*)$$

$$\Downarrow_{\mathcal{R}_t} \qquad\qquad \Downarrow_{\mathcal{R}_t}$$

$$\{(1),(2),(3)\} \vdash \quad (4)$$

Fig. 9. The matching of consequence relations, generated from the parallelism of capturings by $(1^*),(2^*),(3^*)$, and (4^*) and those by $(1),(2),(3)$, and (4).

Now, crucially, the intersection of the collections $\mathcal{G}_{1^*}$, $\mathcal{G}_{2^*}$, and $\mathcal{G}_{3^*}$ is a subset of the collection $\mathcal{G}_{4^*}$. Since the set of tokens supporting a type σ_i is always equal to the set of tokens satisfying at least one member of the collection $\mathcal{G}_i$ captured by σ_i, this immediately means that the set of tokens supporting (4^*) is a subset of the set of tokens supporting all of (1^*), (2^*), and (3^*). That is, the source type (4^*) is a consequence of the set $\{(1^*),(2^*),(3^*)\}$ of source types. Since each of the collections $\mathcal{G}_{1^*}$, $\mathcal{G}_{2^*}$, $\mathcal{G}_{3^*}$, and $\mathcal{G}_{4^*}$ is in a one-one correspondence to its counter-part in $h(\mathcal{G}_{1^*})$, $h(\mathcal{G}_{2^*})$, and $h(\mathcal{G}_{3^*})$, the fact that the intersection of $\mathcal{G}_{1^*}$, $\mathcal{G}_{2^*}$, and $\mathcal{G}_{3^*}$ is a subset of $\mathcal{G}_{4^*}$ transfers to the fact that the intersection of $h(\mathcal{G}_{1^*})$, $h(\mathcal{G}_{2^*})$, and $h(\mathcal{G}_{3^*})$ is a subset of $h(\mathcal{G}_{4^*})$. Hence, the target type (4) is a consequence of $\{(1),(2),(3)\}$, *just as* (4^*) is a consequence $\{(1^*),(2^*),(3^*)\}$. Figure 9 summarizes this situation, where $\vdash$ means a consequence relation.

Thus, we end up with an interesting matching of the consequence relations in the source domain and in the target domain. Due to the consequence on the source part, whenever one expresses the set $\{(1),(2),(3)\}$ of information by creating an $\mathcal{R}_t$-table with the structural properties $\{(1),(2),(3)\}$, the additional condition (4^*) necessarily holds of that $\mathcal{R}_t$-table. Since (4^*) indicates (4) in $\mathcal{R}_t$, this means that the table also expresses the information (4). But due to the consequence of the target domain, this additional information (4) is a consequence of the original set $\{(1),(2),(3)\}$ of information. The automatic expression of the consequence (4) thus takes place.

Generally, if capturings by source types track capturings by target types and the collections captured by some of the source types intersects within the collection captured by another source type, the system has the capacity of on-site inference of consequence for the target types whose capturings are tracked. The system $\mathcal{R}_t$ allows numerous free rides in this way, beside the one we have just seen.

3   Conclusions

In a nutshell, our analysis shows that the opposing functional traits of the system $\mathcal{R}_t$ are rooted in a semantic correspondence between an extensive structure made from source types and another extensive structure made from target types, where those source types and target type are related through basic semantic rules. Specifically, this correspondence is called a homomorphism, and the two extensive structures are called exhaustive sets. It is just like two complex structures are connected by numerous parallel strings. Look back at the picture of the homomorphism in Figure 5.

Now, a capturing is just like setting aside several components in either structure and crushing them together into a single type. Suppose one does the same to those components of the other structure that are connected to the original components with parallel strings. This process is called a tracking of capturings. With these parallel crushings, those parallel strings that used to connect the two sets of components will be also crushed into a single string. This string is a derived meaning relation—it is derived from original parallel strings established by basic semantic rules. Moreover, the system obtains potentials for free rides and over-specificity depending on what components are chosen for these parallel crushings. Thus, the original exhaustive sets connected by numerous strings is the very base on which the system $\mathcal{R}_t$ obtains various functional traits. This is the common semantic mechanism that we have been after.

If so, how much does this account clarify the trade-off phenomenon? Note that the existence of homomorphic exhaustive structures is only the base on which the various functional traits may be constructed; as such it is not a sufficient conditions for them, but only a *near*-sufficient condition. For the system $\mathcal{R}_t$ to obtain those functional traits, there must exist various trackings of capturing relations on that base. Thus questions still remain as to how various types of capturings hold and how capturings in the target domain are paralleled by capturings in the source domain. The first question is, for example, concerned with why we use abstract source types such as (20^*) in addition to more basic source types such as (21^*), or why we use abstract target types such as (20) in addition to more basic target types such as (21). The second question is concerned with why a matching pair of abstract types often exist over the source domain and the target domain, such as the pair of (20^*) and (20).

(20^*) The column labeled "F_1" is all blank.

(21^*) The intersection of the row labeled "C" and the column labeled "F_2" is blank.

(20) No models has the function F_1.

(21) The model C does not have the function F_2.

Strictly speaking, these questions are matters of empirical studies on how humans individuate or construct environmental properties or conditions, but as our analysis of the system $\mathcal{R}_t$ shows, we apparently have a strong capacity for such abstractions. Combined with the homomorphism property of a graphical system, this capacity yields positive functional traits such as expressive richness and inferential efficiency. This gives a competitive edge to the system and makes it preferred and inherited over generations of users.[3] However, the same combination of our strong abstraction capacity and the homomorphism property also yields a negative trait such as expressive inflexibility. This would explain how the opposing traits coexist in many existing graphical systems.

How far can we apply our analysis to other graphical systems then? In our view, the analysis can be directly extended to account for the opposition of functional traits in the system of Euler diagrams, the system of standard geometry diagrams, and many systems of geographical maps. Each of these systems has an extensive semantic correspondence between exhaustive sets in the source and the domain, and various parallel abstractions in these structures give rise to free rides, meaning derivation, and over-specificity.

---

[3] I thank an anonymous reviewer to point out this historical selection process of systems.

This, however, does not mean that the existence of homomorphic exhaustive sets is the only way a system obtains the potentials for meaning derivation, free rides, and over-specificity. Our claim is solely that a homomorphism of exhaustive sets in the source and the target holds *when and where* a graphical system has inferential efficiency, expressive richness, and expressive inflexibility *together*. Thus, there may well be graphical systems with one functional trait without the others, or the degrees of their coexistence may vary. The system of Venn diagrams, for example, is a system with a fairly high capacity for free rides without severe setback from over-specificity. It would be, therefore, an interesting extension of our analysis to investigate various ways a graphical system may have the semantic properties for free rides or meaning derivation without a homomorphism of exhaustive sets, an optimal combination of functional traits.

References

1. Sloman, A.: Interactions between philosophy and ai: the role of intuition and non-logical reasoning in intelligence. Artificial Intelligence **2** (1971) 209–225
2. Stenning, K., Oberlander, J.: A cognitive theory of graphical and linguistic reasoning: Logic and implementation. Cognitive Science **19** (1995) 97–140
3. Barwise, J., Seligman, J.: Information Flow: the Logic of Distributed Systems. Cambridge Tracts in Theoretical Computer Science, 42. Cambridge University Press, Cambridge, UK (1997)
4. Kosslyn, S.M.: Elements of Graph Design. W. H. Freeman and Company, New York (1994)
5. Barwise, J., Etchemendy, J.: Visual information and valid reasoning. In Allwein, G., Barwise, J., eds.: Logical Reasoning with Diagrams. Oxford University Press, Oxford (1990) 3–25
6. Barwise, J., Hammer, E.: Diagrams and the concept of logical system. In Gabbay, D., ed.: What Is a Logical System? Oxford University Press, Oxford (1995)
7. Stenning, K., Inder, R.: Applying semantic concepts to analyzing media and modalities. In Glasgow, J., Narayanan, N.H., Chandrasekaran, B., eds.: Diagrammatic Reasoning: Cognitive and Computational Perspectives. The MIT Press and the AAAI Press, Cambridge, MA and Menlo Park, CA (1995) 303–338
8. Shimojima, A.: Operational constraints in diagrammatic reasoning. In Barwise, J., Allwein, G., eds.: Logical Reasoning with Diagrams. Oxford University Press, Oxford (1995b) 27–48
9. Lindsay, R.K.: Images and inference. In Glasgow, J.I., Narayanan, N.H., Chandrasekaran, B., eds.: Diagrammatic Reasoning: Cognitive and Computational Perspectives. The MIT Press and the AAAI Press, Cambridge, MA and Menlo Park, CA (1988) 111–135
10. Shimojima, A.: A logical analysis of graphical consistency proofs. In Magnani, L., Nersessian, N., Pizzi, C., eds.: Proceedings of MBR 2001 (tentative title). Kluwer Academic Publishers, Dordrecht, Netherlands (Forthcoming)
11. Berkeley, G.: A Treatise Concerning Principles of Human Knowledge. The Library of Liberal Arts. The Bobbs-Merrill Company, Inc., Indianapolis (1710)
12. Hume, D.: A treatise of human nature. In: The Philosophical Works of David Hume. Volume 1. Little, Brown and Company, Boston (1739)
13. Dennett, D.C.: Content and Consciousness. International library of philosophy and scientific method. Routledge and Kegan Paul, London (1969)
14. Pylyshyn, Z.W.: What the mind's eye tells the mind's brain: A critique of mental imagery. Psychological Bulletin **80** (1973) 1–24
15. Shimojima, A.: Derivative meaning in graphical representations. In: Proceedings of the 1999 IEEE Symposium on Visual Languages. IEEE Computer Society, Washington, D. C. (1999) 212–219

Modeling Heterogeneous Systems

Nik Swoboda[1] and Gerard Allwein[2]

[1] DEIMOS Space S.L.
Sector Oficios 34, Tres Cantos, Madrid 28760, Spain
nik.swoboda@deimos-space.com
[2] Indiana University
Bloomington, IN 47408, USA
gtall@cs.indiana.edu

Abstract. Reasoning practices and decision making often require information from many different sources, which can be both sentential and diagrammatic. In such situations, there are many advantages to reasoning with the diagrams themselves, as opposed to re-expressing the information content of the diagram in sentential form and reasoning in an abstract sentential language. Thus for these practices, being able to extract and re-express pieces of information from one kind of representation into another is essential. The main goal of this paper is to propose a general framework for the modeling of heterogeneous reasoning systems and, most importantly, heterogeneous rules of inference in those systems. Unlike some other work in designing heterogeneous systems, our purpose will not be to define just one notion of heterogeneous inference, but rather to provide a framework in which many different kinds of heterogeneous rules of inference can be defined. After proposing this framework, we will then show how it can be applied to a sample heterogeneous system to define a number of different heterogeneous rules of inference. We will also discuss how the framework can be used to define rules of inference similar to the Observe Rule in Barwise and Etchemendy's *Hyperproof* system.

1 Introduction

While navigating the busy streets of London, a small group of tourists find themselves struggling with three different maps to plan a route from the Tower of London to the British Museum. One of these maps is an underground map, another is a detailed book-map, and the last is a tourist map showing icon-like representations of major city buildings. Each of these maps fails to contain information that is essential in the planning of the route. For example, the book map doesn't show the connections between the underground stations, while the underground map doesn't show the location of the British Museum. In deciding which underground lines and roads they will take to reach their destination they will find themselves using information contained in more than one of the maps together. They might even extract information from one map to include it in another, e.g., by sketching the connections between the underground stations in their book map.

M. Hegarty, B. Meyer, and N. Hari Narayanan (Eds.): Diagrams 2002, LNAI 2317, pp. 131–145, 2002.

To plan their trip, these tourists are using an example of a heterogeneous system. They have different instances of three different kinds of map, each expressing diverse pieces of information using various syntactic features. In this paper, we will study such instances of heterogeneous reasoning from a formal logical point of view. To do so, we will first propose a framework for the modeling of heterogeneous reasoning systems. In such systems, the exchange of information between the system's various representations is mediated by heterogeneous rules of inference. Thus we will also show how this framework can be used to define various kinds of heterogeneous rules of inference, which we will generically call Recast Rules. Though the following system can be used to define many different kinds of Recast Rules we will specifically discuss one kind of rule based upon the explicit information content of a diagram.

This is not the first work in modeling a heterogeneous reasoning systems. Notable examples include Barwise and Etchemendy's work on *Hyperproof* [1], Barker-Plumber and Bailin's automated theorem prover &/GROVER [2], and Jamnik's DIAMOND system [3]. However, unlike these projects, our goal will be to provide a framework that can be applied to many different heterogeneous reasoning systems. Also, we hope to provide a framework which is flexible enough to allow the definition of many different kinds of heterogeneous rules of inference.

2   A Framework for Heterogeneous Reasoning

When working with a heterogeneous reasoning system, one of the underlying assumptions is that all of the representations carry information about some common domain. In other words, though each kind of representation can represent different kinds of information in different ways, there is a relationship between the information contained in the various representations. If this were not the case, the system as a whole would be of little use. If there were no connection between the information expressed by the various parts, there would be no point in having combined them into a conglomerate reasoning system.

Thus, we begin with a reasoning system consisting of a common domain, in the form of a collection of mathematical structures, two collections of well-formed representations, and notions of truth relating these representations to the domain.[1] One might think that this alone would suffice for the definition of notions of recasting between these two kinds of representations. For example, we could say that you can recast one representation into a second when all the domain instances in which the first was true were instances in which the second representation was true as well. However, this notion of recasting would be too reminiscent of a notion of logical consequence, e.g., you could recast an inconsistent representation into any representation, and recast any representation into a logically true representation. Unfortunately such a relation does not seem to

---

[1] Simply put, a diagram is true in a domain instance if the information represented by the diagram is contained in the state of the world represented by the domain instance.

capture our goal of modeling the explicit information content of a representation. Thus, to accomplish our goal of being able to define many different kinds of recast relations, including those based upon the explicit content of a representation, we will introduce another collection of structures, information types, into our framework. This collection of types is used to "connect" the representations in the heterogeneous system. It should be noted that the design of this framework was strongly influenced by Barwise and Seligman's work on information theory [4]. Using this collection of types, we can then define three-part type-token relations between these information types, members of the domain, and well-formed representations from both systems. To avoid confusion with the more traditional use of the '$\models$' symbol, we will diverge from Barwise and Seligmen's notation and use a ':' to indicate the type-token relation throughout the rest of this work. Thus $r :^+ t$ will be written when the representation r is known to be of the type t, $r :^- t$ when the representation r is known to not be of the type t and $r :^? t$ when it is unknown whether r is of the type t. An illustration of this system can be found in Fig. 1. Using these type-token relations, we will show in the next section how we can define heterogeneous rules of inference using this framework.

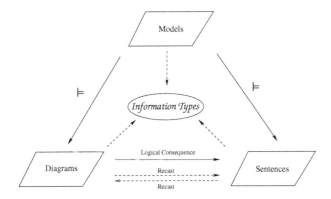

Fig. 1. A Heterogeneous representation system

It should be noted that the careful selection of this collection of information types is a very important to the functioning of the system as a whole. As we will see in the next section, they play an essential role in the definition of the system's heterogeneous rules of inference.

2.1 Information Recasting

Now, we will look at how this framework can be used to define heterogeneous rules of inference, generically called the Recast Rule, that allow the exchange of information between the different representations of the heterogeneous system.

To do this, we will use the proposed framework to define a *recast relation* between each of the system's representations. This relation will hold of two representations when there is a special relationship between the information contained in them.[2] Using this relation we will then say that information in one representation can be re-expressed in another representation if the recast relation holds between those two representations.

Definition 1. *Defining the Recast relation* ($\approx$)
Given two well-formed representations D_1 and D_2, each a token of different representation classifications in a heterogeneous system,

$D_1 \approx^+ D_2$ *if for all types T such that $D_1 :^+ T$, we have that $D_2 :^+ T$.*
$D_1 \approx^- D_2$ *if for all types T such that $D_1 :^+ T$, we have that $D_2 :^- T$.*
$D_1 \approx^? D_2$ *if neither $D_1 \approx^+ D_2$ nor $D \approx^- D_2$ is the case.*

As mentioned earlier, the choice of the collection of information types greatly effects the notion of recasting just presented. More specifically, the selection of a collection of information types highly tailored to the representations at hand will allow a more fine-grained notion of information content to be captured by the type-token relations. This more focused notion of information content will then result in a more specific recast relation. Let's say that, for the information types of one system, we decide to use the collection of models itself, and define the type-token relations based on each homogeneous system's notion of truth. In doing so, one would produce a notion of recasting similar to the notion of logical consequence. By simply moving from full models to a collection of partial models, a much more fine-grained notion of recasting will be produced. For example, using partial models in place of full models would prevent the recasting of any given representation into a trivial representation.

It should be noted that picking just any collection of types and defining sensible type-token relations does not guarantee that the resulting notion of recasting will be sensible. Once defined, this relation's properties should be examined. It is often desirable that the recast relation imply logical consequence, and, in other situations, that the relation be monotonic.

3   A Sample Heterogeneous System

In the last section, the importance of the selection of the collection of information types was discussed. In this section, a simple but concrete example of the effects of the choice of this collection upon the resulting notion of recasting will be presented. For the purpose of this example, we will look at a heterogeneous system consisting of seating arrangement diagrams, age charts, and FOL formulas containing a fixed collection of names and predicates. An example of a seating arrangement diagram and an age chart can be found in Fig. 2.

---

[2] This description is intentionally left vague to allow for the definition of many different kinds of recast relations, and thereby many different kinds of Recast Rules.

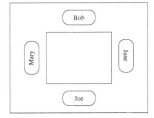 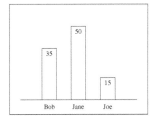

Fig. 2. A seating arrangement diagram and an age chart

3.1   The Syntax of the Three Representations

Seating arrangement diagrams will be interpreted as carrying information regarding the seating of people around a table for a lunch meeting. Well-formed seating arrangement diagrams contain labeled rounded-box tokens, representing people, placed around a rectangle, representing a table, in the center of the diagram. Each of these rounded-box tokens must be labeled with either some person's name (Mary, Bob, Jane, or Joe) or with a '?' (representing an unknown person). Each rounded-box token must be placed adjacent to a unique side of the rectangle, and no two rounded-box tokens can contain the same name (though more than one can be labeled with a '?'). It is not required that each side of the rectangle have a rounded-box token, and the lack of a token is taken to represent that there is no one seated on that side of the table. We will interpret diagrams of this system as carrying information regarding who is "in front of", "to the right of"[3] and "to the left of" other people seated at the table.

Age charts will carry information about the ages of named individuals. Well-formed charts will consist of a base-line with labeled and numbered bars extending upward from the baseline representing the age of the named individual. The labels of the bars must be a valid person's name (Mary, Bob, Jane, or Joe) and the numbers, age labels, must be non-zero natural numbers. Furthermore, we require that within a single chart, the lengths of the bars must be drawn proportionally to one another with respect to their age labels. For example, if one bar is labeled as ten and another as twenty the second bar must be twice as long as the first bar. Note that the bars do not have to be uniquely labeled, thus two bars can exist in the diagram with the same name labels and different age labels thereby resulting in an inconsistent diagram. We will interpret diagrams of this system as carrying information regarding who is "younger than", "older then" and the "same age" as other individuals represented in the diagram.

Lastly, we will describe the syntax of the sentential language that we will use. Atomic wff of the language will consist of the following binary predicates: 'FrontOf', 'LeftOf', 'RightOf', 'Older', 'Younger', and 'SameAge' along with a

---

[3] We will interpret "Mary is to the right of Bob" as saying that, from the point of view of Bob (who is facing the table) Mary is on his right. "Left of" will be interpreted analogously

collection of variables $(x_1, x_2, \dots)$ and names ('Mary', 'Bob', 'Jane', and 'Joe') as usual. Compound wff's will be formed recursively from the atomic wff's and the usual logical connectives and quantifiers: $\neg$, $\wedge$, $\vee$, $\rightarrow$, $\exists$, and $\forall$ in the standard manner.

3.2   The Semantics Based upon a Common Domain

As mentioned before, we begin with the assumption that these three different kinds of representations will carry information about some common domain. We will model this domain with mathematical structures that we will call *worlds*. For our current purposes a world w will consist of the following tuple $\langle People^w, Names^w, Loc^w, Age^w, Const^w \rangle$[4] having the following properties:

- $People^w$ is a set objects $\{o_0, o_1, o_2, o_3\}$ representing four people.
- $Names^w$ is the set $\{Mary, Bob, Jane, Joe\}$ of the names of people about whom we are reasoning.
- Loc^w is a function from $People^w$ to $\{0, 1, 2, 3, \emptyset\}$ representing the locations of the people clockwise around the table. We require that for all o_n and o_m such that $Loc^w(o_n) \neq \emptyset$ and $Loc^w(o_m) \neq \emptyset$ that $Loc^w(o_n) \neq Loc^w(o_m)$, in other words that no two people can occupy the same seat. When $Loc^w(o_n) = \emptyset$ we take that to mean that the person o_n is not seated at the table.
- Age^w is a function from $People^w$ to the positive integers representing the ages of the people.
- $Const^w$ is a one-to-one function from $Names^w$ to $People^w$.

Thus our worlds will consist of four objects representing people and functions representing their state. Each person will have an age and some unique name. Also, we will have information about where individuals are seated around the table with respect to one another. But please keep in mind, it is not required that each person be seated on some side of the table. We will think of each world as representing all the pertinent information in some real world situation. We will now proceed by informally describing when each of the representations of the system can be taken to represent a particular world.

Given a seating arrangement diagram S and a world w, we will say that the diagram holds in the world or represents the world, $w \models S$, if there is some function f mapping the four different sides of the diagram's rectangle to $People \cup \emptyset$ such that:

- for any two different sides s and s' of the rectangle both containing rounded-box icons then $f(s) \neq f(s')$
- if a side s of the rectangle contains no rounded-box icon then $f(s) = \emptyset$, and if it does contain a rounded-box icon then $f(s) \neq \emptyset$
- if a side s of the diagram's rectangle contains a rounded-box labeled with a name n (not a '?') then $f(s) = Const(n)$

---

[4] In the rest of the paper the w superscripts will be omitted when the world to which we are referring is clear from the context.

- if two opposite sides of the rectangle, s_1 and s_2, both contain rounded-box icons then $[Loc(f(s_1)) + 2] \mod 4 = Loc(f(s_2))$
- if two consecutive sides of the diagram's rectangle, s_1 and s_2 with s_2 immediately after s_1, in the clockwise direction, both contain rounded-box icons then $[Loc(f(s_1)) + 1] \mod 4 = Loc(f(s_2))$

Please note that the locations specified by the world's *Loc* function are taken to be relative not absolute locations. Thus a seating arrangement diagram containing four labeled rounded-box icons holds in at least four different worlds each having a different *Loc* function. Another way of thinking of this is that any well-formed seating arrangement diagram can be rotated and result in a diagram representing the same worlds.

Next we will describe the semantics of our age charts. Given some age chart A and a world w we will say that the chart represents the world, $w \models A$, if for each bar named n and labeled with age i in A, it is the case that $Age(Const(n)) = i$ in w.

Finally we will describe the interpretation for the predicates of our sentential language. We will take a variable assignment, Var, to be a function from the collection of variables, $x_1, x_2, \ldots$, to *People*. We then extend this function to include our collection of names by requiring that for all names n, $Var(n) = Const(n)$. Using this function we will define the relation $w \models \varphi[Var]$, that the variable assignment Var satisfies the wff φ in w, recursively on the structure of the formula beginning with atomic wff:

- $w \models \text{FrontOf}(x_1, x_2)[Var]$ iff $[Loc(Var(x_1)) + 2] \mod 4 = Loc(Var(x_2))$
- $w \models \text{LeftOf}(x_1, x_2)[Var]$ iff $[Loc(Var(x_1)) - 1] \mod 4 = Loc(Var(x_2))$
- $w \models \text{RightOf}(x_1, x_2)[Var]$ iff $[Loc(Var(x_1)) + 1] \mod 4 = Loc(Var(x_2))$
- $w \models \text{Younger}(x_1, x_2)[Var]$ iff $Age(Var(x_1)) < Age(Var(x_2))$
- $w \models \text{Older}(x_1, x_2)[Var]$ iff $Age(Var(x_1)) > Age(Var(x_2))$
- $w \models \text{SameAge}(x_1, x_2)[Var]$ iff $Age(Var(x_1)) = Age(Var(x_2))$

Then for complex wff, assuming that φ and ψ are both wff:

- $w \models \neg\varphi[Var]$ iff $w \not\models \varphi[Var]$
- $w \models \varphi \wedge \psi[Var]$ iff $w \models \varphi[Var]$ and $w \models \psi[Var]$
- $\wedge$ and $\rightarrow$ are defined analogously
- $w \models \exists x_1 \varphi[Var]$ iff there is some Var' differing from Var on at most only x_1 such that $w \models \varphi[Var']$
- $w \models \forall x_1 \varphi[Var]$ iff for all Var' differing from Var on at most only x_1 we have that $w \models \varphi[Var']$

3.3   The Recast Relation

In this section, we will propose three different candidates for the collection of information types for the heterogeneous system just presented and examine the resulting notions of recasting that arise.

For our first example, we will use the collection of worlds as our information types and use the above semantic relations to define our type-token relations.

Thus for any seating diagram S we will say that $S :^+ w$ iff $w \models S$ and $S :^- w$ iff $w \not\models S$ (and it will never be the case that $S :^? w$). We define the type-token relations on age charts and formulas of FOL analogously. Using the example diagrams in Fig. 3, we can recast the following sentential information from these diagrams:

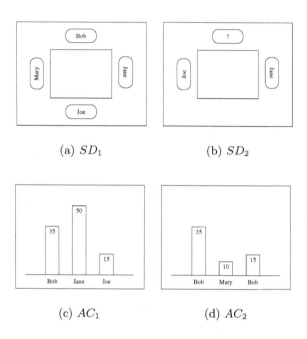

(a) SD_1 (b) SD_2

(c) AC_1 (d) AC_2

Fig. 3. Example seating arrangement diagrams and age chart diagrams

- $SD_1 \approx^+ \mathrm{FrontOf}(\mathrm{Jane}, \mathrm{Mary})$
- $SD_1 \approx^- \mathrm{LeftOf}(\mathrm{Jane}, \mathrm{Mary})$
- $SD_2 \approx^+ \mathrm{LeftOf}(\mathrm{Bob}, \mathrm{Joe}) \vee \mathrm{LeftOf}(\mathrm{Mary}, \mathrm{Joe})$ - Due to the fact that our domain consists of only four persons each who must have a name, the unknown person in the diagram must either be Bob or Mary in all the worlds in which the diagram holds.
- $SD_2 \approx^+ \exists x \, \mathrm{LeftOf}(x, \mathrm{Joe})$
- $SD_2 \approx^+ \forall x, y \, (\mathrm{Older}(x, y) \vee \mathrm{Younger}(x, y) \vee \mathrm{SameAge}(x, y))$ - Any formula that holds in all worlds, logical truths, can be recast from any diagram.
- $AC_1 \approx^+ \mathrm{Older}(\mathrm{Jane}, \mathrm{Joe})$
- $AC_2 \approx^+ \mathrm{FrontOf}(\mathrm{Bob}, \mathrm{Bob})$ - Since there are no worlds in which AC_2 holds, any formula can be recast from it.

As mentioned earlier, by using information types that consist of worlds, or complete models, we end up with a notion of recasting very reminiscent of a notion of logical consequence. Any trivial representation can be recast from any other representation, and an inconsistent representation we can recast into *any* other representation. Furthermore, the inherent structure or regularities of the domain can influence the kinds of information that can be recast. For example, due to the fact that our worlds consist of four named persons, we know that, in this setting, the unlabeled box in SD_2 must represent either Bob or Mary. This information is not contained in the diagram, but rather comes from the constraints put upon the current domain by our choice of worlds. If we wish to capture a notion of recasting more closely related to the explicit information content of the diagram, we must use information types that are more closely akin to the information that our diagrams, in themselves, contain.

We will now look at a second candidate for our information types. This candidate will be similar in spirit to the worlds just presented, but will allow for some degree of uncertainly or unspecificity. A *partial world*, *pw*, will consist of the following tuple $\langle People^{pw}, Names^{pw}, Loc^{pw}, Age^{pw}, Const^{pw} \rangle$ having the following properties:

- $People^{pw}$ is a set objects $\{o_0, o_1, o_2, o_3\}$ representing four people.
- $Names^{pw}$ is the set $\{Mary, Bob, Jane, Joe\}$ of the names of people about whom we are reasoning.
- Loc^{pw} is a *partial* function from $People^{pw}$ to $\{0, 1, 2, 3, \emptyset\}$ representing the locations of the people clockwise around the table. As before we require that no two people be assigned the same location $\{0, 1, 2, 3\}$, and that $\emptyset$ will be assigned when the person is not seated at the table.
- Age^{pw} is a *partial* function from $People^{pw}$ to the positive integers representing the ages of the people.
- $Const^{pw}$ is a *partial* one-to-one function from $Names^{pw}$ to $People^{pw}$.

Note that the difference between a world and a partial world is that the functions *Loc*, *Age* and *Const* are partial in the case of partial worlds and defined for their entire range in the case of worlds. Thus in a partial world, we can have people with unknown names, unknown ages and unknown seating locations. As the type-token relations for these partial worlds we will use notions similar to those used in full worlds except we will use a ":?" when the necessary function is undefined. For complex wff parts of the recursive definition of the type relations for our FOL formulas, we will use appropriate portions of the the standard Kleene three-valued notion of truth. Using these types we then can recast the following information:

- $SD_1 \approx^+ FrontOf(Jane, Mary)$
- $SD_1 \approx^- LeftOf(Jane, Mary)$
- $SD_2 \approx^? LeftOf(Bob, Joe) \vee LeftOf(Mary, Joe)$ - In our partial worlds we do not require that each person have a name, thus though all partial worlds in which this diagram holds have a person to the left of Joe, we don't have to have a name assigned to that person.
- $SD_2 \approx^+ \exists x\ LeftOf(x, Joe)$

- $SD_2 \not\approx^? \forall x, y \ (\mathrm{Older}(x, y) \vee \mathrm{Younger}(x, y) \vee \mathrm{SameAge}(x, y))$ - The partial worlds in which this diagram holds do not have to have an *Age* function defined anywhere, thus it not possible to recast this diagram into this formula nor it is possible to say that this diagram cannot be recast into this formula. The diagram simply doesn't carry the kind of information necessary to make a judgment.
- $AC_1 \not\approx^+ \mathrm{Older}(\mathrm{Jane}, \mathrm{Joe})$
- $AC_2 \not\approx^+ \mathrm{FrontOf}(\mathrm{Bob}, \mathrm{Bob})$ - As before, since there are no partial worlds in which AC_2 holds, any formula can be recast from it.

As you can see, the move to partial structures alleviated a number of the problems that we encountered when working with total structures. Due to their partiality they are able to address more specifically the information content of each of our diagrams. The partial worlds in which a particular seating arrangement diagram holds do not need an *Age* function defined anywhere, thus no age related information can be recast from those diagrams. We also no longer have to worry about being able to always recast any representation into many kinds of trivial representation.[5] Furthermore, we have removed some of the problems associated with our domain structure influencing what we could recast. However undesirable things still occur in the case of inconsistent diagrams. As before, there are no partial worlds in which inconsistent representation hold; thus in our last collection of information types we will attempt to address this issue.

The last collection of information types that we will use is very similar to the collection of partial worlds just presented, except they will also have the ability to carry inconsistent age information. We will refer to this collection as *seating and age information types*. For these types we will use structures similar to the partial worlds just presented, but in place of the Age^{pw} function we will have a finite three-valued relation between *People* and the positive integers representing the ages of the people.[6] Thus in a single seating and age information type one person can have any number of ages (including no age at all). This change will not greatly effect the way in which we define our type-token relations, so we will just briefly describe the new relations. For seating arrangement diagrams, we will use exactly the same type-token relations that we used with partial worlds. For an age chart A and a seating and age information type t we will say that the chart is of the type, $A :^+ t$, if for *each* bar named n and labeled with age i, it is the case that $Age(Const(n), i)$, we will say that the chart is not of a type, $A :^- t$ if there is *some* bar named n and labeled with age i, such that $Age(Const(n), i)$ does not hold, and we will say that it is unknown, $A :^? t$, otherwise. Note that it is unknown whether a chart is of a type when information represented in the chart

---

[5] There are still a few trivial expression that can be recast at times. For example, any seating arrangement diagram can be recast into the formula $\forall x, y \ (FrontOf(x, y) \rightarrow FrontOf(y, x))$. This does not seem to be problematic because such formulas are limited to properties of the kind of information that these diagrams can carry.

[6] Three-valued relations allow some tuples to be in the relation, some to not be in the relation and others to be neither in nor not in the relation. This third possibility allows the relation to be indifferent or uncertain about the status of certain tuples.

is neither supported nor contradicted by the information type. For our formulas, we only have to modify the following clauses of our recursive definition:

- Younger$(x_1, x_2)[Var]$ $:^+$ t when there are two integers $n_1 < n_2$ such that $Age(Var(x_1), n_1)$ and $Age(Var(x_2), n_2)$, Younger$(x_1, x_2)[Var]$ $:^-$ t when there are two positive integers $n_1 \geq n_2$ such that $Age(Var(x_1), n_1)$ and $Age(Var(x_2), n_2)$, and Younger$(x_1, x_2)[Var]$ $:^?$ t otherwise.
- Older$(x_1, x_2)[Var]$ $:^+$ t when there are two integers $n_1 > n_2$ such that $Age(Var(x_1), n_1)$ and $Age(Var(x_2), n_2)$, Older$(x_1, x_2)[Var]$ $:^-$ t when there are two positive integers $n_1 \leq n_2$ such that $Age(Var(x_1), n_1)$ and $Age(Var(x_2), n_2)$, and Older$(x_1, x_2)[Var]$ $:^?$ t otherwise.
- SameAge$(x_1, x_2)[Var]$ $:^+$ t when there are two integers $n_1 = n_2$ such that $Age(Var(x_1), n_1)$ and $Age(Var(x_2), n_2)$, SameAge$(x_1, x_2)[Var]$ $:^-$ t when there are two positive integers $n_1 > n_2$ or $n_1 < n_2$ such that $Age(Var(x_1), n_1)$ and $Age(Var(x_2), n_2)$, and SameAge$(x_1, x_2)[Var]$ $:^?$ t otherwise.

The resulting recast relation is very similar to the one that arose from using partial worlds except in the case of inconsistent diagrams. For example, we now have
$AC_2 \approx^?$ FrontOf(Bob, Bob) since there are seating and age information types in which Bob can have two different ages and no known location at the table. From this diagram we can also say that $AC_2 \approx^+$ Older(Bob, Mary) and $AC_2 \approx^-$ SameAge(Bob, Mary). Hence, as the result of our inconsistent types, only information based upon AC_2's features can be recast from this diagram. A summary of the recast relation arising from the use of the above three different collections of information types is shown in Fig. 4. Though we have only looked at recasting from diagrams to formulas, similar things happen when trying to recast from formulas to diagrams and diagrams to diagrams. Again the more tailored the notion of information type is to the representation at hand, the more fine-grained the notion of recasting that will be produced. Working with worlds can again result in trivial diagrams being recast from formulas or other diagrams, and working with types that cannot express inconsistent information results in a recast relation in which any diagram can be recast from an inconsistent diagram or formula. It should be noted that in the last collection of types presented, we had the ability to express inconsistent information with regard to age, but not with regard to location. Thus, as before, problems arise when recasting from formulas involving inconsistent location information. To remedy this situation the *Loc* function can be changed into a relation in a manner similar to used with the *Age* relation.

One big difference that does occur when recasting into diagrams from other formulas or diagrams is related to their inability to express certain kinds of information. Let's say for example you wish to recast Younger(Bob, Mary) into an age chart. Because age charts must have bars labeled with specific ages there is no way of accomplishing this. We only know that Bob is younger than Mary, we don't know their exact ages thus we cannot produce an age chart. Likewise, it is not possible to recast the same sentence into a seating arrangement diagram;

Diagram	Sentence	Recast Relation Using		
		Worlds	Partial W.	Types
SD_1	FrontOf(Jane, Mary)	$\approx^+$	$\approx^+$	$\approx^+$
SD_1	LeftOf(Jane, Mary)	$\approx^-$	$\approx^-$	$\approx^-$
SD_2	LeftOf(Bob, Joe) $\vee$ LeftOf(Mary, Joe)	$\approx^+$	$\approx^?$	$\approx^?$
SD_2	$\exists x$ LeftOf(x, Joe)	$\approx^+$	$\approx^+$	$\approx^+$
SD_2	$\forall x, y$ (Older(x, y) $\vee$ Younger(x, y) $\vee$ SameAge(x, y))	$\approx^+$	$\approx^?$	$\approx^?$
AC_1	Older(Jane, Joe)	$\approx^+$	$\approx^+$	$\approx^+$
AC_2	FrontOf(Bob, Bob)	$\approx^+$	$\approx^+$	$\approx^?$

Fig. 4. Summary of the recast relation based upon different information types

those kinds of diagrams simply cannot express age related information. In fact, when using seating and age information types it isn't possible to recast any age chart into a seating arrangement diagram and vice-versa. The kinds of information that can be expressed in each does not overlap at all, thus this seems to be desirable outcome.

4 Recasting as the Observation of Explicit Information

One of the basic goals of this section is to explore the question of what information can be observed from a given diagram. There are a range of attitudes that one can take in answering this question. To make the case more dramatic, let's consider the case of photographs. On one end of this spectrum, one might reply that only some set of visual properties of the photograph itself can be observed from the photo. For example, that there is a blue blob on one side. At the other end of the spectrum, the answer to this question would be, any information that follows from the given photograph. Faults can be found in both of these extremes. If one only allows the observation of visual properties of the photo, one gives up the idea that we can observe anything about the subject matter of the photo, what the photo is about. Yet by adopting the other extreme, one is blurring what seems to be an important distinction, what we can see from the photo and what we can infer from what we observe. While diagrams are different from photos in many regards, the same considerations do seem to apply. In our attempt to fight the "beast of observation", efforts will be made to avoid both of these "horns" and aim for a notion of observation somewhere in the middle of these extremes.

4.1 Background

To explain further the notion that we are after, it is convenient to recall the work of Dretske. In *Seeing and Knowing* [5], Dretske attempts to define a notion of epistemic seeing, or seeing that, which has a fundamental relation to a notion of non-epistemic seeing. He takes very seriously the idea that any genuine instance of "seeing that" should have as its basis a visual event. This was done to preserve, as he says, the "visual impact" of "seeing that" and thus to exclude such natural language uses of the phrase "to see that", as "Mary could see that the eggs were completely hard boiled (from the ringing of the timer)", from his notions of seeing. Part of the motivation for doing this was to ensure that the "How?" justification of any information gathered through an act of "seeing that" involves an act of non-epistemic seeing in an essential way. For our purposes, this locution is important because it will help us to characterize information which can be observed, acquired through fundamentally visual means, from a diagram.

Let us briefly recall Dretske's notion of secondary epistemic seeing. According to Dretske, S can see that b is P in the secondary sense when:

1. b is P
2. S can (non-epistemically) see c, and primarily see that c is Q[7]
3. The conditions are such that c would not be Q unless b were P
4. S believes the last condition is the case when considering that b is P

What needs to be emphasized here is that there must be some c which can be non-epistemically seen and from the information that is visually gathered about c, that it is Q, it can be concluded that b is P. Thus an example of a secondary seeing would be "Mary could see that her father's office was on the second floor from the map in the lobby." This is because it is Mary's conviction that what she non-epistemically saw in the map undeniably carries information about the location of the office.

It seems as though something similar goes on with diagrams. Some facts are directly represented in a diagram on the basis of the features of the diagram and the standard interpretation of the diagram alone, while others follow from the diagram, its interpretation, *and* a collection of assumptions or background knowledge which is independent of the interpretation of the diagram. The main motivating goal of this work is to see if we can get a handle on the former and then define a mathematical system consistent with these philosophical intuitions.

4.2 Simple Example

To give a simple example, suppose someone is using the familiar backwards truth-table method to determine whether a given propositional formula is a tautology. He or she starts by filling in F under the main connective and sees if he or she can consistently complete the assignment. So suppose we come upon an incomplete table being used in this process. Perhaps it looks like the following, where p, q and r are complex wffs represented by the atomic letters A, B, and C:

$$\begin{array}{ccc|c} \text{A} & \text{B} & \text{C} & p \to (q \to r) \\ \hline & & & \text{T} \quad \text{F} \qquad \text{F} \end{array}$$

From this table we can *observe* that a certain partial truth assignment has been made: p has been assigned T, $q \to r$ has been assigned F, as has the whole

---

[7] Primary seeing is defined similarly except there is no intermediate c, in other words one must be able to see b itself.

wff. However, even if we observe that r is a conjunct of q, we do not observe from this table that there is no coherent way to finish the truth assignment consistent with the meanings of the connectives. It can be inferred from what is observable, but it cannot be observed because the fact is not justified by the interpretation of the table alone but requires certain principles of logic. What is important to point out is that, while these principles are commonly accepted as given, they in no way follow from the table's use alone. For instance, the above table could just as well be used in a logical system which was multivalued, and in such a situation it would not be the case that there is no way to finish the above table consistently. As a second example, let us consider the Euler/Venn diagram in Fig. 5. From this diagram it seems natural to say:

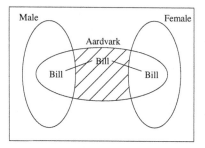

Fig. 5. Bill the aardvark

1. You can observe that Bill is either a Male, a Female, or neither a Male nor a Female, due to the placement of the Bill icons in the diagram.
2. That there are no Aardvarks which are neither Male nor Female can be observed due to the shading of the center region.
3. You can observe that there are no Males which are also Female from the placement of the Male and Female curves.
4. Nothing can be observed regarding Mary since she is not shown in the diagram.
5. That Bill is an Aardvark is observable due to the fact that all the Bill icons are contained in the diagram's Aardvark curve.
6. That Bill is a Male and that Bill is either Male or Female both can be observed to fail from the diagram due to the placement of the Bill icon in the Female curve.

These examples suggest the following principle about observation from diagrams:

Principle of Observation - Immediacy

Given a representation r of a language $\mathcal{L}$, then one can observe r from a diagram if and only if there is a syntactic property p of the diagram that is taken to *explicitly* represent the information conveyed by r.[8]

---

[8] This principle is in the spirit of Dretske's "secondary seeing that" [5]. See also [6].

Though still somewhat vague, we hope that this principle and examples can help to guide further studies into this notion of observation. It would also be interesting to consider more fully the properties of the above notion of observation by looking at the "logic of observation" which arises in a system which uses an observe relation as its basis, i.e., if we think of the observe relation as a kind of counterpart of the normal truth relation, does there arise a natural counterpart of the inference relation? If successful, this raises many other questions including whether in such a setting it make sense to consider counterparts to the properties of soundness and completeness for these relations of observation and "observational inference" and what philosophical significance these properties would have.

5   Conclusion

The general framework for heterogeneous reasoning, presented here, is not specific to just diagrammatic systems, and thus could provide a basis for the design of many kinds of heterogeneous reasoning systems. Examples of some such applications to non-diagrammatic heterogeneous reasoning systems could include: databases with heterogeneous data, simulations of complex physical systems with many different variables and sources of information like meteorological models, and AI reasoning systems which make decisions based upon data in different forms. Though this framework can can be used to define a notion of observation similar to that used in Barwise and Etchemendy's *Hyperproof* [1], and has been applied to a detailed case study on First Order Logic and Euler/Venn diagrams [7,8], further applications are needed to verify its usefulness.

References

1. Barwise, J., Etchemendy, J.: Hyperproof. CSLI Publications, Stanford (1994)
2. Barker-Plummer, D., Bailin, S.C.: On the practical semantics of mathematical diagrams. In Anderson, M., ed.: Reasoning with Diagrammatic Representations. Number FS-97-03 in Technical Reports of the AAAI. AAAI Press (1997) 39–57
3. Jamnik, M.: Automating Diagrammatic Proofs of Arithmetic Arguments. PhD thesis, University of Edinburgh (1999)
4. Barwise, J., Seligman, J.: Information Flow - The Logic of Distributed Systems. Number 44 in Cambridge Tracts in Theoretical Computer Science. Cambridge University Press, Cambridge (1997)
5. Dretske, F.I.: Seeing and Knowing. Chicago University Press, Chicago (1969)
6. Barwise, J., Perry, J.: Situations and Attitudes. MIT Press, Cambridge (1983)
7. Swoboda, N., Allwein, G.: A case study of the design and implementation of heterogeneous reasoning systems. In Magnani, L., Nersessian, N.J., eds.: Logical and Computational Aspects of Model-Based Reasoning. Kluwer Academic, Dordrecht (2002)
8. Swoboda, N.: Designing Heterogeneous Reasoning Systems with a Case Study on FOL and Euler/Venn Reasoning. PhD thesis, Indiana University (2001)

On Diagram Tokens and Types

John Howse[1], Fernando Molina[1], Sun-Joo Shin[2], and John Taylor[1]

[1] School of Computing & Mathematical Sciences
University of Brighton, Brighton, UK
{John.Howse, F.Molina, John.Taylor}@brighton.ac.uk
http://www.it.bton.ac.uk/research/vmg/VisualModellingGroup.html
[2] Department of Philosophy, University of
Notre Dame, Notre Dame, Indiana, USA
Sun-Joo.Shin.3@nd.edu

Abstract. Rejecting the temptation to make up a list of necessary and sufficient conditions for diagrammatic and sentential systems, we present an important distinction which arises from sentential and diagrammatic features of systems. Importantly, the distinction we will explore in the paper lies at a meta-level. That is, we argue for a major difference in meta-theory between diagrammatic and sentential systems, by showing the necessity of a more fine-grained syntax for a diagrammatic system than for a sentential system. Unlike with sentential systems, a diagrammatic system requires two levels of syntax—token and type. Token-syntax is about particular diagrams instantiated on some physical medium, and type-syntax provides a formal definition with which a concrete representation of a diagram must comply. While these two levels of syntax are closely related, the domains of type-syntax and token-syntax are distinct from each other. Euler diagrams are chosen as a case study to illustrate the following major points of the paper: (i) What kinds of diagrammatic features (as opposed to sentential features) require two different levels of syntax? (ii) What is the relation between these two levels of syntax? (iii) What is the advantage of having a two-tiered syntax?

1 Introduction

Many would suppose that there are fundamental differences between linguistic and diagrammatic systems. This assumption justifies a well-accepted classification among existing systems, for example, to call Euler or Venn systems diagrammatic and first-order languages linguistic. However, pinning down what those differences are has turned out to be a daunting task. The difficulty of the task has led some researchers in the area to a skepticism about any demarcation of different types of representation systems.

We find this skepticism vital as a guard against coming up with a quick and easy recipe which would easily ignore important theoretical issues involved in the nature of representation systems. On the other hand, we respect our on-going practice based on the intuition that there are different kinds of representation.

M. Hegarty, B. Meyer, and N. Hari Narayanan (Eds.): Diagrams 2002, LNAI 2317, pp. 146–160, 2002.
© Springer-Verlag Berlin Heidelberg 2002

Recognizing merit in each position, we agree with many researchers that a relation between linguistic and diagrammatic systems should be understood to form a continuum.

Hence, we do not think that it is desirable to seek necessary and sufficient conditions to differentiate various kinds of systems. Instead, the issue of differences among systems should be refocused on clarifying linguistic or diagrammatic *elements* in any given representation system. That is, rather than making a comparison among different systems, we need to make a contrast among different features of a system. Depending on which features become salient in a system, it is called either diagrammatic or linguistic. Hence, strictly speaking, we are talking about *saliently* diagrammatic or *saliently* linguistic when we call a system either diagrammatic or linguistic. Again, we do not claim that we can or should complete a list of diagrammatic or linguistic features of a system, but the longer the list is, the more helpful it becomes.

How, then, can we produce a more comprehensive list. Conducting case studies on various systems is one of the most natural ways to achieve this goal. Exploring our intuitive ideas about various existing systems, this paper argues that it is necessary that a meta-theory of diagrammatic systems has different features from a meta-theory of linguistic systems.

Section 2 starts with the observation that a type-token distinction becomes a more important issue in a diagrammatic system than in a linguistic system. After identifying the main source of this discrepancy in diagrammatic and linguistic elements, we claim that a more fine-grained syntax is needed for diagrammatic representation. In §3, we develop a two-tiered syntax for the Euler system to sharpen our intuition underlying the type-token distinction in the system. With the formalism of §3 in hand, §4 illustrates the utility of two-level syntax, which leads us to the advantages of diagrammatic systems for certain purposes.

2 Tokens and Types

The type-token issue is far from being settled in philosophy. However, there are important assumptions related to the type-token distinction that symbolic logic accepts without much controversy. Consider the following two pairs of symbolic sentences:

1. $(A_3 \wedge A_4)$
2. $(\mathsf{A_3} \wedge\ \mathsf{A_4})$
3. $(A_3 \wedge A_4)$
4. $(A_4 \wedge A_3)$

We may easily make a distinction among different kinds of sameness and difference in the above pairs: sentences (1) and (2) are different tokens of the same sentence type, sentences (3) and (4) belong to different sentence types but with the same meaning. A meta-theory of symbolic logic takes care of these kinds of difference in the following way. A difference among tokens (e.g. sentences (1) and (2)) is ignored at the level of meta-theory, so, syntactically both tokens are treated as the same. The difference between (3) and (4) is reflected at the syntactic level, but semantics guarantees that these two sentences have the same meaning.

These distinctions sound almost trivial, but it is not a trivial question to explore the source of the triviality. Thanks to the conventions of a writing system, we know that neither the size nor the font of a symbol makes a difference to telling whether one string is the same as another. This convention is so ingrained that we almost do not see a visual difference between (1) and (2) or, even if we do, we consider the difference in font and size trivial. Therefore, an ambiguity between sentence-token and sentence-type, if any, is negligible. This is why the syntax of symbolic logic is always a type-syntax, not a token-syntax. On the other hand, every such symbolic system relies on a linear order, which is also related to the conventions of a writing system. Hence, sentences (3) and (4) belong to different types and the meaning equivalence of the two sentences is obtained in the semantics by the definition of the meaning of '∧.'

It is extremely interesting to notice that two features of a sentential system (discussed above) have determined two aspects of its meta-theory. It is well accepted that in a linguistic system neither size nor font of a symbol is a representing fact. Let us call this feature 'token-insensitivity.' The token-insensitive feature justifies a one-level syntax, that is, a type-syntax, of a system. The other feature of a linguistic system is a linear order or 'linearity.' This ingrained feature leads us to an easy agreement on different sentence-types for sentences (3) and (4). However, the price for this straightforward differentiation between (3) and (4) is that the system should have a commutative rule since semantic equivalence is not guaranteed any more for two sentences of different types.

We claim that both token-insensitivity and linearity are linguistic elements of a system and that a saliently diagrammatic system does not have either feature, that is, it is token-sensitive and non-linear. Accordingly, we argue that this difference is and should be reflected at the meta-level. We are making the following proposal: since a diagrammatic system is token-sensitive, it needs two-levels of syntax, that is, token-syntax and type-syntax. Also, we speculate that for a non-linear system a commutative rule is not needed. After illustrating token-sensitivity and non-linearity through examples, we will formalize our proposal and see that our speculation about commutativity is correct, at least for the system formalized.

Consider the three diagrams in Fig. 1. These are three different diagrams whose appearances are clearly distinguishable from one another. Visual differences in these three tokens are hard to ignore, but at the same time we know that these three diagrams are of the same kind in an important sense. In the case of the Venn system [13], the location of contours does not have a representing import. Therefore, we would like to say that these diagrams are the "same" in *some* sense.

On the other hand, we do not want say that the Euler diagrams in Fig. 2 are the same even though they represent the same fact. The semantics of the Euler

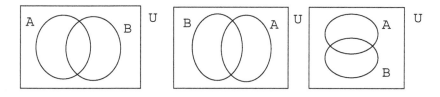

Fig. 1. Equivalent diagrams.

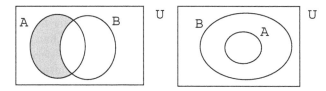

Fig. 2. Semantically-equivalent diagrams.

system tells us that these two diagrams represent the same fact (see §3 for more details of the Euler system we are illustrating). But, there is a clear syntactic distinction between the two diagrams: among other differences, there is shading in one diagram, but not in the other. That is, the difference and sameness in the case of Fig. 2 reflects an intuitive distinction between syntactic and semantic equivalences.

An interesting task is how to capture our intuition about the diagrams in Fig. 1: they are different from one another in one sense, but the same in another sense. First, we would like to decide whether we want to treat the sameness among the three diagrams at a syntactic level or at a semantic level. Again, using our discussions of a symbolic language, we may formulate the question in the following way: are the differences in Fig. 1 similar to the differences between sentences (1) and (2) or the differences between sentences (3) and (4)? We will show that neither of the options is satisfactory.

Suppose that we make no distinction between Fig. 1 and Fig. 2 and conclude that the diagrams in Fig. 1 are syntactically different from one another but semantically the same. That is, the differences among the diagrams in Fig. 1 are treated like the differences between sentences (3) and (4) in the case of symbolic logic. A problem with this approach is that the syntax would become so fine-grained that almost all diagrams would be syntactically different. Hence any equivalence among diagrams is postponed to a semantic level and we would require transformation rules to allow us to commute from one diagram to another in Fig. 1. These rules would be difficult to describe and would complicate the system so much that it would become unusable.

Suppose that we take the other position. The relation among the diagrams in Fig. 1 is just like the relation between sentences (1) and (2). They are syntactically equivalent to one another, period. A problem with this explanation is that the syntax would become too coarse to accommodate differences among obviously present visual properties in diagrams. Interestingly, the reader would more easily notice visual differences among the three tokens of diagrams in Fig. 1 than the visual differences between sentences (1) and (2).

We claim that the discrepancy between syntactically equivalent diagrams and syntactically equivalent sentences is directly related to a difference between diagrammatic and linguistic elements of a system: while a saliently linguistic system adopts conventions of a writing system, a saliently diagrammatic system does not. Hence, there is no convention to rely on to decide that the position of contours is not a representing fact in a given system. It is perfectly possible to imagine a diagrammatic system in which the position of a syntactic object is a representing fact, for example, in many maps, A is to the left of B represents the fact that A is to the west of B. On the other hand, if the size or the font of a syntactic object becomes a representing fact, then we intuitively think that the system has a diagrammatic element.

We present a middle ground solution to the dilemma that these two alternatives present to us. If the syntax of a diagrammatic system is too fine-grained, then the diagrams in Fig. 1 are all syntactically different. But, if the syntax is not fine-grained enough, then we would have to accept a syntax that does not reflect non-trivial visual features of a system. We revive a time-honored distinction between type and token [4,10] and suggest a two-tiered syntax for diagrammatic systems. Hence, the diagrams in Fig. 1 are *different tokens*, but of the *same type*. But, in the case of diagrams, differences in tokens are not negligible unlike with differences in sentence-tokens. Depending on whether we talk about a diagram as a token or as a type, we may attribute different properties to a diagram or different relations among diagrams. Therefore, it is crucial to disambiguate which syntactic status of a diagram is in question, either a concrete token-diagram or an abstract type-diagram. We suggest that the syntax of a diagrammatic system consists of two different levels, that is, a token-level and a type-level. At the token-level the syntax is fine-grained enough to respect our intuition that the diagrams in Fig. 1 are *different*, while the other level of syntax, i.e. type-syntax, lets us say that these diagrams are the *same*.

Thus, our solution seems to incorporate the intuition behind differences and equivalence among diagrams in Fig. 1. However, if the only advantage about making a distinction between type-syntax and token-syntax is to comply with our intuition and the result is to make a more complicated machinery for a system, one might wonder about the utility of the two-level syntax. After presenting a case study of the two-level syntax of the Euler system in the next section, we will return to this issue in §4.

3   Type-Syntax and Token-Syntax of Euler Diagrams

Euler diagrams [1] illustrate relations between sets. This notation uses topological properties of enclosure, exclusion and intersection to represent the set-theoretic notions of subset, disjoint sets, and intersection, respectively. In this paper we have added shading to the notation to increase its expressiveness. Euler diagrams form the basis of more expressive diagrammatic notations such as Higraphs [7], constraint diagrams [3] and some of the notations of UML [9]. This paper is mainly concerned with diagrammatic notations based on Euler diagrams; this is a substantial and important collection in its own right as the above list indicates.

We now give a concise informal description of Euler diagrams. A *contour* is a simple closed plane curve. A *boundary rectangle* properly contains all other contours. Each contour has a unique label. A *district* (or *basic region*) is the bounded area of the plane enclosed by a contour or by the boundary rectangle. A *region* is defined, recursively, as follows: any district is a region; if r_1 and r_2 are regions, then the union, intersection and difference of r_1 and r_2 are regions provided these are non-empty. A *zone* (or *minimal region*) is a region having no other region contained within it. Contours and regions denote sets. Every region is a union of zones. A region is *shaded* if each of its component zones is shaded. The set denoted by a shaded region is empty. In Fig. 2 the zone within A, but outside B is missing from the diagram; the set denoted by such a "missing" zone is also empty. An Euler diagram containing all possible zones is called a Venn diagram.

The **type-syntax** (or *abstract* syntax) of an Euler diagram d is a formal definition that is independent of any concrete visual representation. A concrete **instantiation** of d is a diagram presented on some physical medium (e.g., a sheet of paper, a computer screen, etc) that *complies* with the abstract definition; **token-syntax** describes instantiations. In this section we give the type-syntax of diagrams and a formal definition of the token-syntax. We also give conditions for a concrete diagram to be an instantiation of an abstract diagram and for an abstract diagram to be an abstraction of a concrete diagram.

3.1   Type-Syntax

An *abstract Euler diagram* is a tuple $d = \langle \mathcal{C}, U, \mathcal{Z}, \mathcal{Z}^* \rangle$ whose components are defined as follows:

1. $\mathcal{C}$ is a finite set whose members are called *contour labels*. The element U, which is not a member of $\mathcal{C}$, is called the *boundary rectangle label*.
2. The set $\mathcal{Z} \subseteq 2^{\mathcal{C}}$ is the set of *zones*, while $\mathcal{Z}^* \subseteq \mathcal{Z}$ is the set of *shaded zones*. A zone $z \in \mathcal{Z}$ is *incident* on a contour $c \in \mathcal{C}$ if $c \in z$. Let $\mathcal{R} = 2^{\mathcal{Z}} - \emptyset$ be the set of *regions*, and $\mathcal{R}^* = 2^{\mathcal{Z}^*}$ be the set of shaded regions.

At the abstract level, we identify the concepts of a contour and its label. A zone is defined by the contours that contain it and is thus represented as a set of contours. A region is just a non-empty set of zones.

3.2   Token-Syntax

A *concrete Euler diagram* $\hat{d}$ is a tuple $\hat{d} = \langle \hat{\mathcal{C}}, \hat{\beta}, \hat{\mathcal{Z}}, \hat{\mathcal{Z}}^*, \hat{\mathcal{L}}, \hat{\ell} \rangle$ whose components are defined as follows:

1. $\hat{\mathcal{C}}$ is a finite set of simple closed (Jordan) curves in the plane, $\mathbf{R}^2$, called *contours*. The *boundary rectangle*, $\hat{\beta}$, is also a simple closed curve, usually in the form of a rectangle, but not a member of $\hat{\mathcal{C}}$. For any contour $\hat{c}$ (including $\hat{\beta}$) we denote by $\iota(\hat{c})$ and $\varepsilon(\hat{c})$ the interior (bounded) and the exterior (unbounded) components of $\mathbf{R}^2 - \hat{c}$ respectively; such components exist by the *Jordan Curve Theorem*. Each contour lies within, and does not touch, the boundary rectangle: $\hat{c} \subset \iota(\hat{\beta})$. The set $\hat{\mathcal{C}}$ has the following properties:
   a) Contours intersect transversely (i.e., locally, near the point of intersection, the diagram is a cross).
   b) Each contour intersects with every other contour an even number of times; this, of course, could be zero times.
   c) No two contours have a point in common without crossing at that point.
   d) Each component of $\mathbf{R}^2 - \bigcup_{\hat{c} \in \hat{\mathcal{C}}} \hat{c}$ is the intersection of $\iota(\hat{c})$ for all contours $\hat{c}$

   in some subset X of $\hat{\mathcal{C}}$ and $\varepsilon(\hat{c})$ for all contours $\hat{c}$ in the complement of X:
   $$\bigcap_{\hat{c} \in X} \iota(\hat{c}) \cap \bigcap_{\hat{c} \in \hat{\mathcal{C}} - X} \varepsilon(\hat{c}).$$
2. A *zone* is the intersection of a component of $\mathbf{R}^2 - \bigcup_{\hat{c} \in \hat{\mathcal{C}}} \hat{c}$ with $\iota(\hat{\beta})$. $\hat{\mathcal{Z}}$ is the set of zones. $\hat{\mathcal{Z}}^*$ is the set of shaded zones. Let $\hat{\mathcal{R}} = 2^{\hat{\mathcal{Z}}} - \emptyset$ be the set of *regions*, and $\hat{\mathcal{R}}^* = 2^{\hat{\mathcal{Z}}^*}$ be the set of shaded regions.
3. $\hat{\mathcal{L}}$ is the set of contour labels. The bijection $\hat{\ell} : \hat{\mathcal{C}} \cup \{\hat{\beta}\} \to \hat{\mathcal{L}}$ returns the label of a contour or the boundary rectangle.

Hammer [6] defines an Euler diagram as 'any finite number of closed curves drawn on the page in any arrangement'. This is, of course, a very liberal definition. We have chosen to ban some possible Euler diagrams for the sake of intuitive clarity. For example, the left hand diagram in Fig. 3 in which two contours touch but do not cross is not well-formed by 1(c). Similarly, we ban diagrams with disconnected zones (1(d)), an example of which is given in the middle diagram of Fig. 3, which is attempting to represent $C \subseteq A \cup B$. The zone $\{A, B\}$, i.e., that part of the diagram within A and B, but outside C, is disconnected. Allowing disconnected zones could cause intuitive problems in interpreting the diagram. An alternative way to represent $C \subseteq A \cup B$ is given in the right hand diagram of Fig. 3 in which three contours intersect at a point, but all zones are connected; this is a legal construct.

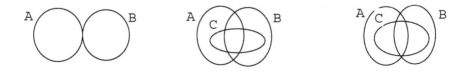

Fig. 3. Touching contours; a disconnected zone; a triple point.

3.3   Mappings between Diagram Types and Tokens

Let $\hat{d} = \langle \hat{C}, \hat{\beta}, \hat{Z}, \hat{Z}^*, \hat{L}, \hat{\ell} \rangle$ be a concrete diagram and let $d = \langle C, U, Z, Z^* \rangle$ be an abstract diagram. Then d is an **abstraction** of $\hat{d}$ if $\hat{L} = C$ and there is a mapping $\mu : \hat{d} \to d$ such that component mappings $\mu : \hat{C} \to C$, $\mu : \hat{Z} \to Z$, are each bijections and satisfy the following conditions:

1. $\forall \hat{c} \in \hat{C} \, \forall c \in C \bullet \mu(\hat{c}) = c \Leftrightarrow \hat{\ell}(\hat{c}) = c$.
2. $\forall \hat{z} \in \hat{Z} \bullet \mu(\hat{z}) = \{ \mu(\hat{c}) \mid \hat{z} \subseteq \iota(\hat{c}) \}$.
3. $\forall \hat{z} \in \hat{Z} \bullet \mu(\hat{z}) \in Z^* \Leftrightarrow \hat{z} \in \hat{Z}^*$.

Such a mapping μ is said to be an *abstraction* mapping.

Similarly, $\hat{d}$ is a (concrete) **instantiation** of d if $\hat{L} = C$ and there is a mapping $\zeta : d \to \hat{d}$ such that component mappings $\zeta : C \to \hat{C}$, $\zeta : Z \to \hat{Z}$, are each bijections and satisfy the following conditions:

1. $\forall c \in C \bullet \hat{\ell}(\zeta(c)) = c$.
2. $\forall z \in Z \bullet \zeta(z) = \bigcap\limits_{c \in z} \iota(\zeta(c)) \cap \bigcap\limits_{c \in C - z} \varepsilon(\zeta(c)) \cap \iota(\hat{\beta})$.
3. $\forall z \in Z \bullet \zeta(z) \in \hat{Z}^* \Leftrightarrow z \in Z^*$.

Such a mapping ζ is said to be an *instantiation* mapping.

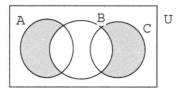

Fig. 4. A concrete Euler diagram.

The Euler diagram in Fig. 4 is an instantiation of the abstract diagram $d = \langle C, U, Z, Z^* \rangle$ where $C = \{A, B, C\}$, $Z = \{\emptyset, \{A\}, \{A, B\}, \{B\}, \{B, C\}, \{C\}\}$, $Z^* = \{\{A\}, \{C\}\}$.

We say that a concrete diagram *complies* with its abstraction. By forgetting the geometric information associated with a concrete diagram we have:

Theorem 1. *For each concrete diagram there is a unique abstract diagram to which it complies.*

However, there are abstract diagrams that have no concrete instantiations. For example, the abstract diagram $d = \langle C, U, Z, Z^* \rangle$ where $C = \{A, B\}$, $Z = \{\emptyset, \{A, B\}\}$, $Z^* = \emptyset$, has no concrete instantiation. The only zones in the diagram are the zone outside all the contours (which must occur in all diagrams) and the zone $\{A, B\}$ which is within both A and B. The only way of representing this

Fig. 5. Illegal and legal diagrams representing $A = B$.

concretely is by equating the two contours as illustrated in the left hand diagram of Fig. 5; this is illegal under 1(a),(c) of the definition of a concrete diagram. The semantic information in this diagram can be represented as a concrete diagram, for instance, as in the right hand diagram of Fig. 5, but this diagram does not comply with d.

3.4   Equivalent Diagrams

Each abstract diagram has zero or many concrete instantiations. For example, the diagrams in Fig. 6 are both concrete representations of the abstract diagram $d_1 = \langle \mathcal{C}_1, U_1, \mathcal{Z}_1, \mathcal{Z}_1^* \rangle$ where $\mathcal{C}_1 = \{A, B\}$, $\mathcal{Z}_1 = \{\emptyset, \{A\}, \{A, B\}, \{B\}\}$, $\mathcal{Z}_1^* = \{\{A\}\}$.

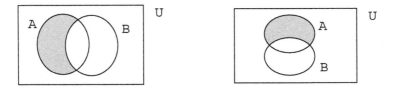

Fig. 6. Type-equivalent concrete diagrams

Two concrete diagrams $\hat{d}_1$ and $\hat{d}_2$ are **type-equivalent** if there exists an abstract diagram d of which each is an instantiation, i.e., if $\mu(\hat{d}_1) = d = \mu(\hat{d}_2)$. The two diagrams in Fig. 6 are type-equivalent. The diagrams in Fig. 1 in §2 are also type-equivalent. Type-equivalence allows us to say that two different looking diagram tokens are equivalent and is probably the most important concept in this paper; most of the points made in §4 on the utility of a two-tiered syntax rely on type-equivalence.

Two concrete diagrams $\hat{d}_1 = \langle \hat{\mathcal{C}}_1, \hat{\beta}_1, \hat{\mathcal{Z}}_1, \hat{\mathcal{Z}}_1^*, \hat{\mathcal{L}}_1, \hat{\ell}_1 \rangle$ and $\hat{d}_2 = \langle \hat{\mathcal{C}}_2, \hat{\beta}_2, \hat{\mathcal{Z}}_2, \hat{\mathcal{Z}}_2^*, \hat{\mathcal{L}}_2, \hat{\ell}_2 \rangle$ are **diagrammatically-equivalent** if $\hat{\mathcal{L}}_1 = \hat{\mathcal{L}}_2$ and there is a mapping $\phi : \hat{d}_1 \rightarrow \hat{d}_2$ that comprises a homeomorphism $\phi : \mathbf{R}^2 \rightarrow \mathbf{R}^2$ which induces a homeomorphism $\phi : \bigcup_{\hat{c} \in \hat{\mathcal{C}}_1} \hat{c} \rightarrow \bigcup_{\hat{c} \in \hat{\mathcal{C}}_2} \hat{c}$ and bijections $\phi : \hat{\mathcal{C}}_1 \rightarrow \hat{\mathcal{C}}_2$, $\phi : \hat{\mathcal{Z}}_1 \rightarrow \hat{\mathcal{Z}}_2$. These mappings satisfy the following conditions:

1. $\forall \hat{c} \in \hat{\mathcal{C}}_1 \bullet \hat{\ell}_2(\phi(\hat{c})) = \hat{\ell}_1(\hat{c})$.
2. $\forall \hat{z} \in \hat{\mathcal{Z}}_1 \bullet \hat{z} \in \hat{\mathcal{Z}}_1^* \Leftrightarrow \phi(\hat{z}) \in \hat{\mathcal{Z}}_2^*$.

Diagrammatic equivalence of concrete diagrams is a topological notion, it captures our "geometric intuition" of equivalence, while type-equivalence is a syntactic notion; however, the two notions are related.

Theorem 2. *If two concrete diagrams are diagrammatically equivalent then they are type-equivalent.*

The proof is straightforward and is omitted.

Type-equivalence is a *label-preserving* equivalence. We now explore a *label-independent* equivalence. Consider the two diagrams in Fig. 7.

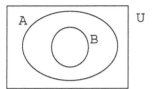   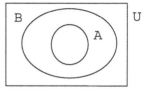

Fig. 7. Structurally equivalent concrete diagrams.

These two diagrams are different tokens of different types; however, if we remove the labels from the diagrams, they appear to be identical. We say that these two diagrams are *structurally equivalent* because their underlying diagrams (without the labels) are the "same."

Formally, two concrete diagrams $\hat{d}_1 = \langle \hat{\mathcal{C}}_1, \hat{\beta}_1, \hat{\mathcal{Z}}_1, \hat{\mathcal{Z}}_1^*, \hat{\mathcal{L}}_1, \hat{\ell}_1 \rangle$ and $\hat{d}_2 = \langle \hat{\mathcal{C}}_2, \hat{\beta}_2, \hat{\mathcal{Z}}_2, \hat{\mathcal{Z}}_2^*, \hat{\mathcal{L}}_2, \hat{\ell}_2 \rangle$ are **structurally equivalent** if there is a mapping $\phi : \hat{d}_1 \to \hat{d}_2$ that comprises a homeomorphism $\phi : \mathbf{R}^2 \to \mathbf{R}^2$ which induces a homeomorphism $\phi : \bigcup_{\hat{c} \in \hat{\mathcal{C}}_1} \hat{c} \to \bigcup_{\hat{c} \in \hat{\mathcal{C}}_2} \hat{c}$ and bijections $\phi : \hat{\mathcal{C}}_1 \to \hat{\mathcal{C}}_2$, $\phi : \hat{\mathcal{Z}}_1 \to \hat{\mathcal{Z}}_2$, such that $\forall \hat{z} \in \hat{\mathcal{Z}}_1 \bullet \hat{z} \in \hat{\mathcal{Z}}_1^* \Leftrightarrow \phi(\hat{z}) \in \hat{\mathcal{Z}}_2^*$.

Now we consider equivalence of abstract diagrams. Abstract diagrams are many-sorted algebras, so the natural concept of equivalence is that of isomorphism.

Two abstract diagrams $d_1 = \langle \mathcal{C}_1, U_1, \mathcal{Z}_1, \mathcal{Z}_1^* \rangle$ and $d_2 = \langle \mathcal{C}_2, U_2, \mathcal{Z}_2, \mathcal{Z}_2^* \rangle$ are **isomorphic** if there is a mapping $\theta : d_1 \to d_2$ such that component mappings $\theta : \mathcal{C}_1 \to \mathcal{C}_2$, $\theta : \mathcal{Z}_1 \to \mathcal{Z}_2$, are each bijections and satisfy the following conditions:

1. $\forall c \in \mathcal{C}_1, \forall z \in \mathcal{Z}_1 \bullet c \in z \Leftrightarrow \theta(c) \in \theta(z)$ and $\theta(U_1) = U_2$.
2. $\forall z \in \mathcal{Z}_1 \bullet z \in \mathcal{Z}_1^* \Leftrightarrow \theta(z) \in \mathcal{Z}_2^*$.

The abstract diagrams of the two diagram tokens in Fig. 7 are $d_1 = \langle \mathcal{C}_1, U_1, \mathcal{Z}_1, \mathcal{Z}_1^* \rangle$ where $\mathcal{C}_1 = \{A, B\}$, $\mathcal{Z}_1 = \{\emptyset, \{A\}, \{A, B\}\}$, $\mathcal{Z}_1^* = \emptyset$ and $d_2 = \langle \mathcal{C}_2, U_2, \mathcal{Z}_2, \mathcal{Z}_2^* \rangle$ where $\mathcal{C}_2 = \{A, B\}$, $\mathcal{Z}_2 = \{\emptyset, \{B\}, \{A, B\}\}$, $\mathcal{Z}_1^* = \emptyset$. Then $\theta : d_1 \to d_2$, where $\theta(A) = B$ and $\theta(B) = A$ is an isomorphism.

Structural equivalence of tokens is a topological notion, while isomorphism of types is an algebraic notion; again the two notions are related.

Theorem 3. *If two concrete diagrams $\hat{d}_1$, $\hat{d}_2$ are structurally equivalent then their abstractions $\mu(\hat{d}_1)$ and $\mu(\hat{d}_2)$ are isomorphic.*

This is just the label-free version of Theorem 2. Structural equivalence of tokens (and hence isomorphism of their abstractions) is important for the classification of diagrams, which is itself important for software tool building and diagrammatic reasoning. However, if two tokens are structurally equivalent, they need not be semantically equivalent. This leads us to yet another level of equivalence of diagrams—*semantic equivalence*. If two tokens are type-equivalent, then they are semantically-equivalent. The two tokens in Fig. 2 are not type-equivalent, but they are semantically equivalent, while the two tokens in Fig. 7 are structurally equivalent, but not semantically equivalent.

4   Utility of Two-Tiered Syntax

In this section we discuss the use of two-tiered syntax in diagrammatic reasoning and in the development of software tools to support such reasoning. Diagrammatic reasoning can take many forms. In this paper we are concerned with self-contained diagrammatic systems involving diagram manipulation.

4.1   Diagrammatic Reasoning

Diagrammatic reasoning is carried out by transforming diagrams, just as we manipulate sentences in order to make inferences in symbolic systems. As we discussed in §2, in the case of symbolic systems, it does not matter whether we mean to manipulate sentence-types or sentence-tokens, since making a type-token distinction does not have important consequences. On the other hand, in the case of diagrammatic systems, we need to make it clear whether transformation rules are being applied to diagram-tokens or to diagram-types.

It is natural to think that diagrammatic reasoning occurs at the token level, since what the user actually manipulates are concrete diagram-tokens not abstract diagram-types. However, we present several cases to illustrate that the relationship between type-syntax and token-syntax can be used to our advantage in carrying out the diagrammatic reasoning process. Thus, we argue that diagrammatic reasoning rules can be stated in terms of token-syntax but formalized and proved valid using type-syntax. As a simple case, consider again the diagrams in Fig. 6. We would like to allow redrawing the second diagram as a copy of the first diagram, or vice versa. The *copy rule* can be stated at the token-level, but with the help of the type-equivalence relation which is defined in the previous section.

Copy Rule: We may transform a concrete diagram $\hat{d}$ to another concrete diagram $\hat{d}'$ if and only if $\hat{d}$ and $\hat{d}'$ are type-equivalent.

A more interesting case is when we consider the reasoning rule, *contour erasure*. Erasing a contour can cause syntactic difficulties at the token level. In Fig. 8 erasing contour C results in shading in part of a zone. This is not a well-formed

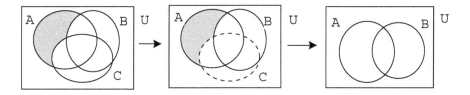

Fig. 8. Erasing a contour.

diagram. To ensure that the resulting diagram is well-formed, the shading must be removed. The reasoning rule can be formulated at the token-syntax level in such a way that this difficulty is overcome:

Contour Erasure Rule (*token-level*): Let $\hat{d}$ be a concrete diagram with at least one contour and let $\hat{d}'$ be the diagram obtained from $\hat{d}$ after erasing a contour and erasing any shading remaining in only a part of a zone. Then $\hat{d}$ can be replaced by $\hat{d}'$.

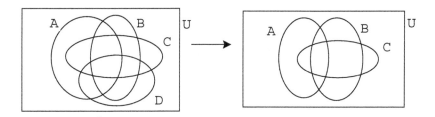

Fig. 9. Problem with erasing a contour.

However, a more serious problem can occur. In Fig. 9 removing a contour results in a non-well-formed diagram because we get disconnected zones in the resulting diagram. This problem was first noticed by Scotto [11] as a flaw in Shin's Venn I system [13]. His solution involves combining the original diagram with the diagram composed just of a boundary contour. From a construction proved by More [8], we can construct a Venn diagram with at least one contour whose erasure would result in a well-formed diagram. Scotto's solution rearranges the contours so that the contour to be removed is this well-behaved contour.

We can offer a different solution. An alternative way of formulating the contour erasure rule is to consider an abstraction d of the concrete diagram $\hat{d}$ and to obtain the abstract diagram d' by removing a contour from d. Then $\hat{d}$ can be replaced by any instantiation of d'. The resulting concrete diagram will be well-formed. The rule at the type level can be stated as follows.

Contour Erasure Rule (*type-level*): Let $d = \langle \mathcal{C}, U, \mathcal{Z}, \mathcal{Z}^* \rangle$ be an abstract diagram with $c \in \mathcal{C}$ and let $d' = \langle \mathcal{C}', U', \mathcal{Z}', \mathcal{Z}'^* \rangle$ be the abstract diagram defined by:

1. $\mathcal{C}' = \mathcal{C} - \{c\}$ and $U' = U$.
2. There exists a surjection $\sigma : \mathcal{Z} \rightarrow \mathcal{Z}'$ defined by $\sigma(z) = z - \{c\}$. Furthermore, $z' \in \mathcal{Z}'^* \Leftrightarrow \forall z \in Z \bullet \sigma(z) = z' \Rightarrow z \in \mathcal{Z}^*$.

Then d can be replaced by d'.

Any token-based solution is problematic in that complicated details of concrete syntax have to be considered, resulting in some very arcane conditions within the rule. The solution that we propose has the advantage of always producing a well-formed diagram in a natural way, because it is an instantiation of a diagram-type which has the required properties.

We suggest that this process extends to diagrammatic manipulation rules in general. What we draw and manipulate are tokens of diagrams. In this sense, diagrammatic reasoning takes place at the token level, but the rules of these operations are stated at the type level, so in another sense diagrammatic reasoning takes place at the type level. This way, we may take advantage of different kinds of flexibility each syntax-level has. At the same time, the mechanisms we developed in the previous section – instantiation, abstraction, type-equivalence, and token-equivalence – provide us with a guide to how to communicate between these two different levels. The following commutative diagram illustrates a general traffic rule: to transform concrete diagram $\hat{d}_1$ under a diagrammatic reasoning rule, we can transform an abstraction d_1 of $\hat{d}_1$ into abstract diagram d_2 and then any instantiation $\hat{d}_2$ of d_2 is the required transformation of $\hat{d}_1$.

$$
\begin{array}{ccc}
d_1 & \longrightarrow & d_2 \\
\mu \uparrow & & \downarrow \varsigma \\
\hat{d}_1 & \longrightarrow & \hat{d}_2
\end{array}
$$

In §2, we have seen that the necessity of a type-token distinction arises from the token-sensitivity and non-linearity of a diagrammatic system. Based on the formalism presented in the above section, in this section we show that this necessary mechanism adds efficiency to the system.

The following two sentences are syntactically different but semantically equivalent.

$$\forall x(Px \rightarrow \neg Qx)$$
$$\forall x(Qx \rightarrow \neg Px)$$

If the semantics is correct, we are able to prove that the two sentences always get the same truth value under the same interpretation. If the system is complete, we have a way to manipulate one sentence to the other.

Let's compare this situation with Fig. 6. Our formalism in the previous section tells us that these two diagrams are type-equivalent. So, we do not need to explore the semantics to see the semantic equivalence. The equivalence is taken care of at a syntactic level. Since they belong to the same type, the copy rule of the system simply allows the manipulation between these two tokens. When a type-token distinction is implemented in syntax, the syntactic operation between these two diagrams comes almost free. Shimojima [12] uses term 'free ride' to refer to an inference in which the conclusion seems to be read off almost automatically from the representation of premises. For further discussion, see [5]. Consider again the diagrams in Fig. 6. Type-equivalence in a syntactic level gives us a free pass for a commutation between these two diagrams without bringing in transformation rules (except the trivial copy rule). Since type-equivalence implies semantic equivalence, we get another free pass for the semantic relation between these two diagrams. Clearly, the relation between "$\exists x(Px \wedge Qx)$" and "$\exists x(Qx \wedge Px)$" involves a more elaborate story.

This efficacy is directly related to our speculation in §2, that is, that a non-linear system might not need the commutative rule. The speculation turns out to be correct, at least for the Euler system, but only when our intuition about a type/token distinction is formalized in type-syntax and token-syntax is the truth of the speculation proven. Interestingly, the 'free ride' commutativity in some diagrams can be problematic; see [3] for a discussion on such problems in the representation of quantifiers in constraint diagrams.

4.2   Software Tools

For diagrammatic notations to be used on a large scale in the software development process, appropriate software tools must be developed. The process of diagrammatic reasoning by hand is difficult; however, with automated support the process becomes much easier and potentially very valuable. Our two-level syntax will play an important role in developing efficient software tools to aid the diagrammatic reasoning process. We claim that each level of ontology has its own merits, and moreover, that the close relationship between them will provide us with a more natural way to implement a diagrammatic system.

Note that the formalization of token-syntax we have given is a mapping to the plane, $\mathbf{R}^2$, and that this mapping itself is an abstraction. Hence, the formalization of token-syntax properly depends on the medium of the instantiation. For a computer instantiation, this would most likely be in terms of pixels. Because of the multiplicity of possible forms of instantiation, it is vital that we have a macro-level of syntax, i.e., the type-syntax, which provides us not only with the basic definition of a diagram but with the relation among different forms of instantiation.

In a given system, the communication established between two levels of syntax (illustrated at the end of the previous subsection) will help us to implement diagrammatic reasoning rules in an efficient way. Continuing the discussion of the contour erasure rule from the last section, it is easy to imagine a computer system in which the most appropriate instantiation of a diagram with a contour erased is produced automatically. In this case, the algorithm for this process must rely on the type-syntax of the diagram. At the same time, our abstraction and instantiation functions do an important part of the work so that a diagram-token on a computer screen is transformed to another diagram-token.

5   Summary and Further Work

We have discussed and formally defined a two-tiered-syntax—type and token—of a diagrammatic system. We have compared this to the situation in linguistic systems and claim that this is a distinguishing feature, at least for diagrammatic notations based on Euler diagrams; this is a substantial and important collection in its own right but it is our belief that our claim extends to other types of diagram. We have shown that it is necessary to consider both forms of syntax in developing "self-contained" diagrammatic inference systems and in their software implementation. The general aim of this work is to provide the necessary mathematical underpinning for the development of software tools to aid reasoning with diagrams. In particular, we aim to develop the tools that will enable diagrammatic reasoning to become part of the software development process. In order for this to happen we need to be able to develop and implement an algorithm that takes a diagram-type and instantiates it as a diagram-token. Work is already underway on this [2].

Acknowledgements. We would like to thank Jean Flower and the anonymous referees for their very helpful comments. Authors Howse and Taylor were partially supported by UK EPSRC grant GR/R63516.

References

1. L. Euler. *Lettres a Une Princesse d'Allemagne*, vol 2. 1761. Letters No. 102-108.
2. J. Flower, J. Howse. *Generating Euler Diagrams*. Accepted for Diagrams 2002.
3. J. Gil, J. Howse, S. Kent. Towards a formalization of constraint diagrams, *Proc Symp on Human-Centric Computing*, Stresa, Sept 2001. IEEE Press
4. N. Goodman. *Languages of Art: An approach to a theory of symbols*. Hackett Publishing Co, INC. 1976.
5. C. Gurr, J. Lee and K. Stenning. Theories of diagrammatic reasoning: Distinguishing component problems, in *Minds and Machines* 8, 533-557, 1998.
6. E. Hammer. *Logic and Visual Information*. CSLI Publications, 1995.
7. D. Harel. On visual formalisms. In J. Glasgow, N. H. Narayan, B. Chandrasekaran, eds, *Diagrammatic Reasoning*, 235-271. MIT Press, 1998.
8. T. More. On the construction of Venn diagrams. J Symb Logic, 24, 303-304, 1959.
9. OMG. UML Specification, Version 1.3. Available from www.omg.org.
10. C. Peirce. *Collected Papers* Vol. 4. Harvard Univ. Press, 1933.
11. P. Scotto di Luzio. Patching up a logic of Venn diagrams. Proc. 6th CSLI WS on Logic, Language and Computation. CSLI Publications, Stanford, 2000.
12. A. Shimojima. Operational constraints in diagrammatic reasoning, in J. Barwise, G. Allwein eds. *Logical Reasoning with Diagrams*. OUP, New York, 27-48, 1996.
13. S.-J. Shin. *The Logical Status of Diagrams*. CUP, 1994.
14. J. Venn. On the diagrammatic and mechanical representation of propositions and reasonings. *Phil.Mag.*, 1880. 123.

Effects of Navigation and Position on Task When Presenting Diagrams to Blind People uUsing Sound

David J. Bennett

Department of Mathematics and Computing
Canterbury Christ Church University College
North Holmes Road
Canterbury, CT1 1QU, UK
+44 (0)1227 767700
djb12@cant.ac.uk

Abstract. This paper questions how we could and should present diagrams to blind people using computer-generated sound. Using systems that present information about one part of the diagram at a time, rather than the whole, leads to two problems. The first problem is how to present information so that users can integrate the information into a coherent overall picture. The second is how to select the area to be presented. This is looked at by using a system that presents graphs representing central heating system schematics. The system presents information by user choice through either a hierarchical split of information and navigation system, or a connection oriented split of information and navigation system. Further, we have a split as to whether a simple system of presenting location of nodes is used, or not. Tasks, classified as being based on hierarchical information or connection-based information, were set using the system and the effect of the different models was recorded. It was found that there was a match of task to navigation system, but that presentation of position had no discernable effect.

1 Introduction

There is much interest in the use of computers as an enabling technology for blind people. This technology relies on substitution of information normally presented visually with information presented so that an alternate sense may be used. Typically, visual information is substituted by either 'auditory' or 'haptic' information, that is, information sensed by the ears or by touch. We are interested in presentation of diagrammatic material using the auditory channel. While the haptic channel is often used for presenting diagrammatic material to blind people, we focus on the possibilities of using sound. Kokjer reports that the bit rate for auditory information is 100 times that for fingertip touch [8]. While this is a simplistic measure by itself, can benefit be had from this rate advantage?

While we may be convinced of the utility of diagrams for sighted people in suitable areas, there is not such strong evidence of their utility for blind people. A typical view, which has been expressed to the author on several occasions, is that diagrams are inherently visual, and therefore are not suitable for presentation in a non-visual manner. We disagree with that view. While the actual presentation of the information

M. Hegarty, B. Meyer, and N. Hari Narayanan (Eds.): Diagrams 2002, LNAI 2317, pp. 161–175, 2002.
© Springer-Verlag Berlin Heidelberg 2002

may be visual and the history of diagrammatic representations inherently influenced by visual presentation, the information content of a diagram may also be represented in other ways. We are not so much interested in whether the information can be presented in alternative ways, but the efficacy of doing so. Is it possible to present diagrammatic information in such a way that makes it useful to blind people?

1.1 Diagrams

If an evolutionary approach to information representation is taken, diagrams are useful because they offer something to users over other types of representations. In their paper, 'Why a Diagram is (Sometimes) Worth Ten Thousand Words', Larkin and Simon [9] attribute three main advantages to diagrammatic materials (for sighted people):

- *"Diagrams can group together all information that is used together, thus avoiding large amounts of search for the elements needed to make a problem-solving inference.*
- *Diagrams typically use location to group information about a single element, avoiding the need to match symbolic labels.*
- *Diagrams automatically support a large number of perceptual inferences, which are extremely easy for humans."* (Page 98)

However, they are keen to point out that these three attributes do not,
> *"insure that an arbitrary diagram is worth 10,000 words". [Ibid.]*

How can we best use technology to enable blind people to gain these benefits of diagrammatic material? Diagrammatic material for blind people has often been presented in tactile form. A number of technologies have been used to create static and dynamic representation. Static forms, such as those created by swell paper [6,13], have no way of being updated unless a new version is created. Dynamic representations may update as representations update, but unfortunately tend to have speed and size restrictions because of engineering limitations [11].

Diagram Systems for Blind People

The use of computers has allowed more use of auditory access to diagrammatic material, and several approaches have been taken.

Rigas developed a system called 'Audiograph' [12] that represents shapes by tracing their outline using *direct sonification*. Direct sonification means the use of sound to represent information in such a way that perceptual features of the sound correspond simply and directly to features of the information to be presented, such that the listener may use them with little need for additional interpretation. Rigas used two scales of notes, played one after the other to indicate the co-ordinates of points on a two dimensional plane. The first scale played on one instrument indicated the x-co-ordinate. The second scale, played on a different instrument, indicated the y-co-ordinate. The origin was at the bottom-left of the diagram. This system could also be used to trace around the outline of a shape, by choosing a starting location, and following an edge in a clockwise direction. Each time the x-coordinate changed, the corresponding note is played using the designated instrument, and similarly for the y-coordinate. The system was found to be effective in test on position and simple shapes, but much less successful for complex images (such as a simple outline of a car), unless hints were given (e.g. it's a mode of transport).

Similarly named, but a different system, Kennel's Audiograf [7] was designed to present diagrams composed of frames, text and connections. These elements could be used in a hierarchical manner, so that frames could contain other frames and text. The system allowed access to diagrams described by the representation through a touch-pad, (synthetic) speech and non-speech sounds. The touch-pad mapped directly location of items on the diagram to location on the touch-pad, so that items on the top-left of the diagram could be accessed by pressing the top-left of the touch-pad.

Blenkhorn and Evans produced a system, 'Kevin' [4] that present data-flow diagrams used in software engineering using earcons [3], synthetic speech and a touch-pad for user input. As the basis for conversion of the information presented to a blind person Kevin used an alternative representation of a data-flow diagram – the N^2 chart. This representation is an incidence matrix by another name, indicating where flows from one diagram element are incident on another. The diagram elements are placed on the diagonal from top-left to bottom-right. Flows are indicated at the intersection square of the matrix horizontally from one element and vertically from the other. The blind user then accesses the N^2 chart by using a touch pad with a grid placed on it to represent the blank matrix. Users press the touch pad at positions on the matrix to hear information about the elements or links indicated at that point. Kevin was evaluated by Obee [10] who found, significantly, that it was easier for the participants to use a tactile (raised line) representation of the data-flow diagram than Kevin for tracing flows. However, participants found it easier to name the diagram elements using Kevin than the tactile representation. This suggests a lack of *computational equivalence* [9, page 67] between data-flow diagrams and the N^2 chart used by Kevin.

The N^2 chart is also lacks *information equivalence* (ibid.) in that it is not possible to determine the original spatial locations of elements in the diagram. This limits the user's ability to use this information to infer relatedness based on spatial location, as put into the original diagram by its designer.

It was decided to look at two limits of the Kevin system. The first limitation was the chosen alternative representational form that removed the original position information. Would a simple method of presenting the position of the elements aid the user's ability to read, understand and work with a diagram.

Secondly, the N^2 chart did not have computational, or near computational, equivalence to the tactile version. One reason for this was that the method of moving focus around the diagram was different to that employed on the original form. It was believed that the way that the user moved focus around the diagram has a significant effect on the ease of performing different types of task, and that this would show a match-mismatch effect.

In order to look at these areas, a diagram form and a way of representing these diagrams in sound to blind people had to be chosen. This representation needed to have more than one method of navigating around the diagram. There was interest in the 'box-and-line' style diagram, where symbols on the diagram are connected using lines to show a relationship. Examples of this type of diagram are common in engineering and computing; circuit diagrams, data-flow diagrams and some visual programming languages use this kind of representation.

One method of navigating around a box-and-line diagram is to allow the user to travel along the lines that link the symbols, much in the same way as a sighted person would follow lines to examine the symbols at the ends. The user could then examine the symbol under the current focus, in the same way that a sighted person would examine a symbol at the focus of their vision.

Recall Larkin and Simon's view of the advantages of diagrams. One of these is that "diagrams can group together all information that is used together". A reader may see groupings at different levels – a group may *contain* smaller groups, while itself being part of a larger group. The groupings could, therefore, form a hierarchical structure. Another way to navigate around a diagram is to see groupings of symbols and use these to form areas in the diagram and move from area to area. Users can then move up and down the hierarchy examining the groupings.

The diagrams used had similar characteristics to the data-flow diagrams used by Kevin, but were based around representation of a house central heating system. Presentation was in sound using a PC based computer program. The sound was a mixture of earcons, Rigas's method for presenting position and digitised, recorded, human speech. The program allowed the experimenter to control the method of navigation and whether the position of elements was presented, or not.

1.2 Participants and Groups

The experiment had eight conditions. The conditions were based around three criteria:
- Navigation model;
- Ordering of questions;
- Presentation of co-ordinates of the components of the diagram parts.

The navigation model criteria had two states:
- Hierarchical;
- Connection-oriented.

The ordering of questions was either:
- Hierarchical questions first;
- Connection-oriented questions first.

The final criteria had participants either:
- Receiving position information of the components using Rigas's method;
- Not receiving any position information at all.

There were sixteen participants in the experiment, eight sighted and eight blind. The participants were randomly assigned to the experimental conditions so that there was one blind and one sighted participant per condition.

Visual Ability
The blind participants all stated that they were 'registered blind', with actual levels of sight varying from no sight at all to having 4/60 vision in one eye only. Two of the participants were congenitally and totally blind, two more had some sight loss since

birth, while the remainder were adventitiously blind. Of these latter four, loss varied from total, to vague colour and shape perception.

The eight sighted participants had excellent sight with or without corrective lenses.

Educational Background

The blind participants had a varied education; four were from Royal National College, Hereford, while the remainder were from local sources in the York area. None of the blind participants had any significant cognitive disability or a disability that prevented them from using computers using a keyboard. The participants from Royal National College were either studying for further education qualifications or undertaking orientation for University study. The participants from the York area varied from one taking a PhD to one who had only received education to secondary level.

All of the sighted participants bar one were undergraduates at the University of York (UK), reading a mixture of arts and sciences. The remaining sighted participant was a DPhil student at the University of York.

Computer Usage by Participants

All the participants had previously used computers, and were confident and competent with them. All participants had used word processors and electronic mail. Other uses included spreadsheets, databases, computer programming, music composition and electronic circuit design.

The blind participants used mostly speech synthesisers to access computers. 'HAL' and 'Jaws' were the most prominent screen-readers. Only one participant used a 'Zoom Text' screen-magnifying program to utilise his remaining sight.

All the sighted participants used a VDU as the standard way to interact with computers.

Haptic Interface Usage

None of the sighted participants had used tactile diagrams, though most were aware of their existence.

All the blind participants had some experience of tactile diagrams. All had some reservations, especially those who had poor tactile ability. None had experience of audio descriptions of diagrammatic material, though one had used audio description of artistic material in conjunction with tactile representations.

Musical Background

Musical background varied. Among the sighted people, one said they had no musical training at all, while two of the blind participants professed to being 'not musical'. All the others had some musical training – there were Music Technology students, a retired piano tuner and a music teacher among the participants.

Match of Blind to Sighted Participants

Ideally there would have been a match between the blind and sighted participants on criteria judged to be important. However, it was not clear what criteria should be used in matching participants. It was decided not to place great importance on comparisons between these two groups.

Similarity within Groups

The second area where there was greater scope for similarity was within the group of blind participants. One prominent difficulty with performing experiments with blind people is finding large groups who are willing and able to take part. In order to obtain eight subjects it is often the case that one must accept all those who volunteer from a number of different sources.

1.3 Materials

An initial questionnaire was used to obtain background on the participants. A training script was developed to be read aloud to the participants, with their interacting with the computer and answering questions to retain their interest and concentration. A computer program was developed so blind people could explore diagrams and is described below.

Computer Program and System

The computer program was designed to enable blind people to access diagrams of central heating systems. It was developed in Borland Delphi and run on a PC Windows platform. Presentation of information was in a two-stage process of 'orientation' and 'reading' as described by Bennett and Edwards [1].

At initialisation for a particular diagram, 'orientation' consisted of giving the diagram name, the number of elements within it and the maximum depth of the hierarchical view. This was intended to give an overall view of complexity.

The 'reading' phase consisted of the user using key presses to elicit information and to navigate around the diagram. Information was presented about the element under the current position 'cursor'. Navigation was by following (heating) pipes in the connection model and by following lines to objects higher or lower on the hierarchical

Table 1. Key Presses and Their Actions

Key	Effect
W	What proportion of diagram has been visited
E	How many pipes are incident to this component (connection)
	How many sub-components has this container got? (hierarchical)
D	Describe far end of 'where-next pointer'
C	Which connection is 'where-next pointer' on? (connection)
	Which sub-component is 'where-next pointer' on? (hierarchical)
X	Give exclusive number for current item under focus
Space	Describe current location
Enter	Silence output
↑	Describe hierarchical location (connection) Move to parent (hierarchical)
↓	Follow 'where-next pointer'
→	Next choice for 'where-next pointer'
←	Previous choice for 'where-next pointer'

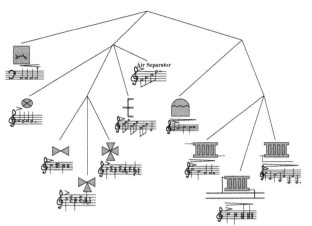

Fig. 1. Earcons Used by the Program to Represent Components of the Central Heating System

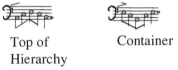

Top of
Hierarchy

Container

Fig. 2. Additional Earcons Used by the Hierarchical Representation

model. Which of the possible choices for moving the cursor from the current location to another would be taken on a move operation was controlled by the concept of a 'where-next pointer'. This pointer was controlled by key presses, allowing users to scroll through the available choices.

The key presses and their actions are given in Table 1. Key presses were chosen on two criteria. The first was so that users could keep their right hand on the cursor keypad for navigation and their left hand on the left-hand side of the keyboard for information gathering. The second was so that some (tenuous) *aide-memoirs* could be used, indicated by **bold** type in the table.

The computer responded to key-presses with a sound response. This used a combination of earcons, direct sonification recorded speech described below.

Earcons, Position Presentation, and Speech

Earcons were designed to be used by the computer program. The principles for earcon design proposed by Brewster [5] and Blattner *et al* [3] were incorporated, as far as possible with the available equipment. The earcons (Fig. 1 and 2) were designed around a hierarchy based on up to three levels: heat effect (generation, transport, output), and then sub-divisions according to type. Earcons were played 'live' on a

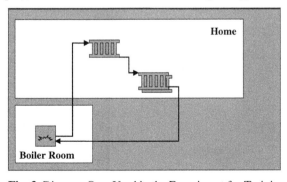

Roland D110 synthe-siser using a trumpet timbre and recorded as '.WAV' files for use in the computer program. The '.WAV' files were then edited to ensure the required volume dynamics were present and all were of a similar per-ceived volume.

Fig. 3. Diagram One, Used in the Experiment for Training

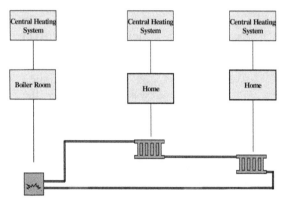

Fig. 4. Connection Model of Diagram One

A method closely resembling Rigas's method of co-ordinate presentation was used. A 16x16 co-ordinate space was presented, rather than Rigas's 40x40 grid. One major difference between Rigas's method and our method was that the last note of the played scale was sustained, as was the eighth or 'octave' note. The scales were again played live on the D-110 and recorded into '.WAV' files. Piano and Organ timbres were used for the x and y co-ordinates respectively.

Speech used in the computer program was human speech recorded into '.WAV' files. This was to provide a certain level of clarity to the speech, so that participants not used to hearing synthetic speech would be able to take part.

The element names and the earcons representing the type of element were combined to form a parallel presentation – the earcon to one speaker and the speech to the other. The timing of the start of the earcon was adjusted so that the sharp attack of the earcon did not interfere with the comprehension of the speech.

Sentence parts and individual words were combined with the earcons and position presentations by the computer program to form complete utterances. These utterances could be interrupted by participants pressing keys to obtain further information.

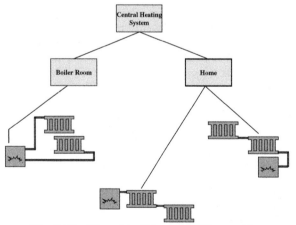

Fig. 5. The Hierarchical Version of Diagram One

Navigation Models

The computer program modelled the diagrams so that they could be accessed in two different ways. The first of these highlighted the pipe connections between components and allowed navigation from component to component along the pipes that linked them (Fig. 4).

At each component a user could have obtained information about the hierarchical structure, by hearing about the parent, grandparent etc. of the component that they are on. However, the Reader cannot traverse up the hierarchy to determine the sub-items for an item higher up the tree from the component.

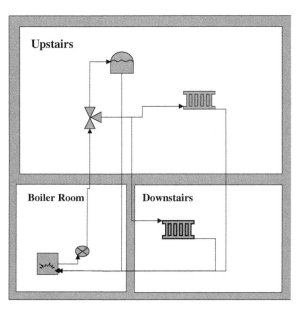

Fig. 6. Diagram Two

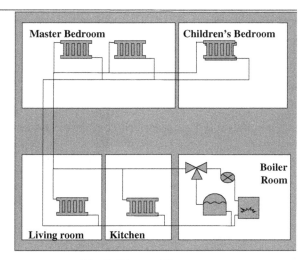

Fig. 7. Diagram Three

The second navigation model highlighted the hierarchical grouping structure and allowed navigation up and down the hierarchy (Fig. 5). It was possible to move up and down the hierarchy and also explore the items that were directly connected to any component, but it was not possible to travel to them in order to reach those items directly connected to them. In this way, a user who wished to traverse the pipe work had to retain information about an item at the far end of a pipe, and then traverse the hierarchical structure to find it and in turn what was connected to that.

Diagrams Used for Testing

Diagram One was used in the training phase of the experiment. The hierarchical and connection oriented navigational equivalent versions of Diagrams Two and Three were used in the main testing phase described below.

1.4 Method

The experiment followed a fixed format: participants undertook a questionnaire and training phase, the main testing phase and a post-test questionnaire phase.

Prior to Testing

Before testing, participants answered a questionnaire and then received training on the diagram representation system and program. The questionnaire asked for basic information about their background and is reported in the 'Participants' section above.

The training was presented in order to bring the level of expertise in the diagram representation up to a basic level of competence.

Testing Phase

The testing phase involved participants interacting with the computer system to complete tasks (answering questions) with Diagram Two, and then with Diagram Three. The interaction sessions with each participant were video recorded for later analysis.

Each session had the participants performing twelve tasks. The task questions were categorised as either connection-oriented or hierarchy-oriented. There were three questions of each type for each diagram. Participants received either the connection- or hierarchy-oriented questions first for each diagram. Participants were timed for each task as they performed it.

After each task participants were asked to rate the task in terms of ease of performance on a numerical scale from 1 to 5, with 1 mapping to an 'impossible task', and 5 to an 'extremely easy or trivial task'. Participants who took longer than 30 minutes, or who indicated that they could not complete a task moved on to the next task, and these were rated 1 automatically.

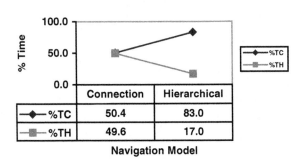

Fig. 8. Percentage Time Spent on Task by Question Type and Navigation Model on Diagram Two.

Post-Testing

After the testing, the participants were asked a series of questions to try to elicit further information about the ease of using the computer program.

2 Results

Statistical differences were looked at using Analysis of Variance at a 95% confidence level. Data were analysed using a four-way split-plot ANOVA where sightedness (2 levels: sighted and blind), navigation model (2 levels: hierarchical and connection) and position information (2 levels: given and not given) are between subject variables and question type (2 levels: hierarchical questions and connection questions) is repeated measure. Results are shown to 3 decimal places.

2.1 Time Taken to Perform Tasks

Time taken to perform tasks was assumed to follow a normal distribution. There were a number of significant effects and interactions in terms of raw time taken to perform the tasks.

For Diagram Two, question type and sightedness combined with navigation model significantly affected time to perform tasks. For Diagram Three, 'question type, position information, question type combined with navigation model', 'question type combined with position information, navigation model, sightedness combined with navigation model' and 'navigation model combined with position information' all significantly affected time to perform task.

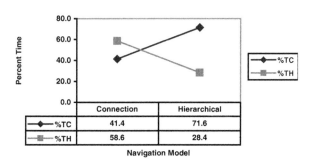

However, emphasis was placed on results pertaining to the percentage of the overall time that people spent on each task. This was done to try to eliminate large individual differences in the time taken to work through the questions posed.

Fig. 9. Percentage time spent on task type by question type and navigation model on Diagram Three

With Analysis of Variance it was found that there was a significant effect on the time spent on task type by questions type and navigation model for both Diagrams Two and Three. For Diagram Two, Analysis of Variance showed Navigation model significantly affected the percentage time taken by the participants [F(1,8) = 15.445], shown in Fig. 8 . For Diagram Three, Analysis of Variance showed navigation model significantly affected the percentage time taken by the participants [F(1,8)= 12.235], shown in Fig. 9.

2.2 Rating of Ease of Performance

For Diagram Two, only question type significantly affected 'ease of performance' [$F(1,8)$=8.128]. For Diagram Three, question type [$F(1,8)$=10.196], position information [$F(1,8)$=7.210] and sightedness combined with navigation model [$F(1,8)$=14.141] all significantly affected ease ratings.

2.3 Correctness

Participants were given one mark for each wholly correct answer given. No changes to the marks were given either for 'correctness of method, but incorrect answer', or 'correct answer, but incorrect method'. For Diagram Two, question type [$F(1,8)$=5.556], sightedness [$F(1,8) = 10.889$] and an interaction between sightedness and position information [$F(1,8)$=5.556] all yielded significant results. For Diagram Three, there were no significant effects.

3 Discussion

3.1 Statistical Results

When reviewing the video recording of participants working on tasks on Diagram Two, it indicated that most participants were still learning how to use the system. This was indicated by the number of times the participants made mistakes in which keys they pressed, the number of times they asked for assistance on keys, earcons or in some cases the method of using the underlying model. By the time participants started on Diagram Three, participants were felt to be more confident and more at ease with using the system.

This would indicate that, on the whole, the training session was too short to deem the participants as being competent at the stage they worked on Diagram 2. With this in mind, it was decided to use evidence from Diagram 2 only as a minor indicator of performance, concentrating on the evidence from Diagram 3, but with extra emphasis on results where differences were found in both Diagrams 2 and 3.

Ceiling or floor effects on measures were looked for and it was felt that the following measures had ceiling effects associated with them:

- Diagram Three – Ease of performing Hierarchical Tasks (mean 4.25 out of a maximum of 5)
- Diagram Three – Number of Hierarchical tasks performed correctly (mean 2.56 [3 s.f.] out of a maximum of 3)

It is felt that this indicates that by the time the participants performed Hierarchical tasks on Diagram Three they had a mature and successful method of performing these, independent of which navigation model they were using. Connection based tasks for the same diagram did not have the same ceiling effect.

Diagram Three had no significant differences related to number of questions answered correctly according to any of the factors, and this may be explained by the generally high level of competence of participants with the system by the time that they reached this stage of using the system.

3.2 Evidence and the Hypotheses

Hypothesis One stated that:
The Navigation Model used by a diagram user significantly affects the model of the diagram they produce internally and hence the ease of performing and the time taken to perform the different types of task with the diagram.

This hypothesis is supported and the Null Hypothesis that Navigation model does not affect the internal representation of the user can be rejected. Although the raw timing information did not show that the Navigation model had a significant effect on Diagram Two, it is felt that the evidence presented by the percentage time analysis shows clearly that for both diagrams Navigation model significantly affected the time spent on performing the tasks of each type. The match-mismatch was evident.

Further, in terms of raw time, sightedness combined with navigation model was significant. It is suspected that the general blind population, because of their background, are less likely to be experienced in the hierarchical view of a diagram and this hinders their use of this efficient form.

The ease ratings were not significantly different for Navigation model solely on either Diagram, but for Diagram 3 sightedness combined with navigation model did show significant differences. These ratings support the information given by the raw time information: that Blind people found the hierarchical model hard to use, while sighted participants found it easier.

Clearly, use of a hierarchical view of a diagram in sound would require ensuring that blind users were sufficiently trained in this model.

Hypothesis Two stated that:
The presentation of Co-ordinate Position of parts of a diagram will significantly reduce advanced guess-ahead and hence ease the learning of a diagram. This will be evident through a higher ease of performing all tasks with the diagrams and an overall reduced time for all tasks.

It is felt that this hypothesis is not supported. There were significant differences in the ease ratings by position information for Diagram Three; there was no contributing evidence from the timings of the tasks. Although there were a number of factors that did include position information as a significant factor in combination with other factors, it was felt that further investigation would be required. What can be said with confidence is that the method of presenting position was not successful in the eyes of the participants and does not give clear evidence of benefit.

3.3 Additional Findings

While reviewing the video recording of the participants it was noted that more than one participant would miss error information while interacting with the system. This was because they were able to 'work-ahead' of the system – interrupting sound output by pressing another key. Usually this was useful; once a user was used to the diagram they could move around rapidly and accurately. However, there were problems when users made errors and error information should have been presented, but was not because it was interrupted by the user's next key press. This meant that the participants either became disoriented or built an incorrect model of the diagram. When producing sound based systems it is useful for sound to be 'interruptible' – long presentations are tedious if the information is not currently being sought. However, error sounds should be classified as being 'non-interruptible' and always played regardless of user interaction with the system.

4 Conclusions

This experimental work raises several further interesting questions. Would a benefit be shown for inclusion of position information if the number of participants had been higher? There was certainly no active and conscious keen interest from the participants; it is possible that the position information did not add enough information to their understanding. Would presentation of more types of position information add sufficient usefulness to make a statistically noticeable difference?

While a match-mismatch was shown for the two different navigation models, the computer program was capable of providing a combined navigation model where both types of navigation were possible. This combined model was not tested with the other two models. It may be that this would be the most useful navigation model to present. However, this was not the *raison d'être* of the experiment. It has been shown that the format of information has an important effect on the way that people can complete tasks and it should be part of an alternative representation designer's work.

Acknowledgements. This work was performed under supervision of Dr. Alistair Edwards of the University of York, sponsored by a studentship from the Engineering and Physical Science Research Council. The work formed part of a DPhil dissertation [2].

References

1.   David J. Bennett and Alistair D. N. Edwards, "Exploration of Non-seen Diagrams". In: ICAD '98: International Conference on Auditory Display, held in Glasgow 1-4 November 1998.
2.   David James Bennett, Presenting Diagrams in Sound for Blind People. Submitted as a DPhil thesis at the University of York, York, U.K.
3.   M. Blattner, D. Sumikawa and R. Greenberg, "Earcons and Icons: Their Structure and Common Design Principles." Human-Computer Interaction Vol. 4 (1) (1989) 11-44.
4.   Blenkhorn and Evans, "A Method to Access Computer Aided Software Engineering (CASE) Tools for Blind Software Engineers". In: Wolfgang L. Zagler, Geoffrey Busby and Roland R. Wagner (eds.): Computers for Handicapped Persons: 4th International Conference. ICCHP '94, (1994) 321-328.

5. Stephen. A. Brewster, "Providing a Structured Method for Integrating Non-Speech Audio into Human-Computer Interfaces". Submitted as a DPhil thesis at the University of York, York, U.K., 1994.
6. John A. Garner, "Tactile Graphics: An Overview and Resource Guide", Information Technology and Disabilities, Vol. 3, Num. 4, 1996. Available at http://bubl.ac.uk/journals/lis/com/itad/v03n0496/article2.htm, last visited 1 November 2001.
7. Andreas Kennel, "AudioGraf: A diagram reader for the Blind". In "ASSETS 96: The Second Annual ACM Conference on Assistive Technologies". The ACM (1996) 51-56.
8. Kenneth J. Kokjer, "The Information Capacity of the Human Fingertips." I.E.E.E. Transactions on Systems, Man and Cybernetics SMC-17(1) (1987) 100-102.
9. Jill H. Larkin and Herbert A. Simon, "Why a Diagram is (Sometimes) worth Ten Thousand Words". Cognitive Science 11, (1997) 65-99.
10. John Obee, "An Evaluation of "Kevin" – a CASE tool for the blind". Technical report from the University of Hertfordshire, Hatfield, UK, October 1995.
11. Scott Orlosky and Deborah Gilden, "Simulating a full screen of Braille". Journal of Microcomputer Applications, 15, (1992) 47-56.
12. Dimitrios I. Rigas and James L. Alty, "The Use of Music in a Graphical Interface for the Visually Impaired". In: S. Howard, J. Hammond and G. Lindegaard (eds.): Proceedings of Interact '97, the International Conference on Human-Computer Interaction, (Sydney). Chapman and Hall, (1997) 228-235.
13. "Maps and Diagrams" page from the RNIB, available at http://www.rnib.org.uk/services/maps.htm, last visited 1 November 2001.

A Fuzzy Visual Query Language for a Domain-Specific Web Search Engine

Christian S. Collberg

Department of Computer Science
University of Arizona
Tucson, AZ
collberg@cs.arizona.edu

Abstract. AλgoVista is a web-based search engine that assists programmers to find algorithms and implementations that solve specific problems. AλgoVista is not keyword based but rather requires users to provide — in a very simple textual language — *input⇒output* samples that describe the behavior of their needed algorithm. Unfortunately, even this simple language has proven too challenging for casual users.

To overcome this problem and make AλgoVista more accessible to novice programmers, we are designing and prototyping a visual language for creating AλgoVista queries. Since web users do not have the patience to learn fancy query languages (be they textual *or* visual), our goal is to make this language and its implementation natural enough to require virtually no explanation or user training.

AλgoVista operates at http://algovista.com.

1 Background

Frequently, working software developers encounter a problem with which they are unfamiliar, but which—they suspect— has probably been previously studied. Just as frequently, algorithm developers work on problems that they suspect have practical applications.

AλgoVista is a web-based, interactive, searchable, and extensible database of problems and algorithms designed to bring together applied and theoretical computer scientists. Programmers can query AλgoVista to look for relevant theoretical results, and theoretical computer scientists can extend AλgoVista with problem solutions.

Unlike most other search engines, AλgoVista is not keyword-based. Keyword-based searching fails exactly in those situations when we are in the most need for accurate search results, namely when we are searching in a new and unfamiliar domain. For example, a programmer looking for an algorithm that solves a particular problem on graphs will not get any help from a keyword-based search engine if she does not know what the problem is called. A Google keyword search for ⌜graph algorithms⌝, for example, returns 300,000 hits that the user has to browse manually.

M. Hegarty, B. Meyer, and N. Hari Narayanan (Eds.): Diagrams 2002, LNAI 2317, pp. 176–190, 2002.

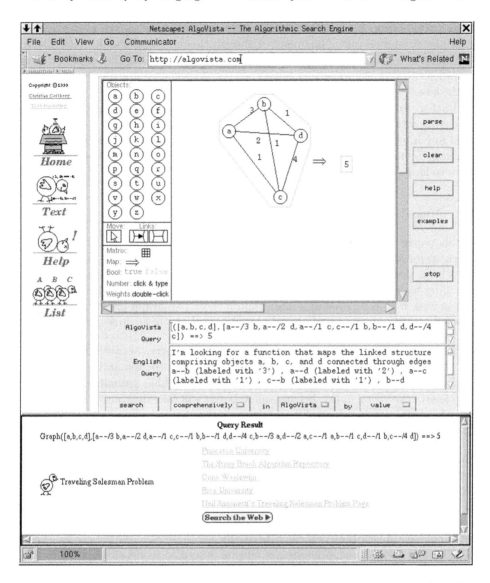

Fig. 1. The AλgoVista user interface. The top right frame is an applet into which the user can enter a graphical query. The applet also displays the textual and English language queries corresponding to the graphical one. The bottom frame shows the result of the search.

Instead, AλgoVista requires users to provide one or more *input⇒output* examples that give a (usually fuzzy and incomplete) description of the problem they are looking for. This technique turns out to be remarkably successful: when looking for links to particular graph algorithms, for example, AλgoVista often returns the requested results in a few seconds.

In previous versions of AλgoVista such *input⇒output* examples were given in a simple text-based query language. Although this language should only take a few minutes to master, most users are too impatient to read the ample on-line documentation or even learn by trying out the canned example queries available at the site.

Rather, immediately after the AλgoVista web-page has been loaded, typical users will enter a few keywords and then hit the SUBMIT button. This will not yield any interesting results since, as we have already noted, AλgoVista's query language is not keyword-based. As a result, the user will get discouraged and leave to look for a different search engine.

1.1   AλgoVista's Visual Interface

In this paper we will describe the design and implementation of a visual query language for AλgoVista. The visual language and its accompanying user interface have been designed to be as self-explanatory as possible. A web user who is unwilling to learn a textual query language will be unwilling to learn a visual one as well, if it means reading more than a short paragraph of documentation. Our main language design strategy is summarized by these three points:

1. Give the user complete freedom in drawing his query, because no web user will take the time to learn complex visual grammars or semantic constraints.
2. Let the user choose herself between different possible semantic interpretations of the visual query, because no web user will take the time to understand why certain parses are valid and others are not.
3. Make each visual query a learning experience.

More specifically: we let the user draw something, we show her the possible interpretations of this drawing, and we allow her to select the most appropriate interpretation. At the same time, we show her the textual query corresponding to the visual one, in the hope that, in time, she will subliminally acquire this language. This may come in handy if, at some later date, she needs to submit a more complex query best described textually.

The textual and visual languages will be described in detail later on in the paper. For now, consider AλgoVista's visual user interface in Figure 1. We note that the interface has a main drawing window in which the user can enter her query by dragging elements from the template window on the left. When a query has been entered the user clicks the PARSE button, the drawing is analyzed, and a textual query is produced in the query window.

At the same time, two things happen to help the user understand the query she just entered:

1. the query is translated into English prose which is shown at the bottom of the screen;
2. convex hulls are drawn around those elements of the drawing that the parser has decided belong together.

If the user is happy about the interpretation of her drawing she clicks the SUBMIT button, and the textual query is sent to the AλgoVista server for processing and search results are returned to the user. If she does not believe the interpretation to be the correct one, she can continue to hit the PARSE button to cycle through all possible interpretations until the desired one is found.

2   AλgoVista — A Search Engine for Programmers

Before we continue our description of AλgoVista's textual and visual query languages, we will briefly motivate the need for specialized search engines for computer scientists. We will also give some examples of how AλgoVista can help a working programmer classify problems and search for algorithms that solve these problems. As we will see, AλgoVista is particularly helpful when you are attempting to classify a problem outside your area of expertise and you have no knowledge of the terminology in this area.

2.1   Interacting with AλgoVista

A programmer will typically interact with AλgoVista by providing *input*⇒*output* samples that describe the problem they are looking to classify. AλgoVista will then search its database of problem descriptions looking for problems that map *input* to *output*. Before we examine the AλgoVista database in detail, we will consider three concrete examples of how a user might query the search engine.

Example 1: Suppose Bob is trying to write a program that identifies the locations for a new franchise service. Given a set of potential locations, he wants the program to compute the largest subset of those locations such that no two locations are close enough to compete with each other. It is trivial for him to compute which pairs of locations would compete, but he does not know how to compute the feasible subset. He starts by trying to come up with an example of how his program should work:

- If there are three locations a, b, c and a competes with b and c, then the best franchise locations are b and c.

If Bob is unable to come up with his own algorithm for this problem he might turn to one of the search-engines on the web. But, which keywords should he use? Or, Bob could consult one of the algorithm repositories on the web, such as http://www.cs.sunysb.edu/ algorith/, which is organized hierarchically by category. But, in which category does this problem fall? Or, he could enter the example he has come up with into AλgoVista at algovista.com.[1] This query expresses:

> "If the input to my program is two relationships, one between a and b and one between a and c, then the output is the collection [b,c]."

Another way of thinking about this query is that the input is a graph of three nodes a, b, and c, and edges a-b and a-c, but it is not necessary for Bob to know about graphs. AλgoVista returns to Bob a link directly to http://www.cs.sunysb.edu/ algorith/files/independent-set.shtml which contains a description of the Maximal Independent Set problem. From this site there are links to implementations of this problem.

Example 2: Suppose that Bob is writing a simple DNA sequence pattern matcher. He knows that given two sequences $\langle a, a, t, g, g, g, c, t \rangle$ and $\langle c, a, t, g, g \rangle$, the matcher should return the match $\langle a, t, g, g \rangle$, so he enters the query[2]

$$([a,a,t,g,g,g,c,t],[c,a,t,g,g]) ==> [a,t,g,g]$$

into AλgoVista which (within seconds) returns the link http:// hissa. nist. gov/ dads/ HTML/ longestcommn.html to a description of the longest common subsequence problem.

Example 3: AλgoVista is also able to classify some simple combinatorial structures. Given the following query

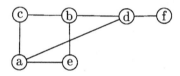

AλgoVista might respond with:

> "This looks like a *complete bipartite graph.* http:// www. treasure-troves. com/math/ Complete Bipartite Graph.html has more information about this structure."

2.2   Program Checking

AλgoVista can be seen as a novel application of *program checking,* an idea popularized by Manuel Blum [1] and his students. The idea is that rather than testing a procedure or attempting to prove it correct, we check, at runtime, that the procedure indeed produces the right output for each given input. The AλgoVista problem description database contains such *program checkers,* and the efficiency and accuracy of these checkers is what makes AλgoVista so successful.

[4] contains an in-depth description of the design of the AλgoVista search engine and the search algorithms it employs.

AλgoVista currently contains some 200 problem descriptions, listed in the table in Appendix A.

---

[1] We invite the reader to visit http://www.algovista.com to try out the queries as they are presented in the paper.

[2] Depending on the nature of the problem, AλgoVista queries can be entered either graphically or textually.

3   The Query Language

The AλgoVista query language was designed to be as simple as possible, while still allowing users to describe complex algorithmic problems. The language primitives include integers, floats, booleans, lists, tuples, atoms, and links. Links are (directed and undirected) edges between atoms that are used to build up linked structures such as graphs and trees. Special syntax was provided for these structures since we anticipate that many AλgoVista users will be wanting to classify graph structures and problems on graphs.

The following grammar shows the concrete syntax of the query language:

$$
\begin{aligned}
S \quad &\rightarrow \texttt{int} \mid \texttt{float} \mid \texttt{bool} \mid \\
& \quad S \; `\texttt{==>}' \; S \mid \\
& \quad \texttt{atom} \, [\, `/' S \,] \mid \\
& \quad \texttt{atom} \, `\texttt{->}' \, [ \, `/' S \,] \, \texttt{atom} \mid \\
& \quad \texttt{atom} \, `\texttt{--}' \, [ \, `/' S \,] \, \texttt{atom} \mid \\
& \quad `[' [ \, S \, \{ \, `,' S \, \}] \, `]' \mid \\
& \quad `(' \, S \, `,' \, S \, `)' \\
\texttt{bool} &\rightarrow `\texttt{true}' \mid `\texttt{false}' \\
\texttt{atom} &\rightarrow `\texttt{a}' \ldots `\texttt{z}' \\
\texttt{int} &\rightarrow `\texttt{0}' \ldots `\texttt{9}' \, \{ \, `\texttt{0}' \ldots `\texttt{9}' \, \} \\
\texttt{float} &\rightarrow \texttt{int} \, `.' \texttt{int}
\end{aligned}
$$

⌜S ==> S⌝ maps inputs to outputs, ⌜(S , S)⌝ represents a pair of elements, and ⌜[S { ,S }]⌝ represents a list of elements. Atoms, ⌜atom [/S]⌝, are one-letter identifiers that are used to represent nodes of linked structures such as graphs and trees. They can carry optional node data. Links between nodes can be directed ⌜atom -> [/S] atom⌝, or undirected ⌜atom -- [/S] atom⌝, and can also carry edge data.

These simple primitives can be combined to produce complex queries. For example, the query

$$[a->b,b->c] ==> [a->a,a->b,a->c,b->b,b->c,c->c]$$

asks which function maps

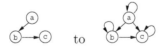

to

(Transitive closure). The query

$$([3,7],[5,1,6]) \; ==> \; [5,1,6,3,7]$$

asks what function maps the lists [3,7] and [5,1,6] to the list [5,1,6,3,7] (List append).

The recursive structure of the grammar allows queries to be deeply nested, although this is fairly uncommon. For example, the query

$$[a->/[1]b,a->/[3,4]c] \; ==> \; [1,3,4]$$

looks for an algorithm that maps a tree to a list of integers, where each tree edge is labeled with a set of integers.

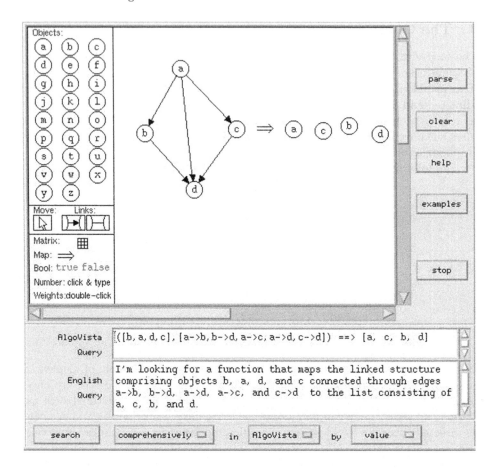

Fig. 2. Searching for topological sorting using a graphical query.

4   The Visual Query Language

AλgoVista's visual query language is closely modeled on its textual counterpart. A user constructs a query by dragging primitive elements from a *template region* on the user interface (Figure 1) onto the drawing canvas. Atoms are modeled by named circles, links by the obvious lines and arrows, booleans and the ==>-arrow by themselves, and numbers are entered by clicking and typing. There are, however, no obvious visual counterparts to the pairs and element-lists of the textual language. These are instead inferred from the positioning of the visual elements.

For example, instead of entering a topological sorting query textually:

$$[a->c,a->b,b->d,c->d,a->d] \implies [a,c,b,d]$$

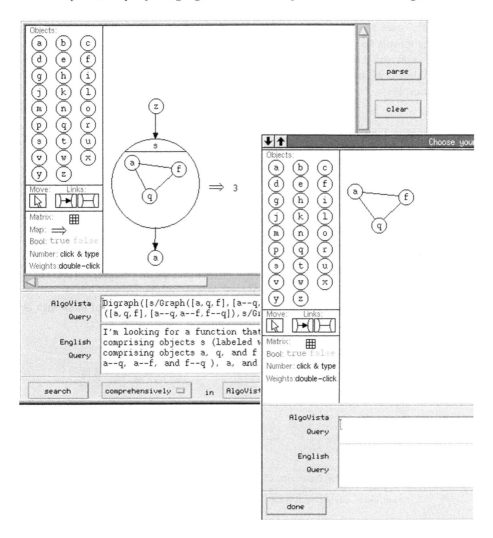

Fig. 3. Example recursive query. The node **s** is labeled with the graph $\ulcorner$[a--f,a--q,f--q]$\urcorner$. The user is editing this graph in the sub-window to the right.

it could instead be drawn like in Figure 2. The textual query is inferred from the drawing, and the English prose query is derived from the textual query.

Recursive queries could be handled in a variety of ways. Many graph editors parse node and edge labels by *proximity*; that is, a graphical element is inferred to be the label of a node a if it is "close enough" to a. This puts a heavy burden on the parser as well as on the casual user who needs to have some understanding of the principles under which the parser operates.

We have instead opted for a much more lightweight solution: if the user double-clicks on an atom or link a new, simpler, drawing window opens up, allowing the user to enter the sub-drawing. This strategy has the advantage of both being simple to implement and trivial to explain to the user. It is now easy to create arbitrarily complex queries, where, for example, the nodes of a graph could be labeled with lists of trees, whose edges are labeled with.... See Figure 3 for an example.

4.1 Parsing Visual Queries

Because of the limited set of graphical elements that AλgoVista supports, parsing visual queries is relatively straight-forward. The ==>-arrow separates inputs from outputs, which means that anything to the left of the arrow is an element of the input, anything to the right belongs to the output.

As we have seen, recursive queries are created by the user explicitly opening up atoms and links (by double-clicking on them) and drawing the label in a sub-editor pane. This limited amount of structure editing greatly simplifies parsing, since it is always clear if an element is the label of an atom or link, rather than a free-standing element.

The main challenge in parsing visual queries is that grouping of elements is not always evident from the drawing. Consider the following query:

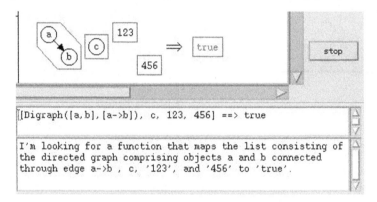

There are four connected components to the left of the ==>-arrow, and it is not clear which of those, if any, form sub-groups. For example, the nodes a, b, and c could form a (dis-connected) graph, or a and b could form one graph and c another.

Many visual parsers use proximity to infer element grouping. In a web-situation where many casual users will visit a site for only a minute or two, there simply is no time to explain what heuristics the parser employs. So, again, we prefer a lightweight and user-centric solution. We will simply guess the user's intentions, and then report back what that guess was. In this case, the parser's first guess was that each connected component is a unique argument in the query. It indicates this by drawing a convex hull (dashed lines) around each component.

If, however, the user's intentions were different, she can simply ask the parser to produce a different parse, by clicking the PARSE-button one or more times.

This will cycle through all the different possible parses of the query, each described visually (using a convex hull), formally (using an AλgoVista query), and informally (using English prose). For example, it could be that the user had planned for the node c to belong to a three-node graph:

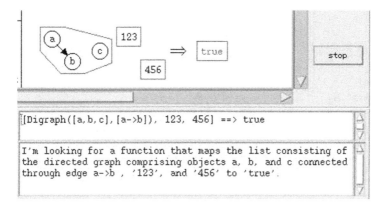

or for the two integer elements to be part of a pair:

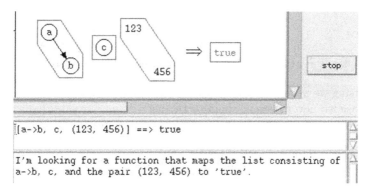

or for the input part of the query to consist of two elements, a graph and a pair of integers:

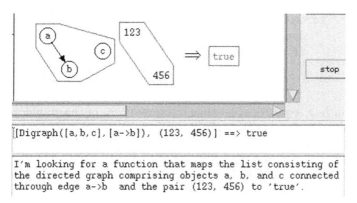

In most cases there are few possible parses and it is immediately clear to the user which is the one she is looking for. To cut down the number of possible parses we can exploit the fact that AλgoVista queries are *typed*. The type system corresponds almost one-to-one to the concrete syntax given in Section 3. The following type assignments map concrete syntax into types:

$$\mathcal{T}[\texttt{int}] \qquad\qquad\qquad = \texttt{Int}$$
$$\mathcal{T}[\texttt{float}] \qquad\qquad\quad = \texttt{Float}$$
$$\mathcal{T}[\texttt{true}] \qquad\qquad\qquad = \texttt{Bool}$$
$$\mathcal{T}[\texttt{false}] \qquad\qquad\quad = \texttt{Bool}$$
$$\mathcal{T}[S_1 \texttt{ `==>' } S_2] \qquad = \texttt{Map}(\mathcal{T}[S_1], \mathcal{T}[S_2])$$
$$\mathcal{T}[\texttt{`(' } S_1 \texttt{ `,' } S_2 \texttt{ `)'}] \qquad = \texttt{Pair}(\mathcal{T}[S_1], \mathcal{T}[S_2])$$
$$\mathcal{T}[\texttt{`['[} S_1 \texttt{ \{ `,' } S_2 \texttt{ \}] `]'}] = \texttt{if } \mathcal{T}[S_1] = \mathcal{T}[S_2] \texttt{ then Vector}(\mathcal{T}[S_1]) \texttt{ else } \bot$$
$$\mathcal{T}[\texttt{atom}/S] \qquad\qquad = \texttt{Node}(\mathcal{T}[S])$$
$$\mathcal{T}[\texttt{atom `->'}/S \texttt{ atom}] \quad = \texttt{DEdge}(\mathcal{T}[S])$$
$$\mathcal{T}[\texttt{atom `--'}/S \texttt{ atom}] \quad = \texttt{UEdge}(\mathcal{T}[S])$$

For example, the query

$$([1,2],[3,4])\texttt{==>}[4,6]$$

has the type

$$\texttt{Map(Pair(Vector(Int),Vector(Int)),Vector(Int)).}$$

During parsing we may find that certain groupings of elements do not typecheck, in which case we never present them to the user.

The above discussion is summarized by the following algorithm:

```
procedure parse(elements)
    sort elements by ⟨x, y⟩ coordinates
    left ← elements left of '==>'
    right ← elements right of '==>'
    input ← connected_components(left)
    output ← connected_components(right)
    for all i←merge(input) & o←merge(output) do
        query ← construct_query(i, o)
        if type_check(query) then
            prose ← query2english(query)
            yield (query,prose)
```

parse generates a sequence of possible interpretations of the graphical elements on the canvas. We first separate the input from the output elements, and then construct a set of connected components for each. We then generate all possible type-correct queries by merging adjacent connected components. Finally, each query is translated to English and presented to the user, along with the textual query and a convex hull around each component.

4.2   Evaluation

AλgoVista's textual interface and its underlying database engine have been operational for some time. The design of the graphical interface was initiated by the fact that our query logs indicated that users had trouble forming correct textual queries. While the graphical interface is operational it has yet to be publically announced so at this point we cannot say whether users will favor one interface over another. Our own experience with the graphical interface, however, has allowed us to make the following observations:

1. For large, disconnected, inputs, the number of possible parses grows exponentially. For example, consider the following query, which a user might have entered to look for information about palindromes:

   This query generates an exponential number of parses, making it time consuming for the user to navigate to the desired one. For most queries this is not an issue, since they tend to contain a small number of connected components. For queries with many components, the textual search interface often works better. For example, the following is an unambiguous palindrome query:

   [m, a, d, a, m]

2. Generally, a query is easy to express if the query language provides the appropriate primitives. For example, since AλgoVista provides node and link primitives it is straight-forward to create queries involving graphs. On the other hand, classifying problems in the Computational Geometry domain is difficult, since AλgoVista currently has no simple way to express points, lines, line segments, circles, etc. We could — and probably will — add such primitives to a future version of AλgoVista, although this will add more complexity to the interface, and hence a steeper user learning curve.

5   Related Work

Rekers [7] provides a nice overview of graphical parsing algorithms. They note that most graph parsing algorithms are worst-case exponential. The paper also presents a new multi-stage graph parsing method with separate phases for determining object locations and spatial relationships, and a final grammar-based rewrite phase. In a web-based visual interface such complex, and potentially slow, parsing methods are unacceptable.

In [6], Liu presents a visual interface to a CASE tool, where boolean queries are constructed in a syntax-directed fashion. Users proceed top-down from a "root" query, iteratively expanding non-terminal nodes until the query is complete.

Novice (as well as expert!) users typically find syntax-directed iterative refinement cumbersome to use. There is a reason why programmers prefer free-form editors like emacs over syntax-directed ones, even though the latter ensures that only correct programs can be constructed. For this reason, AλgoVista is a mostly free-form graph editor, and syntax-directed editing is reserved for recursive edits.

The web ought to present many opportunities for introducing more people to direct-manipulation interfaces. However, we have found few such examples. Marmotta [2], a graphical front-end to online databases, is an exception.

6   Summary

AλgoVista provides a unique resource to computer scientists to enable them to discover descriptions and implementations of algorithms without knowing theoretical nomenclature. However, by monitoring the queries submitted to the web-site we have determined that the textual query language that AλgoVista employs is an impediment to many casual users. It is our belief that the visual language presented here will prove easier to use and faster to learn. To motivate why this will be the case, consider the following two episodes that provided the original inspiration for AλgoVista's principal designers:

Working on the design of graph-coloring register allocation algorithms, Todd Proebsting showed his theoretician colleague Sampath Kannan the following graphs:

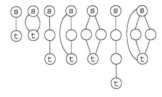

"Do these graphs mean anything to you?" Todd asked.

"Sure," Prof. Kannan replied, "they're series-parallel graphs."

This was the beginning of a collaboration which resulted in a paper in the *Journal of Algorithms* [5].

In a similar episode, the present author showed his theoretician colleague Clark Thomborson the following graph-transformation:

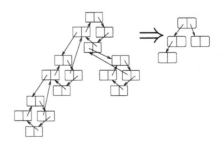

"Do you know what I am doing here?" Christian asked.

"Sure," Prof. Thomborson soon replied, "you're shrinking the biconnected components of the underlying graph."

This result became an important part of a joint paper on software watermarking [3].

It's important to note that in both these episodes the queries were visual in nature, and, in fact, took place while drawing on a white-board. It is our hope that AλgoVista will prove to be a useful "virtual theoretician" that working programmers can turn to with a problem, quickly sketch it out — visually or textually depending on the nature of the problem — and quickly receive a useful answer.

We have stressed throughout this paper that web users are a fickle lot and that speed and simplicity is the key to success for any web-based interface: any code must be small enough to download instantaneously (or the user will go elsewhere), and no user training must be required (or the user will go elsewhere). The AλgoVista visual interface was designed with this in mind: it employs no fancy graph parsing algorithms and any ambiguities are resolved by the user simply cycling through all possible parses.

AλgoVista's visual interface has been implemented as a Java applet. It is available for experimentation at http://AlgoVista.com.

References

[1] Manuel Blum. Program checking. In Somenath Biswas and Kesav V. Nori, editors, *Proceedings of Foundations of Software Technology and Theoretical Computer Science*, volume 560 of *LNCS*, pages 1–9, Berlin, Germany, December 1991. Springer.

[2] Fabrizio Capobianco, Mauro Mosconi, and Lorenzo Pagnin. Progressive http-based querying of remote databases within the Marmotta iconic VQS. In *VL'95*, 1995.

[3] Christian Collberg and Clark Thomborson. Software watermarking: Models and dynamic embeddings. In *POPL'99*, San Antonio, TX, January 1999. http://www.cs.arizona.edu/ collberg/ Research/ Publications/ CollbergThomborson99a

[4] Christian S. Collberg and Todd A. Proebsting. AλgoVista — A search engine for computer scientists. Technical Report 2000-01, 2000. http://www.cs.arizona.edu/ collberg/Research/Publications/Collberg Proebsting2000a.

[5] Sampath Kannan and Todd A. Proebsting. Register allocation in structured programs. *Journal of Algorithms*, 29(2):223–237, November 1998.

[6] Hui Liu. A visual interface for querying a CASE repository. In *VL'95*, 1995.

[7] J. Rekers and A. Schür. A graph grammar approach to graphical parsing. In *VL'95*, 1995.

A   The AλgoVista Database of Problem Descriptions

A Basis for the Null Space of the Matrix, Absolute value, Ackerman's Function, Activity-Selection Problem, Addition of Polynomials, Addition of Rational Numbers, Adjugate Matrix, All pairs shortest path, Arc cosine,

Area of a Circle, Area of a Rectangle, Area of a Trapezoid, Area of a Triangle, Arithmetic Sequence, Articulation points, Average number, Average number of a list of floating point numbers, Average number of a list of integers, AVL Tree, Biconnected Graph, Binary logical and, Binary logical or, Bipartite Graph, Circumference of a Circle, Clique, Clique problem, Combination, Compare less-than, Complete graph, Complex Addition, Connected graph, Continuous Knapsack Problem, Convex Combination, Convex Combination of two points, Cross Product, Cross Product of two 3-D vectors, Derangement of an ordered set, Determinant of a Matrix, Directed Acyclic Graph (DAG), Directed Graph, Directed Linked List, Directed Tree, Distance between two points, Division of two complex numbers, Dot Product of an n-D vector, Euler cycle, Eulerian graph, Eulerian Numbers, Factorial, Finding the closest pair of points, Float Matrix Add, Float Matrix inverse, Float Matrix Multiply, Float Matrix Subtract, Floating-point square root, Full binary tree, Geometric Sequence, Greater integer, Greatest common divisor, Hamiltonian cycle, IEEE remainder, Independent set, Integer Divide, Integer Logarithm, Integer Matrix Add, Integer Matrix inverse, Integer Matrix Multiply, Integer Matrix Subtract, Integer Modulus, Integer multiplication, Integer Partition of n, Integer power, Integer square root, Integer subtract, Intersection of two sets, Invertible Matrix, kth smallest, Landford problem, Langford Sequence, Leaf of directed tree, Least common multiple, less integer, Line Segments Intersect, Linearly Dependent Set of Vectors, Linearly Independent Set of Vectors, List append, List reverse, Longest common subsequence, Lower Triangular Matrix, Matching, Matrix Chain Multiplication, Matrix in Row Echelon Form, Matrix Representing a Consistent Linear System, Matrix Representing an Inconsistent Linear System, Maximal independent set, Maximum bipartite matching, Maximum consecutive subsequence, Maximum element of a list, Median element of a list of integers, Merge lists, Minimum element of a list, Minimum Spanning Tree, Modulus, Multiplication of two complex numbers, Natural logarithm, Nearest Neighbor Search, Palindrome Sequence, Parallel lines, Partition Functions, Perfect matching, Perimeter of a Rectangle, Perimeter of a Trapezoid, Perimeter of a Triangle, Permutation, Perpendicular lines, Philip Hall's Marriage theorem, Polynomial function, Power of e, Proper edge coloring, Proper vertex coloring, Radius of Circle (according to the equation of a circle), Real absolute value, Real add, Real Divide, Real Max, Real Min, Real multiply, Real Power, Real round, Real sub, Relative complement of two sets, Searching, Set Partition of N, Shortest Common Superstring, Single destination shortest path, Single pair shortest path, Single source shortest path, Single source shortest path tree, Sink, Sorting, Spanning Tree, Sparse Matrix, Square Matrix, Steiner System, Strictly Lower Triangular Matrix, Strictly Upper Triangular Matrix, Strongly connected Graph, Subscripts of the Basic Variables of a Matrix, Subscripts of the Free Variables of a Matrix, Subset, Subset Sum, Subtraction of Rational Numbers, Subtraction of two complex numbers, Subtraction of two n-D vectors, Surface area of a cone, Surface area of a Cylinder, Surface area of a Sphere, Symmetric Difference of two sets, The conjugate of a complex number, The Length/Norm of the Vector, The maximum number of incomparable subsets of 1, 2, ..., n, The midpoint of two points, The next permutation, The Not Operator Performed on a Number, The nth Fibonacci number, The Nth Term of an Arithmetic Sequence, The Nth Term of an Geometric Sequence, The number of derangement of a set of n objects, The number of derangements of a set of n objects, The point of intersection of two lines, The Reduced Echelon Form of a Matrix, The Sum of an Infinite Geometric Sequence, The Sum of the First N Terms of an Arithmetic Sequence, The Sum of the First N Terms of an Geometric Sequence, Topological Sort of Directed Acyclic Graph, Transitive closure, Transpose of Matrix, Traveling Salesman Problem, Trigonometric arcsine, Trigonometric cosine, Trigonometric sine, Trigonometric tangent, Undirected Graph, Union of two sets, Unit Vector, Upper Triangular Matrix, Vector in the Column Space of a Matrix, Vector reduction, Vertex Cover, Volume of a Cone, Volume of a Cylinder, Volume of a Sphere, Ways of m choose n, XNOR of 2 Numbers

Diagrammatic Integration of Abstract Operations into Software Work Contexts

Alan F. Blackwell and Hanna Wallach

Computer Laboratory, University of Cambridge
+44 1223 334418
Alan.Blackwell@cl.cam.ac.uk

Abstract.. Most software applications present information to the user in a WYSIWYG form, where the main representation on the screen is made as close as possible to a visual facsimile of the final work product. Wherever possible, users specify required transformations of the product by directly selecting and manipulating areas of this visual facsimile. This brings great usability advantages, but is not adequate for the specification of abstract operations such as generalization and inference commands, which are usually represented linguistically in menus and dialogues. We report a series of experimental implementations exploring alternatives to menus and dialogues. In these six systems, abstract functionality is integrated into the work context through two techniques: diagrammatic interpretation of user's actions (gestures) and diagrammatic overlays superimposed as semi-transparent layers over the visual presentation of the work product. We discuss the diagrammatic justifications and consequences of these alternatives, and present results of preliminary user studies suggesting that both forms of interaction may in future be valuable techniques for exploiting diagrammatic formalisms in software interaction.

1 Introduction

The products of software research regularly include user interfaces based on novel diagrammatic representations. Yet few of these have had widespread influence on human diagram use. The "desktop" on which icons and overlapping windows represent data and applications is one of the few representations that has been sufficiently widely used to become established as a component of public visual literacy. The spreadsheet is another (although it was based on an existing accountancy representation, its behaviour, conventions and applications are so different in degree as to be a new representation). There are also some newly established typographic conventions to indicate ways in which the user can interact with text – cascading menus, folding hierarchy browsers, buttons, and hyperlinks.

However if we consider the representation that is central to the user's task in any given application, most software products simply adopt existing conventions. The word processor imitates a page of text, with all its diagrammatic cues of white space,

M. Hegarty, B. Meyer, and N. Hari Narayanan (Eds.): Diagrams 2002, LNAI 2317, pp. 191–205, 2002.
© Springer-Verlag Berlin Heidelberg 2002

font variation, tabular arrangement and so on. Presentation software such as PowerPoint borrows a different set of typographic conventions from the blackboard and transparency projector. Computer aided design tools adapt the particular diagrammatic conventions that were previously drawn on paper by the mechanical, electrical, architectural or software designers to whom the tool is sold. Image processing programs present the image as a direct facsimile on a notional canvas.

1.1 The Problem with WYSIWYG and Direct Manipulation

All of these design decisions might be justified by the well-established principle of WYSIWYG – what you see is what you get. This argues that the notational representation of the task on the computer screen should be a visual facsimile of the final product. Although appealing as a slogan, WYSIWYG is ultimately an impoverished design criterion. The perennial debate between LaTeX users and users of word processors (not least in the preparation of past proceedings from this conference), demonstrates the inadequacy of WYSIWYG as a simple criterion. The main deficiency of WYSIWYG, and of the equally widely-accepted principle of direct manipulation, is that they do not allow for the abstract nature of computational tasks.

It is precisely in the degree of abstraction that user interfaces become diagrammatic, rather than being directly representational. Diagrammatic representations employ visual notational conventions to express abstract relations. A WYSIWYG depiction of an object (whether painting, document or drawing) on the computer screen ultimately represents that object, not the computational abstractions the user is employing. Direct manipulation [17] emphasizes the modification of the object here and now (which does give some usability benefits) rather than abstract operations over multiple possible classes or situations [3]. An over-emphasis on direct representation or manipulation effectively restricts the abstraction power available to the users – this is the main complaint of LaTeX advocates, as it was previously of those defending the DOS command line against the Windows desktop.

1.2 Representation of Abstract Tools

Of course these software applications do provide their users with the potential for abstract operations in addition to direct manipulation. But the abstract operations are visually separated from the "main" notation of the task representation. They are either represented linguistically by the words on menus and buttons, or visually in the collections of icons representing abstract operations on various toolbars. Both are impoverished uses of the diagrammatic potential of the computer screen. In the case of menus, it is quite natural to represent abstractions by language. Language itself is the cognitive ground of abstraction. But the diagrammatic potential of the bitmapped display is scarcely exploited by the square borders drawn around a menu bar or a button. Iconic toolbars use some small set of diagrammatic relations – exploiting visual similarity between related functions, and grouping the symbols into regions – but once again, we would scarcely recognize the toolbar as being a sophisticated abstract notation.

There is a long history of attempts to create powerful and novel diagrams to represent and manipulate computational abstraction. Visual programming languages have continually been hailed as the wave of the future [4], albeit a wave that never quite breaks and may never do so [7]. A few specialist abstract diagrams have been adopted by scientific or technical communities – charts for project management and workflow, scientific visualization tools – but everyday computer users do not experience the benefits of diagrammatic representation when interacting with the abstract computational aspects of their work.

2 Integrating Abstract Tools through Diagrams

This paper describes a series of experiments in adding abstract diagrammatic representations to conventional computer tools – painting programs, drawing programs, word processors, and computer aided design tools. In each case, the main representation is the product itself, as usual. The experimental modification is that abstract operations specifying transformations of the product have been integrated into the main work area, rather than accessed via verbal menu commands or separate toolbars. Two main diagrammatic mechanisms have been employed. The first is the recognition of gestural components within the work area, by reasoning about the sequence of user actions and assigning abstract diagrammatic meaning to some subset of them. The second is the use of transparent or semi-transparent overlays to superimpose diagrammatic information over the work area.

These two alternatives are directly motivated by theoretical concerns of abstraction use by non-programmers. In an ongoing programme of work, we have been investigating the cognitive factors involved in the use of abstract representations [3,5,6,9]. This is based on a characterization of abstraction as a cognitive investment decision. Creating and using abstractions is an effortful activity, during which a user must concentrate on the abstraction, and is less able to devote attention to regular work. The reason that we use abstractions, however, is to save attention – in direct manipulation, every piece of system behaviour is in response to an action by the user, thus requiring the user's attention. In abstract interaction, the user can reduce the need for future direct manipulation by automating (abstraction over time), or by applying related actions to multiple situations or entities defined by some abstract class (abstraction over situation classes).

Use of abstraction thus represents an investment in which the user undertakes immediate attention costs in order to save attention in future. This is subject to risk assessment as with all investment, and the result describes a wide range of programming activity [5]. We have found empirical evidence for this behaviour, and also implemented a cognitive model that predicts human performance in terms of attention investment decisions [3]. The implications for design of abstract tools are that initial investment cost should be kept low, and that users should be able accurately to assess the risk implications through feedback on the results of abstraction use.

The experimental tools described in the rest of this paper provide for these factors in various ways. The integration of abstraction facilities into the main work area reduces the attentional cost of abstraction use by minimizing the discontinuity in attention required to access and employ the tool. Placing the abstract representation within the context of its effects allows users directly to assess the relationship between abstraction and its effects, thereby minimizing perceived risk.

3   Diagrammatic Integration Using Gestures

We have carried out three experimental implementations of software products that integrate abstract representation into the main task context by monitoring and interpreting the marks that the user makes in that context. In each case, the time course of marking actions is used as a source of data to infer the user's intentions. This gives a gestural character to the interaction. However the gestures themselves are also marks within the main work area, meaning that they act as diagrammatic annotation conventions over the primary task context. This is analogous to the typographic mark-up conventions that are superimposed on a page of text, but refer to another level of abstraction beyond that of the manuscript itself. This move from supplementary markup to interpreted content is a common move in many notational systems, where secondary notation [11] is co-opted as a new layer of semantic structure [10].

3.1   Study 1: Drawing without Toolbars

This experiment extended a conventional object-based drawing editor. In a conventional editor circles, squares and lines can be drawn, selected singly or in groups, then modified by manipulating the selected objects. Normally these editors use direct manipulation to indicate positions and objects within the work area, and menus or toolbars to select the mode of operation or transformation. Our experimental system has no menus or toolbars. Instead it uses a diagrammatic markup vocabulary to interpret some of the marks made by the user within the drawing as referring to abstract operations rather than elements of the drawing itself.

The base drawing modes include lines, circular arcs, ellipses and rectangles. The intended mode is inferred from the shape of the user's mouse movements, as in systems for freehand drawing interpretation [8]. Lines deviating by a certain threshold from the direct path between end-points are taken to be arcs, with curvature then defined by maximum deviation from the line between the end-points. Arcs near 360 degree curvature are interpreted as ellipses. Abrupt changes in the direction of mouse motion are interpreted as right-angled corners, and open corners or lines can be combined to form rectangles, either in a single continuous motion or by consecutive strokes.

Drawing transformations are indicated by markings within the drawing itself, as shown in figure 1. Selection is indicated by drawing two corners that bound the se-

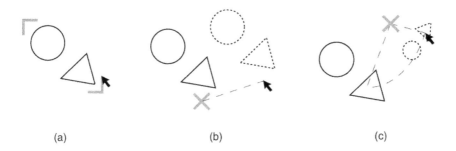

(a)   (b)   (c)

Fig. 1. Use of gestures to transform drawing elements (a) elements are selected by drawing "corners", (b) elements are translated by drawing a cross to show reference point, then moving cursor from that point, (c) elements are rotated and scaled by dragging a reference line drawn between the cursor and the initial reference point.

lected region (selected regions can only be rectangular). Once the selection has been recognized, the corners disappear from the drawing, and the selected objects are highlighted. This is typical of the way that the editor continually identifies and interprets command marks as abstract behaviour. Selections can be transformed by drawing further markup instructions to scale, rotate or reflect selected objects. In each case, the editor first highlights the marks to indicate that they are being interpreted as abstract commands, and then completes the transformation with the indicated parameters. Each markup gesture includes a common vocabulary of a reference point marked with a cross, a direction and magnitude indicated by a line, and a terminal confirmation indicated by an arrowhead. These are interpreted by the editor as an abstract state specification, but appear to the user as a natural command set of marks to be added in the context of the drawing itself.

Structured user trials were conducted in which four users were asked to construct the same drawing, one element at a time, by following a sequence of verbal commands. This was an evaluation, rather than a controlled experiment – the only data recorded was the occasions on which users were unable to complete a particular subtask, or had to ask advice from the experimenter in order to complete it. These trials indicated that the interaction paradigm itself was easily acquired within a short period. The main difficulties arose from the performance of the gesture recognition algorithms. The threshold distinguishing lines from arcs, for example, made it difficult to achieve straight lines over short distances. This in turn could make it difficult to draw the corners of the selection region. If this interaction paradigm were to be widely used, it would be necessary to tune the thresholds of this algorithm, but it might also be necessary to allow explicit control by the user, as noted later in this paper.

3.2   Study 2: Natural Diagrammatic Content in Life Drawing

This experiment was conducted in the life drawing classes run by the Cambridge University faculty of architecture and history of art. The classes take place in a traditional life studio, where 30 or 40 students spend three hours drawing a nude model in pencil or charcoal. As architecture graduates are now expected to work with digital tools as well as traditional pencil and paper, the architecture students in this class were required to complete at least one drawing using a Macintosh drawing program, Fractal Sketcher. Sketcher simulated various paper textures, as well as charcoal, pencil, ink and other traditional drawing instruments. A particularly valuable feature of Sketcher (unfortunately removed in later products from the same company – Sketcher has now been discontinued) is that it can store a journal file of all mouse movements made by the user, and is thus able to "replay" an entire session onto a blank canvas. Students greatly enjoy using this facility to see their drawing reappear in a high speed replay of their own actions.

We exploited the journal facility to investigate the diagrammatic qualities of life drawing conventions. We had already observed that students drawing with Sketcher produced results that were technically very similar to the earlier work that they had produced on paper. Drawings by any given student were easily recognizable as being by the same hand, whether on paper or screen. This fact indicated that the stylistic components of drawing were not developed in reaction to use of Sketcher, but reflected an existing graphical vocabulary. We reverse engineered the format of the Sketcher journal file, obtaining for analysis the complete time sequence and spatial coordinates of marks made by the students in composing their drawings.

Fig. 2. A typical life drawing, and partial boundary inference. The original drawing (on the left) is reproduced with original stroke widths and transparency as specified by the artist. The strokes in the inferred boundary drawing (on the right) are reduced to simple trajectory information. Those strokes which have been identified as regions in which the artist intended to define a boundary are shown with a box drawn around them.

We collected a sample of journal files from 8 different students, and analysed them to find spatially or temporally contiguous patterns that could be used to form abstract inferences about the structure of the drawing. A sample production, with the extracted inferences, is shown in figure 2. This analysis identified several distinctive elements with potentially diagrammatic content. For example, there are several "contour" styles that can be used to identify the points at which students are outlining parts of the body, folds or limb contours. One of these is a long, continuously curving line. Another is a consecutive series of short, aligned, normally overlapping, strokes to indicate a contour. A third is a series of strokes following the same section of a contour to reinforce the line. This last does not necessarily occur in strict sequence, but is often interrupted by strokes elsewhere in the drawing, after which the artist returns to further reinforce the contour. As with contour styles, there are several identifiable "shading" styles that are used to add weight, distinguish figure from ground, or represent shadows cast within the scene. It should be noted that not all drawings could be analysed in these terms. Some students experimented with building form from shading only, or with ambiguous impressionistic representations. These could not meaningfully be analysed.

Life drawing is clearly not "diagrammatic" in the sense that it relies on any formal syntax or semantics of geometric elements. On the contrary, many artists strive to break conventions and employ their tools in unorthodox ways. Nevertheless, in this experiment we did find that many strokes within a drawing could be assigned diagrammatic interpretations related to the abstract structure of the scene and the artist's task. Once again, this can be achieved without requiring that the artist leave the main representational context to define behaviour using menus or toolbars. These observations were exploited in the next experiment.

3.3. Study 3: Diagrammatic Stroke Inference as a Modeling Tool

In this experiment, we created a 3-dimensional human figure modeling tool. Commercial figure modeling tools are included in products such as 3D Studio Max. This type of product requires substantial expertise in abstract solid geometry to construct complex shapes such as human figures. Most users therefore work with predefined figures rather than face the abstract challenges of constructing a figure from scratch. Once again, the objective of our system was to avoid separate abstract commands, either in menus or toolbars. In fact, this system could be used without any user interface or even a screen. Users worked on a CrossPad, a battery powered graphic tablet without any display (essentially a digitizing tablet similar to those used as mouse alternatives, but having internal memory so that it can store several hours' worth of pen motions rather than using a direct link to a computer). A pad of paper is placed on top of the CrossPad, and the user draws on this with a specially instrumented ballpoint pen. The motions of the pen are recorded in the internal memory of the CrossPad, and can later be downloaded to a PC.

This provides very similar data to that we recorded in Study 2 – a time-stamped series of drawing strokes. We used the diagrammatic content of life drawing data, as identified in study 2, to make inferences about implicit 3D structure in a sketch of a human body. If the user is aware of the way her strokes will be interpreted, we found that it is possible for the user to indicate the abstract requirements of 3D structure directly while working in the context of a 2D representation. As in the case of study 2,

this exploits the fact that abstract gestural interpretations are given to a natural vocabulary of drawing strokes. However in this case it is not intended that the user
should be completely free in her choice of drawing strokes – as noted in study 2, many
people make drawings in which they experiment with different representational conventions that this program would not be able to interpret.

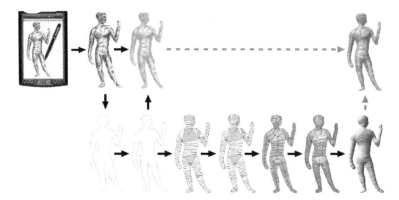

Fig. 3. 3D human figure model generated from diagrammatic content of gestural drawing
strokes. The stroke trajectories (shown at top left) are user to generate a boundary model (bottom row) and blurred to generate a shading model (top row). The boundary model is triangu-
lated in order to generate a generalized cylinder description, which is then lit using a lighting
model that is consistent with the blurred shading.

The system interprets the recorded strokes by classifying them as contours, rein-
forcement strokes, or shading strokes, and identifies an overall boundary contour. If
the user makes strokes that cannot be placed in one of these three categories, this contributes increased error to the interpretation algorithm. The boundary is used to create
a hypothesis for 3D structure using the Teddy algorithm [12], and the shading strokes
are blurred to reconstruct shadows in an inferred lighting model applied to that 3D
structure [18]. An evaluation with a single user, trained to use the expected repertoire
of drawing techniques, resulted in a model that has correctly captured the geometric
structure from this integrated gestural vocabulary, as can be seen in figure 3. This
system has not yet been evaluated with further users, but we note that they would also
require training in order to create drawings with the intended diagrammatic repertoire
of drawing strokes.

4   Diagrammatic Integration Using Transparent Overlays

The first three studies concentrated on providing the user with a single visual representation, in which abstract control information is derived from the shape and sequence of her actions. The diagrammatic content relied on conventions of interpretation for some subset of the user's mouse (or pen) motions. Although not visually separated from the main notation, the three systems all relied on some degree of temporal

separation to identify and group the abstract intentions of the user. Furthermore the user must be aware of the system's expectations, and work within the resulting boundaries (this was not the case in study 2, but the system in study 2 was only able to analyse conventional drawings, and completely failed where students experimented with alternative conventions).

Our second series of studies investigated an alternative. In these systems abstract diagrammatic information is included within the same visual part of the screen as the main representation, but it is superimposed on the main representation as a transparent overlay. This allows users complete freedom within their accustomed representation, while integrating abstract diagrammatic elements into the visible display. User's actions must still be temporally identified as relating either to the abstract diagrammatic layer or to the conventional representation, but this is now done through explicit mode switching by the user rather than through inference by the system.

4.1 Study 4: Overlays for Architectural Constraints

In this experiment we created a tool with the basic functionality of a CAD product for architectural draughting (but in only two dimensions). This was supplemented with an abstract notation allowing users to define required geometric constraints that must be maintained between the elements. Constraint definition is becoming increasingly common in architectural tools for parametric design, where building plans are partially generated by calculating the required dimensions to satisfy constraints on site, materials etc. A recent commercial prototype from Bentley Systems, CustomObjects, allows users either to work directly by manipulating the elements of a plan, or on a neighbouring window showing a diagrammatic network of the constraint relationships between parts of the design [1].

We designed an alternative representation to that used by CustomObjects, differing in that the constraints are not represented in a separate window, but are superimposed directly over the drawing as a semitransparent overlay. This addresses the concern of attention investment theory: that abstract elements of the representation should be visually integrated with directly manipulated elements in order to reduce attentional cost and perceived risk. A constraint on a specific point within the drawing is indicated by a transparent overlay on that point. Constraint relationships over multiple points are indicated by lines between the points. The nature of the constraint (position, length, orientation) is indicated diagrammatically by textures in the transparent overlay as shown in figure 4. The overlay is coloured and rendered in a blurred style so that these visual elements do not interfere with the user's perception of the drawing itself.

This system was evaluated by 15 novice users, all of whom were asked to complete six drawing tasks involving both directly drawn elements and constraints defining required relationships between those elements. The performance of these users was compared to that of an experienced architectural CAD operator undertaking the same tasks with conventional CAD tools, where constraints are defined in an abstract

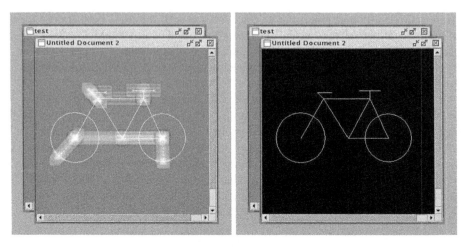

Fig. 4. Geometric constraints using semi-transparent overlays. The display on the left shows a range of constraints that will preserve the bicycle configuration (the display on the right shows the resulting bicycle with the constraint layer turned off). Note that actual constraint representation as it appears on screen employs colouring to render multiple overlay layers more easily distinguished.

notation that is separate from the main drawing. The novice users were able to complete the task just as quickly as an expert user working with her familiar tools, despite the fact that they had never used the experimental system before. (Expert users are more difficult to recruit for experimental participation than novices, and use of a single expert in this study means that no statistical comparison of samples is possible, beyond the observation that the expert was well within one standard deviation of the novice sample mean). This preliminary finding suggests that the transparent overlay approach may bring immediate benefits to existing products that must integrate abstract diagrammatic information into a conventional user interface. Transparency effects are becoming increasingly easy to implement as graphics APIs include alpha channel blending operations, and these support use of transparency at interactive speeds. The next experimental system was created in order to experiment with these facilities.

4.2 Study 5: Overlays for Colouring

This experiment investigated painting programs, rather than the object-based drawing programs of study 1. Each stroke in a conventional paint program replaces the pixels previously in that area, meaning that users only modify the visual appearance – they cannot define abstract operations relating to the intended semantic content of what they draw. Our experimental enhancement was to support semi-transparent strokes that are rendered by alpha blending onto the existing content of the drawing [16]. This allows users to create watercolour-like effects, but also to view the complete history of the painting they have made, to the extent that multiple layers of painting can be viewed through the semi-transparent upper layer.

The principal user abstraction in this program was that of a stack of superimposed painting layers. Each painting stroke was treated as a new layer, blended into the sum of previous layers. A similar facility in Photoshop presents the layers in a separate tool, which must be used to select the current working layer or to delete or reorder layers. We integrated layer management into the main work area, allowing users to select a layer by clicking on a visible stroke that they consider semantically meaningful. This is not completely straightforward, in that the user may be intending to click on a stroke contained in any one of the layers. We made two assumptions. Firstly, we assume that the user will click on the image in a semantically salient area. Rather than click in an area where several strokes are overlaid, the user will tend to avoid ambiguity by clicking in an area where one colour is predominant. Secondly, we identify the predominant colour by sampling the final alpha-blended pixel value at that point, and calculating the relative contribution of all layers to the final colour. If the first assumption is satisfied, it will be possible to identify a single layer that is the best candidate according to the second.

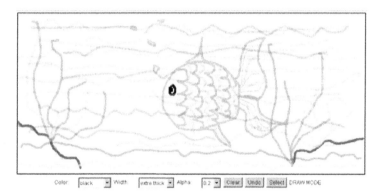

Fig. 5. Sample production from drawing task requiring multiple layered brushstrokes of varying transparency. Experimental subjects were asked to draw a scene with some transparent, some solid, and some semi-transparent elements in order to test their ability to manipulate multiple content layers.

Experimental evaluation of this interface was conducted in an observation study (i.e. no performance measurements were made) of four first-time users of the system, all of whom were asked to draw a scene with a relatively complex set of relationships between visible layers, some of which must be transparent. A sample production is shown in figure 5. Responsiveness of the system was somewhat compromised by the speed of alpha blending calculations (more recent releases of Java include an alpha channel component in the rendering API, meaning that this performance problem could now be avoided). All users found the selection procedure generally usable, although two deficiencies were noted by most – the user's conception of a layer is composed of multiple strokes, whereas the system can only be certain of the homogeneity of individual strokes. In future this should be addressed by area grouping of strokes in the same colour. Secondly, the user is very aware of the selection behaviour being distinct from drawing behaviour. More sophisticated users wished to be able to control this directly. They recognized the behaviour of the inferred selection, but were frustrated by the fact that they did not have direct control over the transition between selection and drawing modes.

4.3   Study 6: Overlays for Text Processing

The final experiment investigated the domain of word processors, especially the case in which a text document contains semi-structured information such as lists of names and telephone numbers, email addresses, financial information and so on. These types of information can often be recognized automatically by the use of regular expressions. (Regular expressions are sequences of text that can also include special control characters such as wild cards, end-of-line markers etc, to define search and matching behaviour. They include options like those described as "Special" and "Wildcard" in the search and replace dialog of Microsoft Word). A sophisticated user can employ a word processor as a reasonably powerful data processing tool by composing macros with regular expressions for search and replace. Several systems have been proposed to allow less sophisticated users to specify this kind of abstract transformation over text documents. These use programming by example techniques so that the user can demonstrate one or more examples of the required transformation, and the system will then work out what operations the user intends by inference over the examples that the user has demonstrated [14,15].

In these previous systems, the abstract inference made by the system is not directly displayed for the user to see. In most cases, the user is able to observe the result of the inferred transformation, and either confirm or reject the inference. Our SWYN system (See What You Need) was constructed to investigate ways in which the user could be provided with sufficient information about the inferred program in order to assess its correctness and modify the program if required. Conventional regular expression notation is not easily readable by non-programmers, so two graphical alternatives were devised, and verified as simpler and less error-prone for non-programmers [3]. An experimental study with 39 participants, each of whom carried out 12 typical text processing tasks, found that the tasks were completed 26% faster when using graphical notations rather than conventional regular expressions, and that participants made 36% fewer errors.

For the purposes of the current paper, the main advantage of one of these alternative notations was that it was suitable for representation as semi-transparent areas, that could be overlaid on a word-processor document while still leaving the document contents partially readable. This means that users can construct their program by direct manipulation, demonstrated in the context of the text they are working on, while also viewing the abstract consequences in the same context. When constructing a regular expression, the user selects strings that should or should not be matched, and a suitable expression is inferred using Hidden Markov Model methods. That expression is

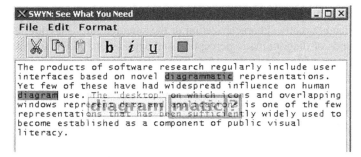

Fig. 6. The SWYN editor, showing interactive construction of a regular expression. The user is selecting examples of text, within the work context, that the regular expression should match. The current state of the inferred expression is displayed in the same visual context as a semi-transparent overlay.

continually updated, and displayed as an transparent overlay on the text, as shown in figure 6. The current version of the system is able to represent complex selection sets within a single window including diagrammatic overlay. A further refinement will use the selection methods developed in experiment 4 to disambiguate manipulation of the diagrammatic regular expression and selections of underlying text – this has proven to be a disadvantage of the current implementation.

5   Conclusions

The experimental systems described in this paper have been developed in order to test a specific theoretical hypothesis of diagrammatic interaction. This hypothesis is that the abstract facilities of software systems can and should be integrated into the primary work area as diagrammatic formalisms, rather than relegated to the principally linguistic representation of menus and dialogues or the impoverished visual representations of iconic commands.

This is an alternative view to the more common belief that the benefits of direct manipulation should be achieved through faithful adherence to the principle of What You See Is What You Get. WYSIWYG has already been strained by application to situations that do not naturally fit this formulation of the user's requirement. In the case where a diagrammatic notation is already established as a conventional intermediary between abstract design activity and physical product, common belief regarding the benefits of WYSIWYG has led to the slightly odd situation in which the conventional notation is treated as a product itself (as in the case of CAD systems), rather than the actual end product "what you get". A further peculiarity in conventional HCI theory is the way in which arbitrary physical analogies are introduced as "metaphors" for abstract operations so that WYSIWYG can be invoked, often without any real usability benefits [2].

These observations regarding the possible limitations of conventional theory are difficult to defend, especially where many borderline theoretical cases seem to have resulted in successful products regardless. This research program has therefore taken the approach that alternative characterization of the situation must be both justified on cognitive grounds of diagrammatic perception and application, and also illustrated by a range of experimental applications rather than simply by implementation of a single system. It is almost always possible to create a research product that improves on current commercial practice in a specific way, but difficult to draw theoretical conclusions from a single example.

The six experimental systems described in this paper are each relatively simple in their functionality and capabilities. Nevertheless, they provide a broad sample in support of our research hypothesis. Some user studies were conducted with these experimental systems, but we do not intend that these be viewed as controlled experiments verifying any one of the prototypes as superior to existing products. The overall argument in our research is that general hypotheses about interaction styles can sometimes be better tested by rapid implementation of a range of very different design options, (each with some degree of confirmation from real users), rather than more laborious and strictly controlled experimental studies.

Our main research hypothesis has been that use of abstraction can be supported by integrating gestural or overlaid diagrammatic content onto a conventional representation of a work product. This combination allows both transient and persistent abstractions to be created with reference to the task domain through local superimposition. It also allows the attentional cost of abstraction to be reduced through overlay of diagrammatic meaning within a uniform attentional context, and reduces the risk associated with investment of attention, through making the effects of abstract operations on the work product directly visible. These are only initial experiments, but they indicate a very promising theoretical basis for future research and product development.

Acknowledgements. The system for drawing without toolbars was implemented by Roman Marszalek, the life drawing analysis system was implemented by Tom Jacques, the system for 3D models from drawing input was implemented by Chris Morgan, the architectural constraint system and the second implementation of the SWYN editor were implemented by Hanna Wallach, and the paint colouring system was implemented by Kate Gardner. This research has been supported by EPSRC grant GR/M16924 "New paradigms for visual interaction". Hanna Wallach's work on the constraint system was rewarded by her recognition in 2001 as computer science student of the year in the UK.

References

1. Aish, R. Custom objects: A model-oriented end-user programming environment. Presented at The Visual End User workshop, IEEE Visual Languages, Seattle (2000)
2. Blackwell, A.F. Pictorial representation and metaphor in visual language design. Journal of Visual Languages and Computing 12 (2001) 475-499

3. Blackwell, A.F. See What You Need: Helping end users to build abstractions. Journal of Visual Languages and Computing 12 (2001) 223-252
4. Blackwell, A.F. Metacognitive theories of visual programming: What do we think we are doing? In Proceedings IEEE Symposium on Visual Languages (1996) 240-246
5. Blackwell, A.F., Green, T.R.G. Investment of attention as an analytic approach to cognitive dimensions. In T. Green, R. Abdullah & P. Brna (Eds.) Collected Papers of the 11th Annual Workshop of the Psychology of Programming Interest Group (1999) 24-35
6. Blackwell, A.F., Hague, R. AutoHAN: An architecture for programming the home. In Proc. IEEE Symposia on Human-Centric Computing Languages & Environments (2001) 150-157
7. Brooks, F.P. No silver bullet – essence and accidents of software engineering. Computer 20(4), (1987) 10-19. Reprinted from Proc. IFIP Congress, Dublin, Ireland, 1986.
8. Do, E.Y-L. What's in a diagram that a computer should understand? In Proc. Sixth Ann. Conference on Computer Aided Architectural Design Futures, Singapore. (1995) 469-482
9. Green, T.R.G., Blackwell, A.F. Ironies of abstraction. Presentation at 3rd International Conference on Thinking. British Psychological Society, London (1996)
10. Green, T.R.G., Blackwell, A.F. Design for usability using Cognitive Dimensions. Tutorial presented at HCI'98, Sheffield UK (1998)
11. Green, T.R.G., Petre, M. Usability analysis of visual programming environments: a 'cognitive dimensions' approach. Journal of Visual Languages and Computing 7 (1996) 131-174
12. Igarashi, T., Matsuoka, S., Tanaka, H. Teddy: A sketching interface for 3D freeform design. Proceedings of ACM SIGGRAPH (1999)
13. Kahneman, D., Tversky, A. Prospect theory: an analysis of decision under risk. Econometrica 47 (1979) 263-291.
14. Mo, D.H., Witten, I.H. Learning text editing tasks from examples: a procedural approach. Behaviour and Information Technology 11 (1992) 32-45
15. Nix, R.P. Editing by example. ACM Transactions on Programming Languages and Systems (1985) 600-621
16. Porter, T., Duff, T. Compositing digital images. Computer Graphics 18 (1984) 253-259
17. Shneiderman, B. Direct manipulation: A step beyond programming languages. IEEE Computer 16 (1983) 57-69
18. Zhang, R.Z., Tsai, P., Cryer, J.E., Shah, M. Shape from shading: A survey. IEEE Transactions on Pattern Analysis and Machine Intelligence 21 (1999) 690-706

Extracting Explicit and Implict Information from Complex Visualizations

J. Gregory Trafton[1], Sandra Marshall[2], Farilee Mintz[3], and Susan B. Trickett[4]

[1] Naval Research Laboratory, Washington, DC 20375-5337, USA
trafton@itd.nrl.navy.mil
[2] San Diego State University
smarshall@sciences.sdsu.edu
[3] ITT Industries
mintz@aic.nrl.navy.mil
[4] George Mason University
stricket@gmu.edu

Abstract. How do experienced users extract information from a complex visualization? We examine this question by presenting experienced weather forecasters with visualizations that did not show the needed information explicitly and examining their eye movements. We replicated Carpenter & Shah (1998) when the information was explicitly available on the visualization. However, when the information was not explicitly available, we found that forecasters used spatial reasoning in the form of spatial transformations. We also found a strong imagerial component for constructing meteorological information.

1 Introduction

How do people comprehend and extract information from a complex graph or visualization? Current models of graph comprehension account very well for simple tasks (e.g., "What is the price of graphium in 1982?"). Similarly, most empirical studies have used simple graphs (e.g., few variables and few data points). For example, within the graph comprehension literature, Lohse (1993) studied some of the most complex graphs, and his standard graphs used 6 variables with a total of 72 data points; Carpenter & Shah (1998) provided participants with graphs that had 2 to 4 variables with 4 to 8 data points (containing interactions); Tan & Benbasat (1990) had participants study graphs that displayed 3 variables with 18 total data points.

In contrast, many domains attempt to display tens of variables and tens of thousands of data points (or more). For example, many scientists use extremely complex visualizations with thousands of data points (Trafton, Trickett, & Mintz, in press; Trickett, Fu, Schunn, & Trafton, 2000; Trickett, Trafton, Schunn, & Harrison, 2001) and weather forecasters routinely examine visualizations with 10 or more variables and thousands of data points over multiple spatial and temporal scales (Trafton, Kirschenbaum, Tsui, Miyamoto, Ballas, & Raymond, 2000). Current wisdom suggests that showing a graph well depends

M. Hegarty, B. Meyer, and N. Hari Narayanan (Eds.): Diagrams 2002, LNAI 2317, pp. 206–220, 2002.
© Springer-Verlag Berlin Heidelberg 2002

on making the variables and patterns explicit (Gillan, Wickens, Hollands, & Carswell, 1998; Tufte, 1983, 1990, 1997). However, when there are an extremely large number of data points, this explicitness rule may need to be relaxed in order to avoid visual clutter.

We believe that as a direct result of the complexity of the graphs, some data will need to be represented more imprecisely (e.g., pressure on Figure 1) and this imprecision may prevent a user from directly reading off information from the graph. The user may need to use a different set of mental procedures than is needed in simpler graphs.

This paper first reviews current models of graph comprehension, suggests augmenting them in a specific manner when the visualization is complex and the task requires information that is implicitly represented, and presents an initial study to explore these issues.

1.1   Current Models of Graph Comprehension

The most influential research on graph and visualization comprehension is Bertin's (1983) task analysis which suggests three main processes in graph and visualization comprehension:

1. Encode visual elements of the display: For example, identifying lines and axes. This stage is influenced by pre-attentive processes and is affected by the discriminability of shapes.

2. Translate the elements into patterns: For example, notice that one bar is taller than another or the slope of a line. This stage is affected by distortions of perception and limitations of working memory.

3. Map the patterns to the labels to interpret the specific relationships communicated by the graph. For example, determine the value of a bar graph.

Most of the work done on graph comprehension has examined the encoding, perception, and representation of graphs. Cleveland and McGill, for example, have examined the psychophysical aspects of graphical perception (Cleveland & McGill, 1984, 1986). Similarly, Pinker's theory of graph comprehension, while quite broad, focuses on the encoding and understanding of graphs (Pinker, 1990). Kosslyn's work emphasizes the cognitive processes that make a graph more or less difficult to read. Kosslyn's syntactic and semantic (and to a lesser degree pragmatic) level of analysis focuses on encoding, perception, and representation of graphs (Kosslyn, 1989). Tracking users' eye movements, Carpenter & Shah (1998) have shown that people switch between looking at the graph and the axes in order to comprehend the visualization. Similarly, Peebles & Cheng (2001a, 2001b) have suggested that in experimental trials, people cycle between looking at the graph and the question they are trying to answer

This scheme seems to work very well when the graph explicitly represents the needed information. Thus, when a meteorologist is asked to provide the wind direction at Pittsburgh on Figure 1, the meteorologist searches for Pittsburgh, finds the wind barb over Pittsburgh, and determines the direction it points (280, or westerly, in this case). Note that it is slightly difficult to see the part of the barb showing wind speed in this figure, which should not affect the perception of wind direction.

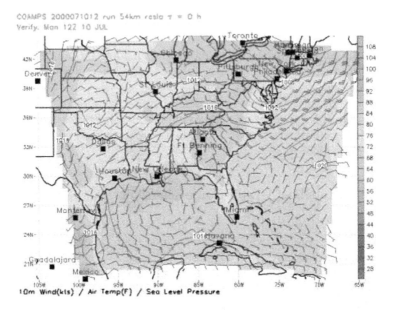

Fig. 1. A typical visualization that meteorologists use. Wind speed and wind direction is shown by the wind barbs; temperature is color coded to the legend on the far right, and pressure is shown by lines connecting the same pressure (1008, 1012, etc.). The original is in color.

1.2   When Information Is Not Explicitly Available

What happens when information is not explicitly available on a graph? For example, in Figure 1, how would the pressure at Pittsburgh be determined? [1]

Current theories either do not deal with how information is extracted when the information is not explicitly available (e.g., Lohse, 1993) or leave the whole process unspecified (e.g., Pinker, 1990; Kosslyn, 1989).

Pinker's (1990) framework, for example, provides the most detailed description of how information is extracted from a graph. He claims that once the graph has been encoded, all inferences and reasoning occur propositionally. Pinker's

---

[1] If a pressure line goes directly through the location for which a forecaster wants pressure, the forecaster simply reads it off (e.g., St. Louis' pressure is 1012). If, however, a pressure line does not pass directly through the desired location, the forecaster must interpolate between two lines (e.g., Pittsburgh's pressure is 1011 because it is between 1008 and 1012 and closer to 1012 than the middle).

"conceptual questions" and "conceptual messages" are his primary method of inferencing and reasoning. Thus, Pinker's theory suggests that people initially encode the aspects of a graph into a propositional representation, and then reason with that propositional representation to answer conceptual questions.[2]

Trafton & Trickett (2001), in contrast, suggested that a great deal of complex visualization reasoning occurs spatially, especially when the information is not explicitly available on the visualization. Trafton & Trickett (2001) showed that when scientists' own complex visualizations did not explicitly show the information they needed, scientists created complex mental representations and manipulated them to help them answer the specific questions they had. Trafton & Trickett suggested that the scientists constructed spatial representations.

Trafton & Trickett (2001) suggested that when people who used complex visualizations needed to extract information that was not available, they used spatial information to create internal mental representations to reason with. We have developed a framework for coding and working with these kinds of graphs and visualizations called *Spatial Transformations* that will be used to investigate these issues. We will argue that spatial transformations are a fundamental aspect of complex visualization usage.

Spatial Transformations are cognitive operations that a scientist performs on a visualization. Sample spatial transformations are mental rotation (e.g., Shepard & Metzler, 1971), creating a mental image, modifying that mental image by adding or deleting features to or from it, animating an aspect of a visualization (Hegarty, 1992) time series progression prediction, mentally moving an object, mentally transforming a 2D view into a 3D view (or vice versa), comparisons between different views (Kosslyn, Sukel, & Bly, 1999; Trafton et al., in press), and anything else a scientist mentally does to a visualization in order to understand it or facilitate problem solving. Also note that a spatial transformation can be done on either an internal (i.e., mental) representation or an external image (i.e., a scientific visualization on a computer-generated image). A more complete description of spatial transformations (along with a mini-experiment to teach interface designers when to use a 2D or a 3D representation) can be found at http://e-lab.hfac.gmu.edu/~trafton/405st.html

Trafton & Trickett (2001) focused on how scientists created mental representations to answer specific questions. Because that study examined scientists working *in vivo* (Baker & Dunbar, 2000; Dunbar, 1995, 1997; Trickett et al., 2001), it was not possible to control what they saw or what information was needed. In this study, we asked experienced meteorologists to provide specific information from weather visualizations like that shown in Figure 1. We also tracked where they were looking with an eyetracker system.

---

[2] There are, of course, a large number of theories that discuss diagrammatic reasoning and suggest that reasoning occurs spatially or via images (Hegarty & Sims, 1994; Larkin & Simon, 1987; Narayanan & Hegarty, 1998; Tabachneck-Schijf, Leonardo, & Simon, 1997). However, most of these theories are concerned with diagram understanding, not graph comprehension.

We had several goals in this study. First, we wanted to provide a baseline measure by showing that when experienced users extracted information explicitly available from a complex visualization, they would extract information based on the canonical graph comprehension model presented earlier; this would also be a partial replication of other graph comprehension studies (Carpenter & Shah, 1998). For a standard meteorological visualization, we predicted that experienced meteorologists would be able to go directly to the desired information and extract it. Second, we wanted to examine how forecasters extract information from the graph that was not explicitly available (i.e., do they use a propositional representation or do they reason with the graph itself). We complete our discussion with some anecdotal evidence of how forecasters remember information that they had seen previously.

2 Method

In order to investigate the issues discussed above, we examined forecasters as they were examining meteorological visualizations.

2.1 Task

Forecasters were presented with a weather visualization (see Figure 1). They were then asked several questions in the following order during the *Graph Comprehension* portion of this study.

- What is the synoptic weather?
- What is the wind speed and wind direction at Pittsburgh, Pennsylvania?
- What is the temperature at Pittsburgh, Pennsylvania?
- What is the pressure at Pittsburgh, Pennsylvania?

Forecasters were then shown a second weather visualization (see Figure 2) and asked two questions:

- What is the wind speed and wind direction at Honolulu?
- What is the relative humidity at Honolulu?

Finally, during the *recall* portion of the study, forecasters were shown a blank screen and asked about wind speed, wind direction, temperature, and pressure at four locations from the first visualization: Pittsburgh, Pennsylvania (Exact), Philadelphia, Pennsylvania (Near), Atlanta, Georgia (Medium), and Houston, Texas (Far). The exact question probed participants' memory for the information they had recently read off. The near, medium, and far conditions allowed us to see how their accuracy changed as the location moved farther away from the area they focused on (Pittsburgh); Philadelphia is very close to Pittsburgh, Georgia is further away, and Houston is quite far from Georgia on the visualization.

It should be noted that all of these question types were based on the types of tasks meteorologists typically do (Lowe, 1999; Trafton et al., 2000).

These questions came in several forms:

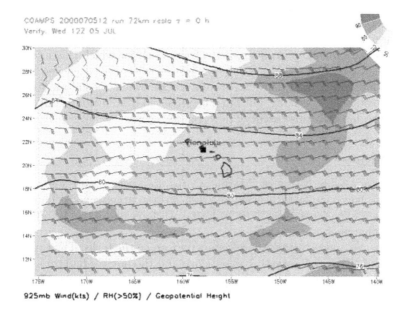

925mb Wind(kts) / RH(>50%) / Geopotential Height

Fig. 2. A typical visualization that meteorologists use. The wind barbs represent wind speed and direction; the black lines represent geopotential height, and the dial in the upper right hand corner shows a legend for relative humidity. The original is in color.

- Questions asking for qualitative information, where the users have to integrate across information. The synoptic weather question is an example of this.
- Questions asking for quantitative information, where the answers (numbers) are explicitly represented (as in traditional graph comprehension studies). The question about temperature and the relative humidity question fall into this category because users only have to map a color to the legend and the numbers are printed on the legend. Also, all locations on the map have an instance of the variable. We expect participants to read-off the information as they need it, just as the graph comprehension literature predicts.
- Questions asking for quantitative information that is imprecisely represented in the graph. In other words, the numbers themselves are not represented explicitly, but there is a symbology associated with them that the user must know in order to extract the needed information. Wind direction and wind speed are part of this question type. If the user has the knowledge about how to interpret the symbology, the graph comprehension theories would predict a read-off strategy.
- Questions asking for quantitative information that is entirely implicit in the graph. Numbers are not represented either explicitly or imprecisely to answer that question, but must be inferred. The question about pressure is an instance of this category. Although pressure is represented explicitly

for some locations, it must be inferred for others (like Pittsburgh). Graph comprehension theories do not make good predictions here, but the spatial transformation theory does suggest a framework for which this information could be extracted.

– Questions asking for quantitative information when the information is no longer available externally, but must be retrieved from memory. All the recall questions fall into this category.

2.2   Participants

All forecasters were Naval or Marine Corps forecasters and forecasters-in-training. All had completed at least the first level weather school. They ranged in forecasting experience from 1 to 10 years. All forecasters had significant operational experience.

Four novices and two expert forecasters performed the tasks. The experts had an average of 10 years forecasting experience and the novices had an average of 2.5 years forecasting experience.

2.3   Setting and Apparatus

Meteorological visualizations were collected from
http://www.fnmoc.navy.mil/. This web site is used a great deal by current Navy forecasters; all forecasters in the study were very familiar with this site and the visualizations that were used. This web site was used extensively by forecasters in a previous study of Navy meteorologists (Trafton et al., 2000)

The experimental sessions took place in a room equipped with a PC and an EyeLink System from SMI. The eyetracker had headmounted optics with three small cameras (left eye, right eye, head compensation). It records eye movements and pupil dilations at 250 Hz (every 4 milliseconds).

2.4   Procedure

Each session began with hooking up the eyetracker and calibrating it to the individual participant. Next, the participant was shown the initial screen. All questions were asked verbally to prevent additional eyetracking to the question (Peebles & Cheng, 2001a, 2001b). After the first set of questions was finished, the next visualization was shown, and the next set of questions was asked. When the participant finished that set, a blank screen was shown for the test questions.

3   Results

How do skilled users extract information from complex visualizations? Is there a difference between extractions that occur when the information is explicitly available on the visualization and when the information is not explicitly available? This results section attempts to answer these questions.

3.1   Overview

All forecasters were able to extract the information from the graphs; they all knew how to read these kinds of meteorological visualizations.

The graph comprehension aspect of this task was not difficult for the participants; experts and novices were very accurate when reading off information from the graph. There were no consistent qualitative or quantitative differences between experts or novices during graph comprehension, so all participants' results were combined.

3.2   Extracting Information That Is Explicitly Available

We first examined how forecasters extracted information that is explicitly on a complex visualization. For several of the questions during graph comprehension, the information was explicitly available on the graph. Thus, for wind speed and wind direction, participants simply had to find the location (e.g., Pittsburgh) and read off the wind speed and wind direction. For other variables (temperature and relative humidity) the forecasters had to match a color code with a legend.

Bertin's model presented earlier suggested that after people encoded a graphical element, they would translate the elements into patterns and map the patterns into labels. Bertin (and others) implied that people would perform this operation serially. Carpenter & Shah (1998) however, found that participants interpreting graphs would cycle back and forth between the graph area and the legend, suggesting a much more iterative aspect to graph comprehension than previously known. We explicitly examined if experienced (both expert and novice) forecasters would cycle back and forth between the graph area and the legend, as Carpenter & Shah (1998) found.

During the graph comprehension stage, we asked two questions whose variables had legends: "What is the temperature at Pittsburgh?" and "What is the relative humidity at Honolulu?" We examined the number of times that participants cycled back and forth between the graph area and the legend. Figure 3 shows a typical subject's eye movements as he finds the answer to the question "What is the temperature at Pittsburgh, Pennsylvania?"

Consistent with Carpenter & Shah (1998), we found frequent switching for both questions. Table 1 shows the average number of cycles for each question.

Table 1. The average number of times that the forecasters looked between the legend and the graph area.

Question	# of glances between graph and legend (avg)
What is the temperature at Pittsburgh?	6.0
What is the relative humidity at Honolulu?	4.5

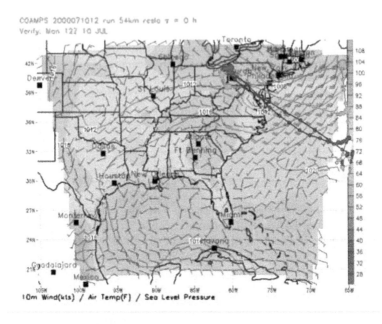

Fig. 3. Eyetracking of a participant after asked the question, "What is the temperature at Pittsburgh, Pennsylvania?" "Dots" are 4 milliseconds apart. There are four glances between the legend and the graph area in this figure.

3.3 Extracting Information That Is Not Explicitly Available

How do experienced forecasters extract information that is not explicitly available on a visualization? There are some propositional theories (e.g., Lohse, 1993; Peebles & Cheng, 2001a; Pinker, 1990) that do not explain the extraction process well if the information is not explicitly available on the graph. At a minimum, these propositional theories leave this process unspecified. Other theories (like Trafton et al.'s (2001) spatial transformation theory) and many theories on diagram comprehension suggest that people use both propositional and spatial reasoning (e.g., Tabachneck et al.'s (1997) CaMeRa theory).

In order to examine the cognitive process of how experienced graph users extract information that is not explicitly available, we asked experienced forecasters what the pressure was at Pittsburgh. Recall that pressure is not represented explicitly for Pittsburgh; forecasters must interpolate between the two pressure lines to determine the actual pressure of Pittsburgh. Eyetracking results show that all participants went through several steps to extract this information:[3]

1. Trace the upper pressure line;
2. Find the pressure itself (the number);
3. Trace the lower pressure line;

---

[3] Note that this was the fourth question asked (the third about Pittsburgh), so participants already knew where Pittsburgh was.

4. Find the pressure itself (the number);

5. Bridging (glancing) back and forth between the two pressure lines.

Table 2 shows the average number of times the forecasters performed each tracing task.

Table 2. The average number of times that the forecasters traced each pressure line and bridging between the two lines.

Tracing type	# of Traces (avg)
Tracing the upper pressure line	2.4
Tracing the lower pressure line	2.6
Bridging between the two pressure lines	7.4

Figure 4 shows one of the forecasters' eye movements as he was bridging between the two lines. We interpret these eye tracking movements to suggest that the forecasters were using the diagram to help them calculate the distance between the two lines. We believe that after the forecasters located the pressure lines, they were mentally drawing a line in between the two lines and dividing up the space between Pittsburgh and the pressure lines. This kind of spatial transformation seems to be more effortful than simply line tracing, as shown by the increased number of eye tracking "bridges" between the two pressure lines as compared to the pressure line tracing, $\chi^2(2) = 19.4, p < .001$, Bonferroni adjusted χ^2s significant at $p < .05$.

3.4   Recalling and Constructing Information

After the graph comprehension part of the task, forecasters were asked for information that they had just extracted (Pittsburgh) or for information that they had seen but not explicitly extracted. This aspect of the task was seen as quite difficult. Several participants were not able to answer some of the recall questions, especially in the far condition. Experts attempted 88% of the questions, while novices attempted only 56% of them, $\chi^2(1) = 7.1, p < .01$. Figure 5 shows the accuracy of those participants who did complete the questions for pressure (wind speed, wind direction, and temperature all showed a similar pattern). While these results must obviously be interpreted with care because of the low completion rate, it is evident that the experts were far more accurate than the novices, especially at the non-recall questions.

A very small number of forecasters showed an interesting pattern while answering questions about temperature (we are missing some eye movement data because several forecasters closed their eyes or looked off the screen). When asked about the temperature for Pittsburgh (an exact question that was being recalled), most forecasters focused on a specific area of the screen. However, when asked about temperature of other areas (near, medium, and far), 67% of the forecasters "looked" to the right to "examine" the legend, even though there

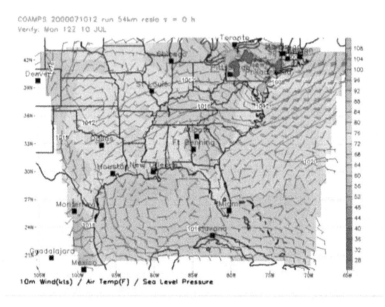

Fig. 4. A forecaster's eye movements as he was bridging between the two pressure lines. This is only a partial display of his eye movements while answering this question (i.e., the tracing the two pressure lines has been omitted.

was nothing on the blank screen to see. We interpreted this kind of glance as evidence for visual imagery.[4]

Figure 6 shows one of the forecasters' eye movements as he made this kind of glance. We have not run enough participants through this type of experiment or even seen enough evidence of this kind of glance, but we know of no other eye movement study that has demonstrated any kind of "examination" of a mental image by eye movements.

4   General Discussion

In this paper, we explored the process that experienced users went through when they extracted information from a complex visualization when the information was explicitly available and when the information was not explicitly available.

We replicated the results from Carpenter & Shah (1998) and showed that even experienced users scan back and forth between a legend and its value.

---

[4] It is interesting that, while looking at the legend was a frequent glance, there were other glances that did not seem to happen. For example, the participants only rarely "looked" at a location that was geographically correct, though there was some change of glance location when questions about a different location were asked. This suggests that the entire representation is not encoded exclusively as either an image or as a proposition.

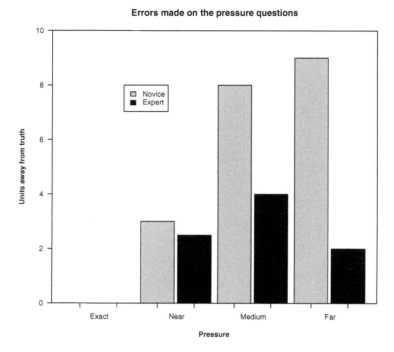

Fig. 5. Accuracy of the forecasters that completed the question for the pressure question across all locations.

We also presented eye movement data that showed that when users need to interpolate between two lines (like pressure lines or isobars), they do not use a propositional representation; rather, they use some spatial or imagerial process to trace a line that allows them to extract the needed information. This extraction literally occurs as a combination between the internal (visualization) and external representations (spatial transformations). This finding also provides further support for the spatial transformation framework (Trafton & Trickett, 2001; Trafton et al., in press).

This study also provided some very preliminary data on differences between experts and novices. We found that both expert and novice forecasters were able to accurately extract information from complex meteorological visualizations. We also found that experts did a better job during the test phase for non-recall items than novices. This finding is consistent with the finding of Trafton et al. (2000) who suggested that experienced forecasters build a qualitative mental model (QMM) and reason with that (rather than simply using climatological knowledge or memory). This finding is consistent with Trafton et al.'s (2000) view that expert forecasters are able to generate complex quantitative relationships by extracting primarily qualitative information from a complex weather visualization.

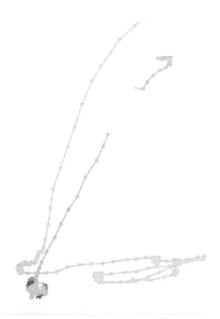

Fig. 6. Eyetracking of a participant after asked the question, "What is the temperature at Philadelphia, Pennsylvania?" This was the "Near" question during the recall stage. The screen itself was blank; it showed nothing but white, so there was literally nothing for the forecaster to look at.

We also suggested that some forecasters use visual imagery to generate information they can not recall. Other researchers have suggested that visual or spatial imagery is used when reasoning with complex diagrams (e.g., Hegarty & Sims, 1994; Tabachneck-Schijf et al., 1997), but most theories of graph comprehension do not. One of our main goals in this paper was to provide some evidence for combining some of the visual and spatial research from diagrammatic reasoning into the graphical reasoning research.

Acknowledgments. This research was supported in part by grants N00014-00-WX-20844 and N00014-00-WX-40002 to the first author from the Office of Naval Research and by grant number N00014-00-1-0331 from the Office of Naval Research to the second author. The views and conclusions contained in this document are those of the authors and should not be interpreted as necessarily representing the official policies, either expressed or implied, of the U. S. Navy.

References

Baker, L. M., & Dunbar, K. (2000). Experimental design heuristics for scientific discovery: the use of "baseline" and "known standard" controls. *International Journal of Human Computer Studies, 53*(3), 335–349.

Bertin, J. (1983). *Semiology of graphs.* Madison, WI: University of Wisconsin Press.

Carpenter, P. A., & Shah, P. (1998). A model of the perceptual and conceptual processes in graph comprehension. *Joural of Experimental Psychology: Applied, 4*(2), 75–100.

Cleveland, W. S., & McGill, R. (1984). Graphical perception: theory, experimentation, and application to the development of graphical method. *Journal of the American Statistical Association, 79,* 531–553.

Cleveland, W. S., & McGill, R. (1986). An experiment in graphical perception. *International Journal of Man-Machine Studies, 25,* 491–500.

Dunbar, K. (1995). How scientists really reason: Scientific reasoning in real-world laboratories. In R. J. Sternberg, & J. E. Davidson (Eds.), *The nature of insight,* (pp. 365–395). Cambridge, MA: MIT Press.

Dunbar, K. (1997). How scientists think: Online creativity and conceptual change in science. In T. B. Ward, S. M. Smith, & S. Vaid (Eds.), *Creative thought: An Investigation of Conceptual Structures and Processes,* (pp. 461–493). Washington, DC: APA Press.

Gillan, D. J., Wickens, C. D., Hollands, J. G., & Carswell, C. M. (1998). Guidelines for presenting quantitative data in HFES publications. *Human Factors, 40*(1), 28–41.

Hegarty, M. (1992). Mental animation: Inferring motion from static displays of mechanical systems. *Journal of Experimental Psychology: Learning, Memory and Cognition, 18*(5), 1084–1102.

Hegarty, M., & Sims, V. K. (1994). Individual differences in mental animation during mechanical reasoning. *Memory and Cognition, 22*(4), 411–430.

Kosslyn, S. M. (1989). Understanding charts and graphs. *Applied Cognitive Psychology, 3,* 185–226.

Kosslyn, S. M., Sukel, K. E., & Bly, B. M. (1999). Squinting with the mind's eye: Effects of stimulus resolution on imaginal and perceptual comparisons. *Memory and Cognition, 27*(2), 276–287.

Larkin, J. H., & Simon, H. A. (1987). Why a diagram is (sometimes) worth ten thousand words. *Cognitive Science, 11,* 65–99.

Lohse, G. L. (1993). A cognitive model for understanding graphical perception. *Human Computer Interaction, 8,* 353–388.

Lowe, R. K. (1999). Extracting information from an animation during complex visual learning. *European Journal of Psychology of Education, 14*(2), 225–244.

Narayanan, N. H., & Hegarty, M. (1998). On designing comprehensible interactive hypermedia manuals. *International Journal of Human-Computer Studies, 48,* 267–301.

Peebles, D., & Cheng, P. C.-H. (2001a). Extending task analytic models of graph-based reasoning: A cognitive model of problem solving with cartesian graphs in ACT-R/PM. In *Proceedings of the 4th International Conference on Cognitive Modeling,* (pp. 169–174). Mahwah, NJ: Lawrence Erlbaum Association.

Peebles, D., & Cheng, P. C.-H. (2001b). Graph-based reasoning: From task analysis to cognitive explanation. In *Proceedings of the Twenty Third Annual Conference of the Cognitive Science Society,* (pp. 762–773). Hillsdale, NJ: Lawrence Erlbaum Association.

Pinker, S. (1990). A theory of graph comprehension. In R. Freedle (Ed.), *Artificial intelligence and the future of testing,* (pp. 73–126). Hillsdale, NJ: Lawrence Erlbaum Associates, Inc.

Shepard, R. N., & Metzler, J. (1971). Mental rotation of three-dimensional objects. *Science, 171,* 701–703.

Tabachneck-Schijf, H. J. M., Leonardo, A. M., & Simon, H. A. (1997). CaMeRa: a computational model of multiple representations. *Cognitive Science*, *21*(3), 305–350.

Tan, J. K. H., & Benbasat, I. (1990). Processing of graphical information: A decomposition taxonomy to match data extraction tasks and graphical representations. *Information Systems Research*, *1*(4), 416–438.

Trafton, J. G., Kirschenbaum, S. S., Tsui, T. L., Miyamoto, R. T., Ballas, J. A., & Raymond, P. D. (2000). Turning pictures into numbers: Extracting and generating information from complex visualizations. *International Journal of Human Computer Studies*, *53*(5), 827–850.

Trafton, J. G., Trickett, S. B., & Mintz, F. E. (in press). Connecting internal and external representations: Spatial Transformations of scientific visualizations. *Foundations of Science*.

Trafton, J. G., & Trickett, S. B. (2001). A new model of graph and visualization usage. In J. D. Moore, & K. Stenning (Eds.), *Proceedings of the Twenty Third Annual Conference of the Cognitive Science Society*, (pp. 1048–1053). Mahwah, NJ: Lawrence Erlbaum Associates.

Trickett, S. B., Fu, W., Schunn, C. D., & Trafton, J. G. (2000). From dipsy-doodles to streaming motions: Changes in representation in the analysis of visual scientific data. In *Proceedings of the Twenty Second Annual Conference of the Cognitive Science Society*, (pp. 959–964). Hillsdal, NJ: LEA.

Trickett, S. B., Trafton, J. G., Schunn, C. D., & Harrison, A. (2001). "That's odd!" How scientists respond to anomalous data. In J. D. Moore, & K. Stenning (Eds.), *Proceedings of the Twenty Third Annual Conference of the Cognitive Science Society*, (pp. 1054–1059). Mahwah, NJ: LEA.

Tufte, E. R. (1983). *The visual display of quantitative information*. Cheshire, CN: Graphics Press.

Tufte, E. R. (1990). *Envisioning information*. Cheshire, CN: Graphics Press.

Tufte, E. R. (1997). *Visual Explanations*. Cheshire, CN: Graphics Press.

Visual Attention and Representation Switching During Java Program Debugging: A Study Using the Restricted Focus Viewer

Pablo Romero, Richard Cox, Benedict du Boulay, and Rudi Lutz

Human Centred Technology Group,
School of Cognitive & Computing Sciences
University of Sussex, Falmer, Brighton,
East Sussex, BN1 9QH, UK
juanr@cogs.susx.ac.uk
http://www.cogs.susx.ac.uk/

Abstract. Java program debugging was investigated in programmers who used a software debugging environment (SDE) that provided concurrently displayed, adjacent, multiple and linked representations consisting of the program code, a functional visualisation of the program, and its output.

A modified version of the Restricted Focus Viewer (RFV)[3] - a visual attention tracking system - was employed to measure the degree to which each of the representations was used, and to record switches between representations. Other measures included debugging performance (number of bugs identified, the order in which they were identified, bug discovery latencies, *etc.*).

The aim of this investigation was to address questions such as 'To what extent do programmers use each type of representation?' and 'Are particular patterns of representational use associated with superior debugging performance?'.

A within-subject design, and comparison of performance under (matched) RFV/no-RFV task conditions, allowed the use of the RFV as an attention-tracking tool to be validated in the programming domain.

The results also provide tentative evidence that superior debugging using multiple-representation SDE's tends to be associated with a) the predominant use of the program code representation, and b) frequent switches between the code representation and the visualisation of the program execution.

1 Introduction

When trying to perform a programming activity in everyday settings, programmers normally work with a variety of external representations as well as the program code. Some of these external representations are used in debugging packages, prototyping and visualisation tools in software development environments, or are included as part of internal and external documentation. Therefore, programming normally requires the co-ordination of multiple representations.

M. Hegarty, B. Meyer, and N. Hari Narayanan (Eds.): Diagrams 2002, LNAI 2317, pp. 221–235, 2002.
© Springer-Verlag Berlin Heidelberg 2002

Probably the most typical case, at least for beginner programmers, of co-ordination of external representations in programming is working with debugging packages, a common example of a visualisation tool. Novice programmers often spend a good amount of their learning time attempting to understand the behaviour of programs when trying to discover errors in the code. To perform this task, novices normally work with both the program code and the debugger output, trying to co-ordinate and make sense of these representations. Yet studies of program comprehension have not, to the best of our knowledge, addressed the issue of how multiple external representations are used for this kind of programming task.

We believe that the investigation of the co-ordination of multiple external representations in programming can be effectively supported by visual attention tracking methods, and that a tool like the Restricted Focus Viewer (RFV)[3] can be used for this purpose. The use of this experimental tool allows us to analyse the *process* of representational use in program debugging by addressing questions such as 'How much time do users spend using each representation?', 'Under what circumstances do programmers switch between representations' and 'Are particular patterns of representational use associated with superior debugging performance?'.

1.1   Coordination of Multiple External Representations in Programming

Two important aspects to consider regarding the co-ordination of multiple representations in programming are *modality* and *perspective* [9]. The term 'modality' is used here to mean the representational forms used to present or display information, rather than in the psychological sense of sensory channel. A typical modality distinction here is between propositional and diagrammatic representations. Thus, the first aspect refers to co-ordinating representations which are basically propositional with those that are mainly diagrammatic. It is not clear whether co-ordinating representations in the same modality type has advantages over working with mixed multiple representations or whether including a high degree of graphicality has potential benefits for performing the task.

Modality. Although programmers normally have to coordinate representations of different modalities, there has not been much research on these issues in the area of programming. One of the few examples is the GIL system [15], which attempts to provide reasoning-congruent visual representations in the form of control-flow diagrams to aid the generation and comprehension of LISP, a functional programming language which employs mainly textual representations. In [15], it is claimed that this system is successful in teaching novices to program in this language; however, this work did not compare co-ordination of the same and different modalities.

Work in the algorithm animation area ([5]) has found advantages for the use of multiple representations of mixed modality. In [5], it was found that students might benefit from the dual coding that results from presenting a graphical visualisation of the program together with a textual explanation of it.

Other studies in the area have been concerned with issues related to the format of the output of debugging packages [16,18]. Those studies have offered conflicting results about the co-ordination of representations of different modalities. In [18], it was found that subjects working with representations of the same and different modalities had similar performance, while in [16], it was reported that the ones working with different modalities showed a poorer performance than those working with the same modality. In both cases, participants worked with the program code and with the debugger's output. The debugger notations used by both of these studies were mostly textual. The only predominantly graphical debugging tool used by these studies was TPM [10]. While the performance of the participants of the former study [18] was similar for the textual debuggers and TPM, the subjects of the latter study [16] found working with TPM more difficult. One important difference between these two studies is that while the former used static representations, the latter employed a visualisation package (dynamic representations). The additional cognitive load of learning and using a multi-representational visualisation package may explain the difference in findings.

Perspective. The second aspect refers to co-ordinating representations that highlight either the same or different programming information types. Computer programs are information structures that comprise different types of information [20], and programming notations usually highlight some of these aspects at the cost of obscuring others [12]. Experienced programmers, when comprehending code, are able to develop a mental representation that comprises these different perspectives or information types, as well as rich mappings between them [19]. Some of these different information types are: function, data structure, operations, data-flow and control-flow. It is an open issue whether co-ordinating notations that highlight different information types will be more beneficial to programmers than working with those that highlight the same ones.

Java debugging. To date, there have been numerous investigations of debugging behaviour across a range of programming languages [4,11,22,25] and previous research has also examined the effect of representational mode upon program comprehension [13,15,16,18].

However, these studies were performed mainly in the context of procedural or declarative computer languages. It is not clear whether the results will generalise to the (currently popular) Object-Oriented paradigm. Research in program comprehension for Object-Oriented languages suggests that these kinds of language highlight functional information [6,26]. However, it is not clear whether novice programmers working with medium size programs find comprehending function in Object-Oriented languages an easy task [27], specially because as program size increases, functional information tends to become diffuse.

Furthermore, debugging studies have not tended to employ debugging environments that are typical of those used by professional programmers (*i.e.* multi-representational software debugging environments (SDE's)). Such environments typically permit the user to switch rapidly between multiple, linked, concurrently displayed representations. These include program code listings, data-flow and control-flow visualisations, output displays, etc. In [23], a comprehensive survey of external representations employed in object-oriented programming environments is provided.

1.2   Aims

The aim of this work was to conduct a small-scale, exploratory study as a first step towards the development of a descriptive model of representational behaviour in program debugging.

To date, the RFV has been validated in the context of reasoning about simple mechanical systems via the inspection of static diagrams [3]. A secondary aim, therefore, was to validate the RFV for use in an active debugging context (in which users make frequent switches between multiple, heterogeneous, interactive representations).

Additional aims were:

- to employ a multi-modal debugging environment; for multi-modality is a characteristic typical of those used in professional practice;
- to investigate debugging in the context of the Java programming language - a modern, Object-Oriented and widely used programming language;

2   Method, Materials, and Procedure

2.1   The Experimental Debugging Environment

The Java SDE enabled participants to see the program's code, its output for a sample execution and a visualisation of this execution in terms of the program's functionality. A screen shot of the system is shown in Figure 1. Participants were able to see the several program modules in the code window, one at a time, through the use of the side-tabs ('coin', 'pile', 'till'). Also, the visualisation window presented a functional visualisation of the program's execution similar to those found in code animation systems [14]. Functional representations were selected in preference to other program perspectives because research in Object-Oriented program comprehension has suggested that function is an important information type for these languages (see Section 1.1).

The SDE was implemented on top of a modified version of the Restricted Focus Viewer (RFV). The RFV has been reported in [3] as an alternative to eye-tracking devices. This program presents image stimuli in a blurred form (but note that none of the images in Figure 1 are so blurred). When the user moves the mouse to an image, a section of it around the mouse pointer becomes focused. In this way, the program restricts how much of a stimulus can be seen clearly.

It allows visual attention to be tracked as the user moves an unblurred 'foveal' area around the screen. Use of the RFV enabled moment-by-moment representation switching between concurrently displayed, adjacent representations to be captured for later analysis. Here, we used a modified version of the original RFV to track visual attention and representation switches during program debugging.

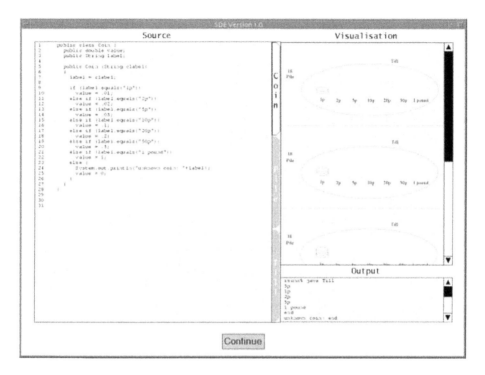

Fig. 1. The debugging environment used by participants (with RFV switched off)

The original RFV was modified for use in this study in several ways. First, stimulus images can be presented in a scroll or a tab pane. This allows us to present big images or more than one image in a specified display area. Second, the focused ('foveal') spot no longer follows the movement of the mouse. In our modified version, participants may click a mouse button to set the focused spot in the desired place. Every window image 'remembers' where its focused spot is, so when the user returns to that window the region in focus is the one that was set by the previous mouse click performed on that window image. This feature makes switching between stimulus images easier, because participants do not have to re-establish the place where they were looking at every time they switch their attention from one window image to another. Also, this change allowed us to distinguish between two kinds of mouse-usage - *i.e.* using the mouse to navigate among images versus using it to position the focused region.

In order to assess the effect of using the RFV itself upon debugging performance, each participant also debugged an equivalently-matched program using the SDE with the RFV disabled. Experimental program versions were counterbalanced in RFV/non-RFV conditions across participants.

The data collected by the RFV consists of a log of mouse and keyboard actions, as well as the times taken by participants to perform the debugging sessions[1]. Additional data recorded included the number of program errors (bugs) identified by subjects, their description and the order in which they were identified.

2.2   Participants and Procedure

The experimental participants were four DPhil students and one professional programmer. All of them knew Java and the four students were using it in coursework projects. Table 1 gives details of the participants programming experience.

Table 1. Programming experience of participants

Participant	Java experience (in months)	General programming experience (in years)
1	6	7
2	24	12
3	3	5
4	1.5	6.5
5	36	6

Participants performed three debugging sessions. The first one was a warm-up session and it was performed under the restricted focus condition. The two main sessions followed — one with and the other without the restricted focus condition (order was counterbalanced across participants). Participants were allowed as much time as they needed in each of the sessions. They were instructed to find as many errors as they could in the programs. The debugging sessions consisted of two phases. In the first phase participants were presented with a specification of the target program. This program specification consisted of two paragraphs describing in plain English the problem that the program was intended to solve, the way it should solve it (detailing the solution steps, specifying which data structures to use and how to handle them), together with some samples of program output (both desired and actual). When participants were clear about the task that the program should solve and also how it should be solved, they moved on to the second phase of the session.

---

[1] These data are also used as input to a screen movie capture mode for 're-plays' post-session.

```
import    java.io.*;
public class Till {
    private MoneyPile[] piles;

    public Till () {
        piles = new MoneyPile[7];
        piles[0] = new MoneyPile("1p",.01);
        piles[1] = new MoneyPile("2p",.02);
        piles[2] = new MoneyPile("5p",.05);
        piles[3] = new MoneyPile("10p",.1);
        piles[4] = new MoneyPile("20p",.2);
        piles[5] = new MoneyPile("50p",.5);
        piles[6] = new MoneyPile("1 pound",1.0);
    }

    public void add(Coin c) {
        for (int i=0; i<piles.length; i++) {
            if (c.label.equals(piles[i].coin_type))
                piles[0].add(c);
        }
    }

    public void count() {
        double total = 0;
        double pile_total;

        for (int i=0; i<piles.length; i++) {
            pile_total = piles[i].n_coins * piles[i].coin_value;
            System.out.println(piles[i].n_coins+" "+ piles[i].coin_type+
                            " coins is "+ pile_total+ " pounds");
        }
        System.out.println("The total is: "+ total+" pounds");
    }

    public static void main(String args[])  throws IOException {
        Till myTill = new Till();
        boolean end_of_coins = false;
        BufferedReader in = new BufferedReader
            (new InputStreamReader(System.in));

        while (!end_of_coins) {
            String coin_type = in.readLine();
            if (coin_type.equals("end"))
                end_of_coins = true;

            Coin coin = new Coin(coin_type);
            myTill.add(coin);
        }
        System.out.println("Counting the till contents: ");
        myTill.count();
    }
}
```

Fig. 2. Code for the Till class.

In the second phase they were presented with three windows containing the program code, a sample interaction with the program and a visualisation which illustrated this interaction graphically. They were instructed to identify as many errors as possible in this program. When subjects reported that they thought they had detected all of the errors they moved on to the next debugging session.

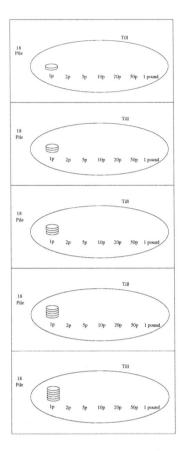

```
rsunx% java Till
5p
1p
2p
5p
1 pound
end
unknown coin: end


Counting the till contents:
5 1p coins is 0.05 pounds
0 2p coins is 0.0 pounds
0 5p coins is 0.0 pounds
0 10p coins is 0.0 pounds
0 20p coins is 0.0 pounds
0 50p coins is 0.0 pounds
0 1 pound coins is 0.0 pounds
The total is: 0.0 pounds

rsunx%
```

Fig. 3. Output from a sample execution session of the *Till* program.

Fig. 4. Functional visualisation of a sample execution session of the *Till* program.

The target programs consisted of three short Java programs. The 'warm-up' session program detects whether a point is inside a rectangle, given the co-ordinates of the point and the vertices of the rectangle. The first experimental program prints out the names of the children of a sample family. The second experimental program ('Till') counts the cash in a cash register till, giving subtotals for the different coin denominations. Some of the code, output for a sample execution session and a functional visualisation to this execution for the *Till* program are shown in Figures 2, 3 and 4 respectively.

The programs of the two main debugging sessions were seeded with three errors, and the 'warm-up' session's program was seeded with two errors. The errors of the main debugging sessions programs can be classified as 'functional', 'control-flow' and 'data structure'. In this classification, functional errors are those that occur in the line or lines in which the main computation of the program is performed [21]. For example, in the *Till* program of Figure 2, the functional error can be found in the method count, where the grand total of the money being counted is not computed.

Control-flow errors have to do with the execution of the program not following a correct path. For example, the control-flow error in the *Till* program is located in the two last lines of the *while* loop of its *main* procedure. These two lines should be included within an *else* structure, so that the execution of the program either acknowledges an end-of-coins case or adds the new coin to the till, but never follows both paths at the same time.

Data structure errors normally have undesired consequences for the program data structures. For the *Till* program of Figure 2, the data structure error is located within the only instruction of the *if* structure of the *add* method. This error consists of every coin added to the till being sent only to the first money pile, regardless of its type. In this way, the money pile receiving all coins is one which should only accumulate coins of a one-pence denomination.

3   Results

3.1   Debugging Performance

The results of the experiment in terms of debugging performance are presented in Table 2. It can be seen that although the restricted view condition slowed down the debugging performance of some participants, the number of errors found did not seem to be affected by this experimental condition (participants tended to spot more errors working under the restricted view condition). This result to some extent replicates that of [3], which compared eye-tracking with the RFV and found that response times were generally slower for the RFV, but other aspects of performance were less affected. The results reported here represent a more controlled validation of the RFV than that of [3], since in that study RFV data from a diagram inspection task were compared to eye-tracking data of an earlier study conducted by another investigator. In the current investigation a within-subject design ensured that participants served as their own controls and parallel forms of the tasks were counterbalanced across subjects for the RFV/non-RFV conditions.

Table 2 shows that in terms of number of errors spotted and for the two main sessions, the most successful participant was number 2 (5 errors found), then participants 3 and 4 (4 errors spotted each), then 5 with 3 errors located and finally participant 1 with only one error found.

3.2   Debugging Behaviour

The global experimental results in terms of debugging behaviour are shown in Table 3. This table presents the percentage of time that participants spent looking at each representation when working in the restricted view condition. Additionally, this table also presents the average number of switches per minute between the SDE representations.

Table 2. Number of errors found and time taken for the three debugging sessions for each participant (the warm-up program was seeded with two errors and the other two with three errors)

Participant	Warm-up		RFV on		RFV off	
	Errors found	Time	Errors found	Time	Errors found	Time
1	0/2	19.53	1/3	5.9	0/3	10.22
2	2/2	19.24	3/3	36.28	2/3	6.28
3	2/2	29.58	3/3	17.23	1/3	9.56
4	1/2	22.46	2/3	33.42	2/3	21.26
5	2/2	12.25	2/3	17.34	1/3	13.12

Table 3. Percentage of time spent in each representation and average number of switches per minute for each participant.

Participant	Code	Visualisation	Output	Switches per minute
1	82.5	6.3	11.2	1.55
2	92.2	5.9	1.9	1.54
3	95.5	2.5	2.0	1.44
4	87.5	7.6	4.9	2.19
5	80.8	7.1	12.1	1.93

The more successful participants (2,3 & 4) spent a longer amount of time focusing on the code representation compared to the less successful ones (1 & 5). No clear pattern seems to exist relating the average number of switches per minute to debugging performance in this representation of the data. However, when the averages are represented as annotations to cyclic directed graph depictions of switches between representations, a distinctive pattern emerges.

Table 4. Number of switches per minute for each switch type.

Participant	Code to visualis.	Visualis. to code	Code to output	Output to code	Visualis. to output	Output to visualis.
1	1.54		1.54	2.31	0.77	
2	1.98	1.98	0.99	1.1	0.11	0.11
3	2.07	2.07	0.23	0.46	0.46	0.23
4	2.84	3.08	1.19	1.07	0.24	0.48
5	2.05	2.05	1.82	1.37	0.23	0.23

The different types of switches considered for this analysis are: a switch from the code representation to the visualisation, from visualisation to code, code to output representation, output to code, visualisation to the output and output to visualisation. Table 4 presents the results in these terms. For each participant, the number of switches per minute is reported. Figures 5, 6, 7, 8 and 9 present these data in the form of cyclic directed graphs for each participant, with the frequency of switches normalised. In each of the figures, the frequency of code-visualisation, visualisation-output, and code-output switches are represented by annotated directional arrows. From these Tables and Figures it can be observed that the most successful participants (2, 3 & 4) performed more frequent switches between the code and the visualisation than the less successful ones (1 & 5).

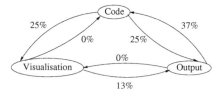

Fig. 5. Number of switches per minute for each kind of switch for participant 1.

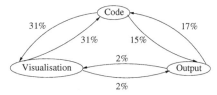

Fig. 6. Number of switches per minute for each kind of switch for participant 2.

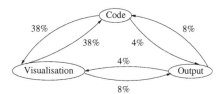

Fig. 7. Number of switches per minute for each kind of switch for participant 3.

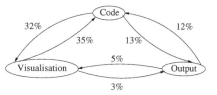

Fig. 8. Number of switches per minute for each kind of switch for participant 4.

4   Discussion

This investigation aimed to relate debugging performance to representation use in a multi-representational, multi-modal debugging environment similar to those found in commercial software development environments and software visualisation packages [23]. These sorts of environment are characterised by having several concurrently displayed representations of the program. There is a central representation, the program code, and a series of secondary representations that support it (program output and execution visualisations). Because software debugging environments are an important tool for novice programmers, modeling the process of representation use in this sort of environment can be of central relevance for educational purposes.

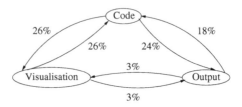

Fig. 9. Number of switches per minute for each kind of switch for participant 5.

Several hypotheses provide a basis for an initial descriptive model. First, better performing participants quickly identified the code representation as the central one and devoted a high percentage of inspection time to it. Secondly, the global frequency of switching *per se* does not seem to be related to good debugging performance. Successful and unsuccessful performance share similar levels of switching frequency among all the available representations. This differs from experimental results obtained in studies of representational behaviour in the domain of analytical reasoning [8,7]. Those studies found that poor performers switched more frequently than successful ones. However, there are several differences between those studies and the one reported here. Although analytical reasoning as a cognitive task might be remarkably similar to program comprehension, the analytical reasoning studies encouraged participants to build their own representations. Therefore, switching representations represented 'a strategic decision by the subject to abandon the current external representation and construct a new one' [8]. In the present study, representations were complementary rather than alternative, therefore, switching did not represent discarding one representation for another, but complementing the information of one with another. Also, representations in the analytical reasoning study were presented serially, while the ones of the study reported here were displayed concurrently.

Our third hypothesis is related to the frequency of switching but it considers the different types of switches between the representations. A relatively high frequency of switches between the code and visualisation representations seems to be related to good debugging performance. This seems to be in agreement with findings in the program comprehension area that suggest that experienced programmers, when comprehending a program, are able to develop a mental representation that consists of different program perspectives, as well as rich mappings between them [19]. It also suggests that efficient debugging performance using SDE's is associated with the exploitation of representations that differ maximally in terms of their modality and expressiveness [24], *i.e.* in other words, good software bug diagnosticians tend to be heterogeneous reasoners [2].

The finding that performance, in terms of number of bugs detected, was superior in the RFV condition (compared to the non-RFV condition) was unexpected. This finding might be explained by at least two different causes (or by a combination of them). The first is that such a result might be explained by two features of the modified RFV which may have provided a degree of additional user-support.

The amount of visual search performed by users in the RFV condition may have been reduced in two ways: a) by the RFV's blurring of unused parts of the display - thus reducing visual clutter, and, b) by the RFV's window-by-window location memory for the user's 'last-attended-to-position' that re-instated the 'fovea' on each switch back to a particular representation. The second explanation assumes that restricting the visual focus of the screen made local action more effortful. As suggested in [17], increasing the cost of performing the task could have increased the level of planning, which in turn could have enhanced the task performance.

5   Conclusions

This study investigated Java program debugging performance and behaviour through the use of a software debugging environment that provided concurrently displayed, adjacent, multiple and linked representations and that allowed visual attention switches of participants to be tracked. The experimental results allowed us to propose several hypotheses which can be considered as a first step in building a preliminary descriptive model of the process of representation use for program debugging. Also, the modified version of the RFV was validated for use in the program representation domain².

The preliminary hypotheses about representation use in program bug diagnosis need to be further tested with a larger sample population and an experimental design that takes into account several important issues. For example, this investigation has only considered functional visualisations, while commercial software development environments also offer visualisations of data structure and controlflow [23]. Considering different kinds of visualisations would allow us to relate debugging performance of bug type to type of representation.

Another issue concerns the co-ordination of uni-modal versus multi-modal representations. Some evidence suggests that co-ordinating representations of different modality seems to be more complicated than co-ordinating those of the same perspective [1]. However, it is not clear how modality differences between the representations affect representational behaviour. These questions will be investigated in further work.

Acknowledgments. This work is supported by the EPSRC grant GR/N64199.
The authors would like to thank the referees for their helpful comments and the participants for taking part in the study.

---

² Like the original RFV, our modified version has also been placed in the public domain for use by the diagrammatic research community. It can be downloaded from `http://www.cogs.susx.ac.uk/projects/crusade/`

References

1. S. Ainsworth, D. Wood, and P. Bibby. Co-ordinating multiple representations in computer based learning environments. In P. Brna, A. Paiva, and J. Self, editors, *Proceedings of the 1996 European Conference on Artificial Intelligence on Education*, pages 336–342, Lisbon, Portugal, 1996.
2. J. Barwise. Heterogeneous reasoning. In G. Allwein and J. Barwise, editors, *Working papers on diagrams and logic*. Visual Inference Laboratory, Indiana University, 1993.
3. A. Blackwell, A. Jansen, and K. Marriott. Restricted focus viewer: a tool for tracking visual attention. In M. Anderson, P. Cheng, and V. Haarslev, editors, *Theory and Application of Diagrams. Lecture Notes in Artificial Intelligence 1889*, pages 162–177. Springer-Verlag, 2000.
4. D. Bergantz and J. Hassell. Information relationships in PROLOG programs: how do programmers comprehend functionality? *International Journal of Man-Machine Studies*, 35:313–328, 1991.
5. M. D. Byrne, R. Catrambone, and J. T. Stasko. Evaluating animations as student aids in learning computer algorithms. *Computers & Education*, 33:253–278, 1999.
6. C. L. Corritore and S. Wiedenbeck. Mental representations of expert procedural and object-oriented programmers in a software maintenance task. *International Journal of Human Computer Studies*, 50:61–83, 1999.
7. R. Cox. *Analytical reasoning with multiple external representations*. PhD thesis, University of Edinburgh, Edinburgh, Scotland, U.K., 1996.
8. R. Cox and P. Brna. Analytical reasoning with external representations: Supporting the stages of selection, construction and use. *Journal of Artificial Intelligence in Education*, 6(2/3):239–302, 1995.
9. T. de Jon, S. Ainsworth, M. Dobson, A. van der Hulst, J. Levonen, and P. Reimann. Acquiring knowledge in science and mathematics: The use of multiple representations in technology-based learning environments. In M. W. van Someren, P. Reimann, H. P. A. Boshuizen, and T. de Jon, editors, *Learning with Multiple Representations*, pages 9–40. Elsevier Science, Oxford, U.K., 1998.
10. M. Eisenstadt, M. Brayshaw, and J. Paine. *The Transparent Prolog Machine*. Intellect, Oxford, England, 1991.
11. D. J. Gilmore. Models of debugging. *Acta psychologica*, 78(1):151–172, 1991.
12. D. J. Gilmore and T. R. G. Green. Comprehension and recall of miniature programs. *International Journal of Man-Machine Studies*, 21(1):31–48, 1984.
13. J. Good. *Programming Paradigms, Information Types and Graphical Representations: Empirical Investigations of Novice Program Comprehension*. PhD thesis, University of Edinburgh, Edinburgh, Scotland, U.K., 1999.
14. S. P. Lahtinen, E. Sutinen, and J. Tarhio. Automated animation of algorithms with eliot. *Journal of Visual Languages and Computing*, 9:337–349, 1998.
15. D. C. Merrill, B. J. Reiser, R. Beekelaar, and A. Hamid. Making processes visible: scaffolding learning with reasoning-congruent representations. *Lecture Notes in Computer Science*, 608:103–110, 1992.
16. P. Mulholland. Using a fine-grained comparative evaluation technique to understand and design software visualization tools. In S. Wiedenbeck and J. Scholtz, editors, *Empirical Studies of Programmers, seventh workshop*, pages 91–108, New York, 1997. ACM press.
17. K. P. O'hara and S. J. Payne. Planning and the user interface: the effects of lockout time and error recovery cost. *International Journal of Human Computer Studies*, 50:41–59, 1999.

18. M. J. Patel, B. du Boulay, and C. Taylor. Comparison of contrasting Prolog trace output formats. *International Journal of Human Computer Studies*, 47:289–322, 1997.
19. N. Pennington. Comprehension strategies in programming. In G. M. Olson, S. Sheppard, and E. Soloway, editors, *Empirical Studies of Programmers, second workshop*, pages 100–113, Norwood, New Jersey, 1987. Ablex.
20. N. Pennington. Stimulus structures and mental representations in expert comprehension of computer programs. *Cognitive Psychology*, 19:295–341, 1987.
21. R. S. Rist. Schema creation in programming. *Cognitive Science*, 13:389–414, 1989.
22. P. Romero. Focal structures and information types in Prolog. *International Journal of Human Computer Studies*, 54:211–236, 2001.
23. Romero, P., Cox, R., du Boulay, B. & Lutz, R. A survey of representations employed in object-oriented programming environments. (in press) *Journal of Visual Languages and Computing*.
24. K. Stenning and J. Oberlander. A cognitive theory of graphical and linguistic reasoning: logic and implementation. *Cognitive Science*, 19(1):97–140, 1995.
25. I. Vessey. Toward a theory of computer program bugs: an empirical test. *International Journal of Man-Machine Studies*, 30(1):23–46, 1989.
26. S. Wiedenbeck and V. Ramalingam. Novice comprehension of small programs written in the procedural and object-oriented styles. *International Journal of Human Computer Studies*, 51:71–87, 1999.
27. S. Wiedenbeck, V. Ramalingam, S. Sarasamma, and C. L. Corritore. A comparison of the comprehension of object-oriented and procedural programs by novice programmers. *Interacting with Computers*, 11:255–282, 1999.

Guiding Attention Produces Inferences in Diagram-Based Problem Solving

Elizabeth R. Grant[1] and Michael J. Spivey[2]

[1] Cornell University, Department of Human Development, Ithaca, NY 14853, USA
erg6@cornell.edu
[2] Cornell University, Department of Psychology, Ithaca, NY 14853, USA

Abstract. Many eye-tracking studies have shown that visual attention patterns during diagram-based problem solving, measured by eye movements, reveal critical aspects of the problem solving process that traditional measures like solution time and accuracy cannot address. In our first experiment (n = 14), we use this method during the solution of a widely-studied high level reasoning problem, Duncker's (1945) radiation problem, to show that differences in visual attention to a particular diagram feature corresponds with correctly solving the problem. We then extend these findings in a second experiment (n = 81) to evaluate cognitive sensitivity to perceptual changes in the diagram. We show that problem solvers are highly sensitive to the diagram structure, and that the shifts in attention that result from subtle perceptual changes in the diagram appear to have a dramatic positive effect on reasoning.

1 Eye Movements and Problem Solving

When solving academic and everyday problems, we often rely on external visual stimuli to help us analyze and infer the answer. This behavior requires the interaction between attention to an external visual representation and the highest levels of thought [31]. If we assume that eye movements reflect visual attention to a diagram, then we can analyze how people look at diagrams and how their looking patterns correlate with inference-making. This theory already has some evidence in its favor [6]; [8]; [15]; [26]. In this work, we extend this method to a widely-studied reasoning problem, Duncker's [7] radiation problem, to see how attention to a schematic diagram corresponds to reasoning accuracy. Then, we ask a novel question that evaluates the reverse effect: how shifts of visual attention to critical diagram features might themselves affect inference-making. We know that cognition often directs attention, but can attention sometimes direct cognition? In other words, if we lead one's eyes to fodder, can we make one think?

The usefulness of diagrams to support reasoning has recently received much attention across domains, particularly since Larkin and Simon's [19] comparison of diagrammatic and sentential representations of information. Domains of study have included mathematical reasoning [2]; [3]; [29], logical reasoning [4], analogical transfer [5]; [9]; [10]; [11]; [24], and insight problem solving [18].

M. Hegarty, B. Meyer, and N. Hari Narayanan (Eds.): Diagrams 2002, LNAI 2317, pp. 236–248, 2002.
© Springer-Verlag Berlin Heidelberg 2002

Early studies of eye movements during diagram-based problem solving reported some rudimentary relationships between reasoning strategies and eye movements [16]; [20]; [23]. Results from more recent eye movement studies have also shown that visual attention is related to the inferential process during reasoning. Studies have found evidence that task-specific eye movement patterns reflect strategic cognitive phases across the time course of a problem solution during geometric reasoning [8] during the Tower of London problem, used to test impairments following frontal lobe damage [15], and during insight problem solving [18]. In the realm of mechanical reasoning, [13] found that participants tended to look at diagram components important to the causal sequence referred to in a corresponding text; [26], found that visual scanpaths reflected mechanical diagrams' causal sequence and correlated with solution accuracy. [14] have linked use of a diagram in problem solving to the complexity of the problem. Most recently, [18] have established a connection between eye movements to elements of physically-depicted insight problem and achieving insight after an impasse.

Given this strategic reliance on visual representations during the problem solving process, how sensitive are humans to slight changes in representational structure? According to [19], humans are sensitive to whether information is spatially or linguistically presented. Diagrammatic representations are computationally faster to use than sentential representations because they use spatial location to depict relationships between elements of a problem. One explanation for this benefit is that diagrams represent information in an analogous way to cognitive and perceptual representations in the mind. This would make relevant information easy to access and would lessen memory demands by providing an efficient on-line reference [17]; [21]; [22]. In contrast, sentential or propositional representations preserve the "temporal or logical sequence" between components of a problem ([19], p. 66). Thus, they guide the reader through the components of the whole to make constrained inferences in order to reach the solution, whereas the diagrams we refer to in this paper do not, when they depict the relations between components of real or abstract objects (e.g., maps, anatomical depictions, or geometric shapes) or systems (e.g., gear-and-belt or rope-and-pulley physics problems). These types of spatial diagrams serve as unified representations without guiding attention to any particular components in sequence, unlike hierarchical diagrams and flowcharts.

If we find that humans are indeed sensitive to the way information is structured, then the subtle guidance of attention within a visually-represented problem could provide an inferential advantage. In a follow-up study to Gick & Holyoak (1983), [24] animated Gick & Holyoak's source analogue priming diagrams and significantly improved participants' performance. Our goal was to see if a very different kind of animation, of perceptual diagram features rather than conceptual ones, could also improve participants' performance.

1.1 Duncker's Radiation Problem

Karl Duncker's radiation problem, a typical and fairly difficult reasoning problem, was used in the following two experiments to evaluate the interaction between perceptual feature salience, attention, and inference-making during diagram-based problem solving. Figure 1 depicts the diagram participants viewed while generating solutions to the following instructions (diagram and instructions adapted from [7]),

> Given a human being with an inoperable stomach tumor and lasers which destroy organic tissue at sufficient intensity, how can one cure the person with these lasers and, at the same time, avoid harming the healthy tissue that surrounds the tumor?

The solution requires firing multiple weak lasers from several angles outside the healthy tissue so that they intersect at the tumor, where their combined intensity is enough to destroy the tumor. The diagram is divided into four areas

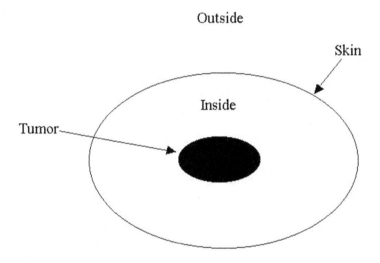

Fig. 1. The diagram that participants viewed while solving Duncker's (1945) radiation problem. Diagram feature labels were not shown; they were verbally explained.

of relevance to the solution. The center black oval represents the tumor and is the target for the lasers. The white area outside the center oval and within the outside oval represents the healthy tissue, which should not be damaged. The outside black oval represents the skin that surrounds the healthy tissue inside it, through which the lasers must innocuously pass. The white area outside the outer oval represents the outside area from which multiple lasers must be positioned and fired at different angles in relation to the tumor. We have called these four areas, respectively, the Tumor, the Healthy Tissue, the Skin, and the Outside.

The lasers are not depicted in the diagram; their sources and trajectories must be imagined.

Contrary to many diagram-based problems, Duncker's radiation diagram reduces the confound between visually-interesting and task-informative areas. The diagram is simple and idealized. Eye movements that are nonrandom and those that extend beyond the boundaries of the visual stimulus would implicate the influence of task-relevant cognition rather than purely stimulus-bound visual processing. At the same time, all three depicted components (tumor, healthy tissue, skin) and the one non-depicted component (outside) are meaningful within the context of the problem. Interestingly, solving this problem requires attention outside the scope of visual depiction, as only the lasers, positioned in the blank white outside area, can be adjusted in position and intensity to solve the problem correctly. Solutions that attempt to manipulate the tumor itself, the healthy tissue, or the skin, by slicing at it, inserting tubes, adjusting the tumor's position within the healthy tissue, for example, are incorrect.

In Experiment 1, we monitored participants' eye movements while they solved the radiation problem. The goal of this study was to identify any locus of visual attention that correlated with a successful solution.

This study tested the following two hypotheses:

1) that looking patterns would be non-random, and

2) that successful and unsuccessful problem solvers would show different looking patterns.

2   Experiment 1

2.1   Method

Participants. Fourteen Cornell University students with normal or corrected-to-normal vision participated for course credit. Individuals who were familiar with Duncker's radiation problem did not participate. One additional subject never reached the solution and was excluded.

Apparatus. Participants were seated approximately 30 cm away from a vertical white marker board. The board was affixed with a clear overhead transparency sheet printed with a two-dimensional static diagram whose area equaled approximately 30x30 degrees of visual angle. Participants' eye movements were monitored using a lightweight ISCAN headband-mounted eyetracker, which allowed participants' heads to move naturally. Viewing was binocular, and eye position was recorded from the left eye with an accuracy of about 0.5 degrees, sampled at 60 Hz. Eye position, verbal protocols, drawings, and solution times were recorded on a Sony Hi-8 VCR with 30 Hz frame-by-frame playback.

Materials and Design. The diagram from Duncker's radiation problem was used for all participants (see Figure 1).

Procedure. Participants were tested individually by the same experimenter in a laboratory with controlled lighting. The eye tracker was placed on each participant's head and was calibrated before the task began by having each participant look sequentially at a grid of 8 black dots surrounding the diagram, at the inside oval of the diagram, and at one or more points along the outside oval of the diagram. After calibrating the eye tracker, which lasted approximately 5-8 minutes, participants were allowed to move their heads normally.

Participants were asked to give a verbal protocol of their solution. They were also asked to draw their solutions with dry-erase markers on the diagram in order to confirm the accuracy of their spoken solutions (e.g., placement of laser sources at appropriate angles). The experimenter then read the problem instructions aloud and explained how the elements of the diagram correspond to the elements of the problem. During the solution attempt, the experimenter answered direct questions participants asked about the problem. The task ended after 10 minutes if the correct solution had not been produced, although, in order to create an equivalent time segment for coding eye movements immediately prior to the solution, unsuccessful participants were read one or more hints after 10 minutes had elapsed in order to allow them to reach the solution. Hint 1 read, "What if you could adjust the intensity of the lasers?" Hint two read, "What if you had more than one laser?" The task was ended and hints were given before 10 minutes only if participants repeatedly stated that they could not generate any further solutions. During long pauses in the protocols, the experimenter asked participants to "Think aloud."

Coding. Accuracy was defined as a binomial variable. Participants who spontaneously inferred the solution before the task ended (within 10 minutes) composed the Successful group. Participants who failed to reach the solution in 10 minutes and solved the problem only with hints, composed the Unsuccessful group. Subjects who received hints and still failed to solve the problem were excluded.

Eye movements were analyzed by coders who were blind to participants' solution times and accuracy (intercoder reliability ranged from 90%-100% across time). Two 900-frame time segments were coded for each participant: the 30 seconds after they heard the instructions (beginning time segment) and the 30 seconds before they stated and drew the correct solution (end time segment). This moment of insight was clearly identifiable, typically marked by an intake of breath and a comment like "Aha!" "Oh, I know", or "OK, I have it," followed by simultaneously drawing and explaining the correct solution. Although they may have drawn prior to this, those drawings did not depict the correct solution.

Coders assigned the position of each participant's gaze during each video frame (sampled on the VCR at 30 Hz) of these two 30-second time segments to one of five mutually-exclusive and exhaustive diagram location codes: Tumor, Healthy Tissue, Skin, Outside, or Irrelevant, where Irrelevant denoted eye-position crosshairs either absent from the screen or focused on an object other than the diagram, such as the participants' hand or the marker.

2.2   Results and Discussion

Accuracy. Thirty-six percent (n = 5) of participants solved the problem successfully, whereas 64% (n = 9) of participants were unsuccessful.

Eye Movements. One participant from the successful condition was excluded from the eye-movement analysis due to an audio recording failure. Analyses were conducted to compare the eye movement patterns of successful problem solvers with unsuccessful problem solvers. The goal was to discover if successful problem solvers' eye movements could be characterized by a higher proportion of time spent looking at particular areas of the diagram.

Using the coding system specified above, proportions of time spent looking at each area of the diagram were calculated for each participant for both the beginning and end time segments. Proportions were normalized to eliminate frames coded as Irrelevant. Figure 2 shows the mean proportion of time that successful and unsuccessful participants spent looking at each area of the diagram at the beginning and end of the task. For the beginning period during the first 30 seconds after the instructions, t-test comparisons of the mean proportions of time the successful and unsuccessful groups spent looking at each area of the diagram indicated no significant differences. However, during the end period during the 30 seconds before inferring the solution, t-test comparisons of mean looking times indicated that successful participants spent significantly more time looking at the Skin area of the diagram than unsuccessful participants did; $t(12) = 2.734, p < .02$. No other significant differences between unsuccessful and successful groups were found in the proportions of time spent looking at the Tumor, Healthy Tissue, or Outside areas of the diagram. Results from Experiment 1 indicated that a higher proportion of frames spent looking at the Skin area of the diagram characterized the eye movements of successful versus unsuccessful problem solvers. Given this, we considered this Skin area to be a critical diagram feature that corresponded with inferring the solution to the problem.

Thus, the eye movement results in Experiment 1 provided the specific information we needed to further explore the relationship between attention and inference-making. Based on these results, our goal in Experiment 2 was to test the sensitivity of problem solvers to changes in the structure of the visual representation and the possibility of capitalizing on this sensitivity by manipulating attention within the diagram to produce inferences. Specifically, we evaluated the hypothesis that participants who use a diagram that increases the perceptual salience of the critical diagram feature would produce more successful solutions than those who use a static diagram or a diagram that highlights a noncritical feature. It could be that increased perceptual salience of the critical diagram feature has a bottom-up influence that, in interaction with the visual system, increases the likelihood of generating the correct inferences to solve the problem.

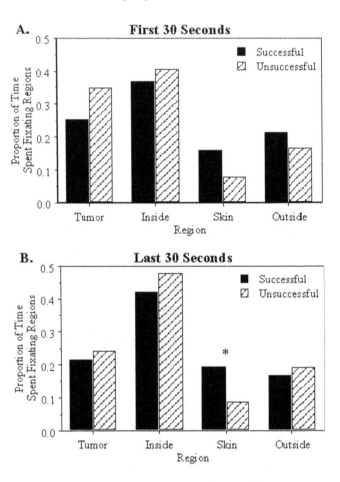

Fig. 2. Mean normalized proportion of looking time by area during the first thirty seconds of viewing (panel A) and the last thirty seconds of viewing (Panel B) in Experiment 1. The asterisk (*) indicates a significant difference between successful and unsuccessful problem solvers.

3   Experiment 2

In the baseline condition of this experiment, one third of the participants faced Duncker's radiation problem with the static diagram from Experiment 1. Another third of the participants faced the problem using an animated diagram that highlighted the critical feature we discovered in Experiment 1, the oval perimeter that represents the skin. As an additional control condition, the final third of participants faced the problem using an animated diagram that highlighted the tumor, a non-critical area based on results from Experiment 1. This study tested the hypothesis that using a diagram that drew attention to the critical feature in the diagram would increase the frequency of correct solutions compared to using a static diagram or a diagram that drew attention to a non-critical diagram feature.

3.1   Method

Participants. Eighty-one Cornell University undergraduates with normal or corrected-to-normal vision participated for course credit, 27 in each of the three conditions. As in Experiment 1, participants were screened for familiarity with Duncker's radiation problem. The following additional subjects were excluded: 5 never reached the solution in the static condition, and 7 never reached the solution in the blinking skin condition. Data for excluded subjects in the Blinking Tumor condition are not available.

Apparatus. Participants were seated approximately 30 cm away from a Macintosh computer with a 20" display that depicted the tumor diagram. The diagram was identical to that in Experiment 1 but in this case was depicted on the computer screen. The computer screen was covered with a sheet of transparency plastic so participants could draw on the diagram. Participants' solutions were videotaped using the apparatus described in Experiment 1, but eye movements were not recorded.

Materials and Design. Participants were randomly assigned to one of three conditions: Static diagram, Animated Skin diagram, and Animated Tumor diagram. In the Static condition (n = 27), the diagram appeared fixed, as in Experiment 1. In the Animated Skin condition (n = 27), the diagram's Skin area subtly "pulsed" by increasing and decreasing its thickness by one pixel three times per second. In the Animated Tumor condition (n = 27), the diagram's Tumor area subtly "pulsed" in the same pattern as in the Animated Skin condition, by increasing and decreasing its outer edge by one pixel three times per second.

Procedure. Procedures were identical to those in Experiment 1, except that eye movements were not recorded.

3.2   Results and Discussion

Figure 3 shows the percentage of successful and unsuccessful participants in the three conditions of Experiment 2 compared with accuracy results from Experiment 1. Highly similar to the participants in Experiment 1, who also used a static diagram, 63% (n = 17) of participants in the Static condition were unsuccessful and 37% (n = 10) were successful (in Experiment 1, 64% were unsuccessful and 36% were successful). In the Animated Tumor condition, 67% (n = 18) were unsuccessful and 33% (n = 9) were successful. However, in the Animated Skin condition, 33% (n = 9) were unsuccessful and 67% (n = 18) were successful. Chi-square tests indicated significantly more successful solutions produced by participants in the Animated Skin condition than in the Static condition [$\chi^2(1, N = 54) = 4.747, p < .05$] and than in the Animated Tumor condition [$\chi^2(1, N = 54) = 6.000, p < .05$]. We found no significant differences between success rates in the Static and Animated Tumor conditions. Using a diagram that drew

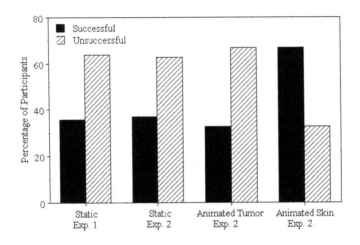

Fig. 3. Percent of successful and unsuccessful problem solvers who used Static diagrams in Experiments 1 and 2 and Animated diagrams in Experiment 2.

attention to the critical feature identified in Experiment 1 reversed the accuracy pattern shown by those in the Static conditions of Experiments 1 and 2 and in the Animated Tumor condition of Experiment 2. Using a diagram that subtly highlighted the critical area significantly increased the frequency of inferred solutions.

4   General Discussion

In Experiment 1, we found that while participants dealt with Duncker's radiation problem, their eye movements were nonrandom and discriminated successful from unsuccessful problem solvers. In Experiment 2, we demonstrated that a subtle increase in perceptual salience of a critical diagram component produced a higher frequency of accurate solutions. We review possible explanations and propose several directions for further research to illuminate this striking context effect on high level cognition.

Concerning results from Experiment 1, eye movements during this task were non-random and seemed to reflect task-related thought. Participants' eye movements frequently roved beyond the scope of the visual stimulus to blank areas of the screen; thus, fixations were not solely driven by the depicted visual stimulus. As suggested by related eye-movement work on memory and imagery [25]; [27], it seems as though eye position may be used to coordinate spatial locations in the visual field with internally represented elements from the mental model. In addition, both the successful and unsuccessful groups spent varied amounts of time looking at each diagram area; eye movements were not uniformly distributed between the three depicted diagram areas or even between those and the additional blank Outside area. Indeed, compared to those who failed to solve the problem, those who solved the problem spent more time looking at one particular diagram feature: the Skin. Based on these results, we propose that eye movement patterns correlated with the problem solving process.

Beyond a correlation, however, our results go a step further and suggest that attentional shifts to a critical diagram feature can actually cause correct inference-making. The results of both experiments suggest that attending to the Skin characterized correct solutions and that attracting attention to this area indeed helped people solve the problem. What valuable information or looking pattern does the Skin represent that could so dramatically create inferences? Or, why is the critical area critical? We propose two possible explanations and encourage further research to elucidate this strong context effect.

One possible explanation for our results is that the Skin contains information necessary to solve the problem and that drawing attention to it helped subjects recognize this information. While one third of the subjects recognized this without the help of an animated diagram, an additional third only recognized it when we drew their attention to it via animation. The Skin area represents the relationship between the outside and the healthy tissue and is the point at which the lasers can begin to harm healthy tissue. If the Skin area was inherently informative, however, one might expect it to have captured a higher relative percentage of looks relative to the other diagram areas than it actually did for both the successful and unsuccessful groups, particularly because eye movements were coded for the time segment immediately preceding the solution for both groups. In fact, the skin area captured the smallest percentage of looks in the unsuccessful group and was approximately equal to looks to the Outside in the successful group. This is not compelling evidence that subjects looked at the Skin because it provided them with important information, per se, but our results do not rule out this possibility. If this were the case, then we could argue that the diagram that animated the critical area served as a hybrid information representation that both spatially depicted relationships between elements of the problem and, at the same time, created a hierarchy of informativeness among the individual diagram features that the static diagram lacked. However, the small percentage of looks to the Skin suggests that participants were not simply staring at the Skin to absorb whatever information it could provide. Rather, the Skin was necessarily crossed during attentional transitions between the Tumor and the

Outside. This attentional pattern, and the failure of the feature informativeness hypothesis to explain our results, leads to a second possible explanation that requires further investigation.

Our data allow the possibility that the shift in attention between the tumor or healthy tissue and the outside could have produced problem solving insights. Spivey and Geng [27] have found that the eyes will simulate directional movement in a mental model even when no visual stimulus is present. It could be that animating the Skin drew attention away from the tumor and healthy tissue, which are unchangeable in the context of the problem, and toward the outside, from which the lasers must be fired. Our qualitative observations suggest that eye movements of successful participants tended to reflect such attentional patterns in frequent triangular "in-and-out" eye movement patterns from a point Outside to the Skin, then to the Tumor, and back out to a different point Outside, which corresponds directly to the correct solution, which requires firing two lasers from different points Outside so that the lasers intersect at the Tumor. However, further experiments are needed to explore this hypothesis.

An obvious next step to speak to an attentional mechanism would be to record eye movements in a protocol similar to Experiment 2 and to compare them to verbal solution protocols. We did not record eye movements in Experiment 2 due to time considerations and the large number of subjects, and, in Experiment 1, we found that aggregate data were more reliable than subjectively-coded patterns of eye movements. We would encourage the collection of data that would discriminate fixations on the Skin versus saccades that cross the Skin while shifting attention between the Tumor or Inside and the Outside in order to further illuminate the attentional and visual mechanisms that supported inferring the correct solution.

While we have discussed this potential explanation of our results in terms of shifts in attention, based on our data we cannot rule out the more speculative but fascinating hypothesis that the triangular "in-and-out" eye-movement pattern may have served as a physical mechanism that jump-started a perceptual simulation [1] of multiple incident rays, and wound up supporting the inference that multiple lasers could be fired (at low intensities) from different points outside the diagram. This could be investigated by an experiment that requires a fixed gaze position at one location while requiring attention in another location.

In general, we would encourage further experimental investigations concerning how attention and eye movements mediate the relationship between external and internal information representations to support reasoning.

In this work, we have proposed intelligent attentional guidance as a possible way to improve reasoning in a problem solving task that relies on a diagram. Although we typically experience our eye movements as under our internal or mental control, mounting evidence in the eye movement literature shows just how intertwined the interaction between even subtle aspects of the visual environment, attention, and mental operations might be. Particularly now, as the cognitive sciences are increasingly acknowledging that cognition is an interaction between internal mental process and situated environmental constraints (e.g., [12]; [28]; [30]), this knowledge can be applied to building representational structures and interfaces that exploit our reliance on the processes that mediate the relationship between our intelligible and sensible worlds.

Acknowledgments. This research was supported by a National Science Foundation Graduate Fellowship and a Cornell University College of Human Ecology Research Grant to the first author, and by a Sloan Foundation Fellowship in Neuroscience and a grant from the Consciousness Studies Program at the University of Arizona to the second author. Thanks to Daniel Richardson, Melinda Tyler, Sybil Montas, and Antonio Javier for their assistance.

References

1. Barsalou, L. W. (1999). Perceptual symbol systems. Behavioral and Brain Sciences, 22, 577-660.
2. Barwise, J., & Etchemendy, J. (1996). Visual information and valid reasoning. In G. Allwein & J. Barwise (Eds.), Logical reasoning with diagrams (pp. 3-26). Oxford: Oxford University Press.
3. Barwise, J., & Hammer, E. (1996). Diagrams and the concept of logical system. In G. Allwein & J. Barwise (Eds.), Logical reasoning with diagrams (pp. 49-78). Oxford: Oxford University Press.
4. Bauer, M. I. & Johnson-Laird, P. N. (1993). How diagrams can improve reasoning. Psychological Science, 4, 372-378.
5. Beveridge, M., & Parkins, E. (1987). Visual representation in analogical problem solving. Memory and Cognition, 15, 230-237.
6. Demarais, A. M. & Cohen, B. H. (1998). Evidence for image-scanning eye movements during transitive inference. Biological Psychology, 49, 229-247.
7. Duncker, K. (1945). On problem solving. Psychological Monographs, 58, Whole No. 270.
8. Epelboim, J. & Suppes, P. (1997). Eye movements during geometrical problem solving. Proceedings of the 19th Annual Conference of the Cognitive Science Society. (p. 911). Mahwah, NJ: Erlbaum.
9. Gick, M. L. (1985). The effect of a diagram retrieval cue on spontaneous analogical transfer. Canadian Journal of Psychology, 39, 460-466.
10. Gick, M. L. (1989). Two functions of diagrams in problem solving by analogy. In H. Mandl & J. R. Levin (Eds.), Knowledge acquisition from text and pictures (pp. 215-231). Amsterdam: North-Holland.
11. Gick, M. L., & Holyoak, K. J. (1983). Schema induction and analogical transfer. Cognitive Psychology, 15, 1-38.
12. Greeno, J. G. (1998). The situativity of knowing, learning, and research. American Psychologist, 53, 5-26.
13. Hegarty, M. (1992). The mechanics of comprehension and comprehension of mechanics. In K. Rayner (Ed.), Eye movements and visual cognition: Scene perception and reading (pp. 428-443). New York: Springer-Verlag.
14. Hegarty, M. & Just, M. A. (1993). Constructing mental models of machines from text and diagrams. Journal of Memory and Language, 32, 717-742.
15. Hodgson, T. L., Bajwa, A., Own, A. M., & Kennard, C. (2000). The strategic control of gaze direction in the tower of London task. Journal of Cognitive Neuroscience, 12, 894-907.

16. Hunziker, H. W. (1970). Visuelle Informationsaufnahme und Intelligenz: Eine Untersuchung ueber die Augenfixationen beim Problemloesen. Schweizerische Zeitschrift fuer Psychologie, 29, 165-171.
17. Johnson-Laird, P. N. (1983). Mental models: Towards a cognitive science of language, inference, and consciousness. Cambridge, MA: Harvard University Press.
18. Knoblich, G., Ohlsson, S., & Raney, R. E. (in press). An eye movement study of insight problem solving. Memory & Cognition.
19. Larkin, J. H., & Simon, H. A. (1987). Why a diagram is (sometimes) worth ten thousand words. Cognitive Science, 11, 65-99.
20. Lenhart, R. E. (1983). Conjugate lateral eye movements and problem solving ability: Or, where to look for the right answer. Psychophysiology, 20, 456.
21. Mayer, R. E. (1976). Comprehension as affected by structure of problem representation. Memory and Cognition, 4, 249-255.
22. Mayer, R. E. (1983). Thinking, problem solving, cognition. NY: W. H. Freeman and Company.
23. Nakano, A. (1971). Eye movements in relation to mental activity of problem-solving. Psychologia, 14, 200-207.
24. Pedone, R., Hummel, J. E., & Holyoak, K. J. (2001). The use of diagrams in analogical problem solving. Memory & Cognition, 29, 214-221.
25. Richardson, D. C. & Spivey, M. J. (2000). Representation, space, and Hollywood Squares: Looking at things that aren't there anymore. Cognition, 76, 269-295.
26. Rozenblit, L., Spivey, M., & Wojslawowicz, J. (in press). Mechanical reasoning about gear-and-belt diagrams: Do eye-movements predict performance? In B. Meyer (Ed.), Diagrams and spatial reasoning. Springer.
27. Spivey, M. J., & Geng, J. J. (in press). Oculomotor mechanisms activated by imagery and memory: Eye movements to absent objects. Psychological Research.
28. St. Julien, J. (1997). Explaining learning: The research trajectory of situated cognition and the implications of connectionism. In D. Kirshner & J. Wilson (Eds.), Situated cognition: Social, semiotic, and psychological perspectives. (pp.261-280). Mahwah, NJ: Erlbaum.
29. Wheatley, G. H. (1997). Reasoning with images in mathematical activity. In L. D. English (Ed.), Mathematical reasoning: Analogies, metaphors, and images (pp. 281-297). Mahwah, NJ: Erlbaum.
30. Young, M. F. & McNeese, M. D. (1995). A situated cognition approach to problem solving. In P. Hancock & J. Flach (Eds.), Local applications of the ecological approach to human-machine systems. (pp. 359-391). Hillsdale, NJ: Erlbaum.
31. Zhang, J. & Norman, D. O. (1994). Representations in distributed cognitive tasks. Cognitive Science, 18, 87-122.

ViCo: A Metric for the Complexity of Information Visualizations

Johannes Gärtner[1], Silvia Miksch[2], and Stefan Carl-McGrath[3]

[1] Institute of Design and Assessment of Technology
Vienna University of Technology, Argentinierstraße 8/E 187, A-1040 Vienna, Austria
johannes.gaertner@tuwien.ac.at
http://time.iguw.tuwien.ac.at/JOG

[2] Institute of Software Technology and Interactive Systems
Vienna University of Technology, Favoritenstraße 9-11/E 188, A-1040 Vienna, Austria
silvia@ifs.tuwien.ac.at
http://www.ifs.tuwien.ac.at/~silvia

[3] Otto-von-Guericke University, Department of Simulation and Graphics
P.O. Box 4120, D-39016 Magdeburg, Germany
stefan.carl@student.uni-magdeburg.de
http://www.uni-magdeburg.de/carl/

Abstract. Information Visualization produces a visual representation of abstract data in order to facilitate a deeper level of understanding of the data under investigation. This paper introduces *ViCo*, a metric for assessing Information Visualization complexity. The proposed metric allows for the measurement of Information Visualization complexity with respect to tasks and users. The algorithm for developing such a metric for any chosen collection of visualizations is described in general and then applied to two examples for purposes of illustration.

1 Introduction

Within the field of visualization, Information Visualization aims for supporting individuals in understanding and detecting the relevant features of a field of interest. Information Visualization is the use of computer-supported, interactive, and visual representations of abstract data to facilitate cognition. The goal of Information Visualization is to ease understanding, promote a deeper level of understanding of the data under investigation, and foster new insight into underlying processes. The fields of application may vary from scientific tasks to everyday purposes.

Important contributions to the field of Information Visualization have come from various directions. In his seminal books on visualization Edward Tufte ([16], [17], [18]) discussed a number of interesting visualizations. He also introduced a number of recommendations for the design of such graphics (e.g., removing elements that do not contain information, minimizing gray) and a number of interesting concepts to enhance visualization design and analysis (e.g., macro and micro reading, which enables readings of a visualization at various levels of abstraction and detail).

M. Hegarty, B. Meyer, and N. Hari Narayanan (Eds.): Diagrams 2002, LNAI 2317, pp. 249–263, 2002.
© Springer-Verlag Berlin Heidelberg 2002

Information Visualization covers a broad field of visualizations and a number of books try to present relevant knowledge on dos and don'ts of specific design elements (e.g., [3], [14], [15]).

In many cases there is broad consensus on whether a specific visualization is good or not good. However, there is little theory to support such judgment. One way to deal with this situation is to develop benchmarks for the evaluation of visualizations where standardized sets of data and tasks are visualized in various ways [8]. it is difficult to quantitatively measure visualizations and to understand when to apply one visualization compared to another. To address some of these limitations, we are focusing on the development of an appropriate metric, called *ViCo,* by which to judge visualization and Information Visualization in particular. Our approach takes into account the tasks to be accomplished and the users' knowledge and needs on the one hand, and on the other hand, the difficult procedure of quantifying qualitative information concerning what we call here cognitive elements or operations. Hence, the metric *ViCo* can be seen as an algorithm that allows a quantitative comparison of the relative complexity of a set of visualizations for any given situation.

For example, assume we have two visualizations and let #N be the number of items being represented. Then, keeping all other parts being equal, the representation that makes it necessary to read #N items once is substantially better than one that necessitates reading #N * #N items. We want to facilitate the development of such formulas and comparisons.

In the next section, we develop the conceptual fundamentals of the *ViCo* algorithm. The algorithm itself is then introduced in the third section. The fourth section illustrates the algorithm through two examples. Finally, we discuss related issues and present concluding remarks.

The authors have experiences in task-specific approaches of Information Visualization, which range from visualization for software development and management consulting [5], to visual representations for various monitoring data and processes of patients in intensive care units ([10], [12]) and for the design of shift-rotas in various industries [6].

2   Conceptual Fundamentals

Here we present central conceptual fundamentals and definitions, which are needed to proceed with our approach and then are heavily exploited in the later algorithm we develop in section 3. These definitions include (1) reading and writing, (2) comparisons and calculations, (3) tasks and users, (4) complexity, and (5) the metric.

- **Reading and Writing**

Berg [2], inspired by actor-network theory and work within Computer Supported Cooperative Work (CSCW), tries to circumvent technological-determinist as well as social-constructivist accounts in discussing the changes brought about by the use of artifacts. He aims for a relational conceptualization of what such tools do, without attributing the activities exclusively to the tool itself or to the person working with it. He conceptualizes the activities associated with information technology in work practices as *reading* and *writing* of artifacts. This enables a consistency of approach in analyzing the paper-based and computer-based technologies. For his field of analysis – electronic patient records – he describes the generative power of artifacts as accumulating inscriptions and coordinating activities, thus making the handling of more complex work tasks possible.

Transferring this conceptualization of computer artifacts to the field of Information Visualization, a first element of complexity comparisons will relate to such reading and writing of visualization. Specifically, how many things do users have to read or to write for a given visualization?

- **Comparisons and Calculations**

Expanding on Berg's [2] approach (i.e. considering the use of artifacts in terms of the cognitive activities of users) one has to consider other activities that might be of relevance for Information Visualization. Two additional activities are considered here: *comparisons* and *calculations*. Comparisons deal with comparing one or more elements of Information Visualizations with respect to specific features. Calculations may influence the task or problem processing in two ways: first, that something can be computed (compare [2]) and, second, the effort of computing may vary [5].

- **Tasks and Users**

Two critical elements are missing so far: *tasks* and *users*.

It is impossible to discuss the amount of reading, and writing, comparing, and calculating that is necessary without specifying a task and supposing a user up front. Only when it is clear whether a task is completed or not can one discuss the amount of reading, writing, comparing, or calculating that is necessary.

Information Visualization complexity can only be discussed with respect to the same tasks. Similar constraints are described in the field of designing maps. For example, MacEachren [11] argues that there cannot be a discussion of how good or bad a map is without knowledge on the various ways of its use.

Users are to be considered too. The analysis of visualization complexity cannot be conducted without some reference to the users of a given visualization. Reason is that the information users can gain by using an Information Visualization depends also on their general and task-specific knowledge (e.g., to interpret graphics on various accounting measures, one must understand the categories of accounting; to understand the tableau of chemical elements, one must know something about chemistry).

- **Complexity**

We conceptualize the complexity of visualizations in terms of the operations – or cognitive elements - needed to accomplish the tasks by users. This approach relates strongly to the field of computational complexity [13], a part of computer sciences.

The proposed metric of complexity will not deliver a single number but will describe a function with various variables (e.g., number of items to be compared). For example, a simple algorithm for finding the median of n items uses $k*n \ log \ n$ comparisons, (where k is a constant of proportionality). Here "n" is a "*variable*". In computing, it is common to use the "size of the input" as the main variable. In our case, we use variables to denote the different dimensions of input, which are relevant to comprehend the visualization. Additionally, complexity analysis in computer science provides both *upper* and *lower bounds*. For example, median finding has a lower bound of $2n$ comparisons (a proof that any algorithm for median finding must make at least this many comparisons), and an upper bound of about $3n$ (worst-case runtime of the best algorithm for median finding) [1]. For our approach, it would also be interesting to consider upper and lower bounds for visualization tasks.

In our case, the necessary variables may be difficult to identify and the number of variables considered is expected to spread over time, as the analysis of a specific field of Information Visualization matures and deepens. Though a function is more difficult to handle than a single number, a function seems an appropriate way for the comparison of visualization. For instance, researcher and designers can gauge which visualization to use under what circumstances. Furthermore, it is not unusual to work with functions to describe complexity. Again, computational complexity within computer sciences works strongly with such elements.

- **The Metric**

Science distinguishes a number of ways to compare or describe features of objects of interest. From a mathematical point of view, the highest level of such comparisons leads to scalar, absolute values. On a level lower, observers would agree on the ordering and relative distances of complexity (e.g., 1-2-4; 3-6-12), or even weaker ordering function (e.g., A > B, B > C).

As mentioned in the previous section we do think that the computation of complexity relies on defining tasks to be accomplished by users. It would be too much to expect the metric (and its procedures) to guarantee that its users of the metric reach consensus on which tasks and user groups to take as the starting point. However, we consider it plausible – and will discuss it later on – that it should be possible to come up with a list of relevant tasks in close to all situations and to articulate reasonable assumptions with respect to the users. Afterall, Information Visualization typically makes use of information that already refers to such tasks and user groups.

Under the condition of shared assumptions regarding users and tasks, the metric we develop will be able to compute the complexity of Information Visualizations on a particular level.

3   Our Approach: A Metric for the Complexity of Visualizations

In the following we will describe the proposed algorithm to develop the metric of complexity for a chosen set of visualizations, called *ViCo* (Visualization and Complexity). We first describe the steps of the algorithm and then show their application on two examples.

The algorithmic steps of *ViCo* are:
1. Analyze the tasks to be accomplished by the use of a set of given visualizations and select those tasks to be taken as the basis of measurement.
2. Define minimal reading, writing, comparing, and calculating operations with respect to users' groups and variables of the data set to be visualized.
3. Develop the functions that describe the number of such operations needed to accomplish such a task.

We make the assumption here that the visualizations under consideration include all the information necessary to complete the tasks at hand. Though similar visualizations ([6], [10]) may vary substantially in what tasks they allow one to work on, this line of inquiry shall not be pursued here, because we are focusing on approaches which stay as simple as possible to communicate complex data and information in diagrammatic form.

3.1   Tasks

The first step of *ViCo* is to define the tasks that are the basis of the later measurement. In many cases this selection will be straightforward. For example,

– Understanding differences between object A and object B,
– Finding an object, or
– Being able to decide whether something is true or false.

In other cases, with a large number of tasks, a selection process may be needed. In most cases it should be possible to come up with a reasonable number of the most relevant tasks or at least relevant examples of tasks. However, if developers of an Information Visualization have no idea about possibly relevant tasks that users will try to accomplish with such Information Visualization, we would recommend to do more exploration in that direction, before starting the work of visualization design and analysis.

After selecting tasks, a further refinement is needed. A task is defines as such for our further analysis if (and only if) we are able to determine whether it is completed or not, and this typically calls for further refinement:

– Understanding differences between object A and object B with respect to pre-defined quantity of features (e.g., all, some, a percentage, etc. of the features),
– Finding a particular object (e.g., the street within a map), or
– Being able to decide whether statements A, B, C are true or false.

Again we assume that such refinements should be possible in most or all practically relevant situations.

3.2   Reading & Writing, Comparing & Calculating, and Users & Variables

As mentioned in the second section, Berg [2] focused on reading and writing operations. As long as we deal with visualizations drawn by the computer there is little user-writing involved. However, if we take into account interactive parts of the Information Visualization process, then the tasks of writing and typing become a crucial part of the complexity analysis as well.

Besides reading and writing, we consider comparisons and calculations as separate operations. It seems possible to develop additional categories as well that might help to focus better on further activities (e.g., group processes). Our metric is open to such extensions.

In the following we explain how the reading, writing, comparing, and calculating operations are defined. At first glance this might look rather tricky. However, it is so only to some degree, as we go for *relative complexity* of visualizations and not for *absolute complexity*. We do not attempt to develop a metric that covers all possible visualizations, for all possible tasks for all possible user groups. We go for a smaller objective: We want to be able to compute the complexity for any given set of visualizations with given tasks and given assumptions regarding the user group. This allows for incremental enlargement for any specific field but avoids the pitfalls of a universalistic approach.

Looking closer at reading, writing, comparing, and calculating, the question arises at what level to measure these activities. It is possible to conceptualize these operations in extremely complex ways. Again, we go for a smaller aim. We try to find the simplest possible operations for a given set of visualizations.

When looking at simple conceptualizations, possible types of such operations could look like the following. The conceptualization of reading, comparing, and calculating, can be seen in analogy to the various levels of perception (see for example [7]). Writing we conceptualize as straightforward activity:

At least three levels of reading operations can be distinguished:
1. Operations with the eye (e.g., finding a legend)
2. Basic operations for reading a letter or a word; or finding the next row, etc.
3. Cognitive operations (e.g., memorizing)

At least two levels of comparison operations can be distinguished:
1. Direct comparison
2. Comparison with memorizing

At least two levels of calculation operations can be distinguished:
1. Actual calculation
2. Cognitive processes in order to develop a way how to calculate

It is not always necessary to work with the operations on the visual level. Dropping such measurement seems reasonable if no relevant differences can be expected between the visualizations to be analyzed. This might be the case if the operations defined (e.g., finding the start of row, finding a column) do not vary strongly between the visualizations at hand. If high differences between visualizations can be expected then measurement of eye movement should be done. Techniques to measure and compare such eye movement are used within usability labs (e.g., eye tracking), and the results of such measurements depict the time needed for a task or operation depending on relevant variables (e.g., number of columns). Statistical measures would then apply here.

The simplest measures – and those this article tries to exploit as far as possible – are simple operations of reading a letter or a word, comparing two lines, etc. Such operations should be selected on the highest possible level with respect to the visualizations under investigation. For example, if two visualizations both rely on bars and make it necessary to compare them, such basic operations could be: (a) Find pairs of bars that shall be compared, and (b) Compare two bars.

Using cognitive operations [7] as a foundation of measurements may sound unusual from the perspective of computer scientists. However, they are not as bad as one might expect. For reasons of measuring complexity, we can simplify dramatically by again defining basic cognitive elements (e.g., reading a word). These basic elements can be used without further clarification as long as they are used in the same way for all visualizations of interest in a situation.

For example, if the comparisons of numbers have to be made, such a "comparison" would be a cognitive minimal element. It would not make sense to go into further detail (e.g., understanding all the processes involved in such a comparison) as long as the minimal element meets the following requirements:
1. It is used consistently with respect to the visualizations at hand (*consistent*).
2. It does not vary internally in relevant ways (e.g., words in visualization D are dramatically shorter than words in visualization E) (*invariant*).
3. It does not overlap with other operations – either within or in between tasks or visualizations (i.e., if two operations are used that somehow overlap in their utility, they have to be split up in smaller operations) (*irreducible*).

If a cognitive element does not meet the above criteria then further refinement is needed. Such refinement typically brings in features of *users* (e.g., users do or do not know how to read a specific element of a representation) or additional *variables* (e.g., length of words). These variables may refer back to the task or to other features of the process, the visualization, etc.

Whenever decisions have to be made regarding the level of knowledge of users one can expect, this either brings in an additional variable or an assumption regarding the users that holds true for all visualizations under investigation. Knowledge of users may refer to general knowledge and capabilities or task-specific, situation-specific knowledge. It is important to understand that this does not call for a complete collection of all user knowledge. Only if a reading operation depends (in its feasibility or complexity) on specific knowledge will a decision have to be made about whether to assume that expected users will have that knowledge or to make a variable out of it (compare the explanations about variables used in computational complexity in section 2). The first approach simplifies the function but limits its applicability. The second approach increases the scope of applicability of a comparison to more user groups. However, this comes at the price of higher complexity of the function. We are aware that such a set of assumptions regarding users can be increased indefinitely. From a practical point of view however, the number of elements to be added will depend on the interactions of those persons involved in developing the metric. Therefore, the list should be limited, but open for later amendments.

Again, if designers of visualizations do not have an idea about their users, it seems worthwhile to think about this. In most cases however it should be clear. If different basic operations lead to different complexity results, this indicates weaknesses or differences in these definitions.

Summing up, after defining a set of basic operations, the variables to be considered, and (some) assumptions regarding the knowledge of the expected users, we can start with the calculation. The variables of the complexity function are a side product of the above analysis.

3.3   Develop the Functions to Compute the Number of Such Operations

After defining basic operations and variables one should be able to describe the complexity of reading, writing, comparing, and calculating of Information Visualizations in terms of software programs. Such programs finish when the corresponding task is fulfilled.

Correspondingly, it is necessary to develop an algorithm that accomplishes the task with the operations defined. Then – with standard techniques of computer sciences – one can compute the complexity as a function of the variables introduced in a reproducible way.[1]

---

[1] We are aware of the fundamental limitations in this field (e.g., the question whether an algorithm is the simplest possible algorithm for the task to be accomplished cannot be solved in general). However, we expect most actual algorithms to be simple, because the building blocks of the algorithms – the operations – are complex. The complexity is in the operations and not in the overall algorithm. For example, it is very difficult for us to code good algorithms for face recognition, but people do this with ease. "Simple for the human brain" does not mean "simple for us to code as an algorithm".

The results of this approach should be rather stable. Algorithms should not vary too strongly between applicants. A change of basic operations should only lead to a change in the resulting function if it introduces a new operation or a new variable. Both of these options are consistent with the metric.

The complexity of the algorithm (building upon well-defined tasks and well-defined operations) then is also the complexity of the visualization. To facilitate visualization comparisons, it may make sense to further simplify the functions describing the complexity. The basic operations used in these algorithms have constant time (they do not depend on variables!). Correspondingly, one operation can be described as a multiple of the other `Op1=a*Op2` by using scalars `a`, `b`, etc.

Summing up, a complexity analysis builds upon the elements listed in Table 1 (all necessary definitions were given in the previous section):

Table 1. Definitions of the various elements of the metric *ViCo*.

$$
\begin{aligned}
Vis &= \left\{ Visualization_A, \ Visualization_B .... \right\} \\
Task &= \left\{ Tasks \ to \ be \ achieved \ with \ Vis \right\} \\
User &= \left\{ Assumptions \ about \ the \ users \ e.g., \ knowledge \right\} \\
Var &= \left\{ Variables \ used \ in \ at \ least \ one \ operation \right\} \\
Op   &= \left\{ \begin{array}{l} operations\,(Var) \\ \forall v \in Vis \ \exists \ algorithm \ to \ accomplish \ \forall t \in Task \ building \ upon \ these \ operations \\ operations\,(Var) \ is \ consistent \ \land \ invariant \ \land \ irreducible \end{array} \right\}
\end{aligned}
$$

Using the described algorithm it is possible to develop the metric *ViCo* for any chosen set of visualizations and correspondingly compute the relative complexity of a set of visualizations. This measure of complexity relies on reasonable definitions of tasks, reasonable assumptions regarding users, well-defined operations, and variables describing features of the problems at hand that are considered in the assessment. In the next section, we illustrate *ViCo* with two examples.

4 Examples

In this section we explain how our complexity metric, *ViCo,* is applied to the following two examples, (1) Tasting Whisky and (2) Visualizations of some issues regarding the Challenger's Disaster.

4.1 Example 1: Tasting Whisky

Tasting whisky is a very complicated task, which is done principally with the nose, then by the tongue, etc. The taste of Whisky can be graded in 10 categories on the scale of 0-3 for each (3 being the highest). If you use a star plot [4] (also called a wheel) each category corresponds to a spoke of the wheel. When you finish the grading and join up the lines, a particular shape of wheel appears, which reflects the characteristics of the Whisky. Figure 1, shows two examples: on the left-hand side the star plot of "The Balevenie, 12 years old" and on the right-hand side "Glenfiddich" (taken from http://www.scotchwhisky.com/).

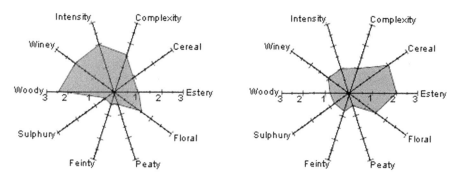

Fig. 1. Visualizing the taste of two types of Whisky. On the left-hand side, the star plot of "The Balevenie, 12 years old" and on the right-hand side, "Glenfiddich" (taken from http://www.scotchwhisky.com/).

In the following Table 2, the visualizations (Vis), the Tasks (T), the assumptions regarding the users (Users), the variables (Var) and then Operations (Op) are defined. After that the algorithms for accomplishing two tasks (compare Table 2) with the operations are described. Building upon that the complexity functions are developed.

Table 2. Example 1: Tasting Whisky, defining the elements needed to proceed with the complexity analysis

Kind	Name	Explanations
Vis1	Star Plot	See above Figure 1
T1	Highly Similar?	Determine whether the whiskys under consideration have highly similar features
T2	Identify Differences	Find and identify main differences of the whiskys
Users		• are able to understand and read star plots • are familiar with the 10 features of whiskey
Var	#C	• number of categories/spokes
Op1	Read shape	• read and comprehend the overall shape of a star plot
Op2	Compare two shapes	• compare two shapes and decide whether they are highly similar
Op3	Read scale value	• read and comprehend a scale value
Op4	Find corresponding scale	• after having read a scale name or value, find the corresponding scale in another picture
Op5	Compare two scale values	• compare two scale values and decide whether they are identical
Op6	Read scale name	• read and comprehend the scale name

Task 1: **Highly Similar**	For TWO star plots
The Algorithm	Read shape A (Op1)
	Read shape B (Op1)
	Compare two shapes A + B (Op2)
The complexity	2*Op1 + Op2
Task 2: **Identify Differences** *The Algorithm*	For TWO star plots for EACH Scale Read scale value (Op3) Find corresponding scale (Op4) Read scale value B (Op3) Compare two scale values (Op5) Read scale name (Op6)
The complexity	#C * (2*Op3 + Op4 + Op5 + Op6)

The result of the task 1, which checks for highly similar features of whiskys is 2*Op1+Op2 and the result of the task 2, which identifies the main differences of the whiskys is #C * (2*Op3 + Op4 + Op5 + Op6).

The complexity metric *ViCo* could be easily expanded to consider further issues (e.g., Var2=#W, which covers the number of whiskys to be compared) or it could be refined (e.g., Var3= #Identical counts the scales that do not show substantial differences).

Looking at an additional visualization that shows differences of corresponding features (see Figure 2), the complexity and the savings can be easily computed. Simplifying, if we assume that no additional operation is needed for finding the next scale with a difference (which only holds true for small numbers) and *#C* is the number of spokes and *#Identical* is the number of identical strokes, then the complexity function would be (#C - #Identical) * (Op3+Op6).

If we compare this result with the result of Table 2 (Task 2), them we can easily recognize that the computational complexity of the second visualization applying the same operators is much easier than the complexity of the other visualization.

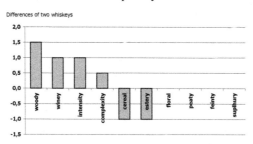

Fig. 2. Visualizing the differences of the two whiskys of Figure 1.

Further possible expansions of the analysis might consider, for example, how users would deal with large numbers of comparisons. Then also user interactions (e.g., selecting two whiskys and switching to another visual representation) might become relevant basic operations. A closer look at the basic operations might also lead to refinement of the metric and to a better understanding of the visualization. E.g., to what number of scales is a reading and comparison of shapes as a single operation reasonably possible? To what precision is the reading of scale values possible?

4.2 Example 2: Visualizations of Some Issues Regarding the Challenger's Disaster

Within the field of Information Visualization, scatter plots are another important class of diagrams and visual aid. Such diagrams can lead to great insight, but also to its occlusion. As an example for this Tufte [18] cites the accident of the space shuttle Challenger.

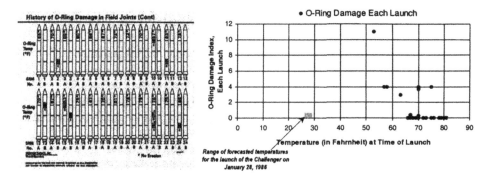

Fig. 3. Visualization of the shuttle's disaster showing the original diagram used by the NASA and the booster rocket manufacturer.

Fig. 4. Visualization of the shuttle's disaster showing the final re-visualization by Tufte [18].

There was the question about whether the shuttle should be launched on a cold day (January 27, 1986). The decision depended on whether the temperature would make the O-Ring that sealed the sections of the booster rocket unsafe.

The Figure 3 reprints one of the diagrams used in making the decision by the booster rocket manufacturer. Based on that diagram, NASA decided to launch the shuttle and the O-Ring was damaged and the shuttle crashed. The next Figure 4 shows the re-visualization by Tufte [18]. It uses a simple scatter plot depicting the relation between the two major variables of interest. Different types of damage are combined into a single index of severity. The proposed launch interval of temperature is also put on the chart to show it in relation to the data. The new diagram tentatively indicates a pattern of damage below 70° or 60°.

In the following we discuss these two visualizations in terms of *ViCo* for the same task of understanding whether there is a relation between temperature and O-Ring damage. Table 3 shows the use of *ViCo* in detail.

In the case of Figure 3, to achieve that task, 'normal' users will need a lot of processing (including ordering and calculating appropriate measures, e.g., calculating the averages of damage/no damage launches). In the case of Figure 4, users can start

Table 3. Example 2: Tufte's visualization of some issues regarding the Challenger Disaster, defining the elements needed to proceed with the complexity analysis

Kind	Name	Explanations
Vis2	Diagrams	See above Figure 3 and 4
T	Relate Damage and O-Ring	Is there a relation between O-Ring damage and temperature?
Users		• are able to understand and read scatter plots
Var	#N #O	• number of shuttle's starts • number of O-Rings
Op1	Read Damage	• read and comprehend severity of O-Ring damage
Op2	Read Temperature	• read and comprehend the temperature
Op3	Write Data	• write down in corresponding column the damages of O-Rings and temperatures
Op4	Calculate Measures	• compute the average and the measures needed to compare the data series
Op5	Read Shape & Decide	• read and comprehend the overall shape/distribution of the data points • make decision whether there is a very clear relationship
Op6	Read Data Points	• read and comprehend data points
Task: **Relate Damage and O-Ring with Fig. 3** *The Algorithm*		For EACH start (#N) 　For EACH O-Ring (#O) 　　Read Damages (Op1) 　　Read Temperature (Op2) 　　Write Data (Op3) 　　Calculate Measures (Op4)
The complexity of Fig 3.		#N * (#O * Op1 + Op2 + Op3 + Op4)
Task: **Relate Damage and O-Ring with Fig. 4** *The Algorithm*		Read Shape & Decide(Op5) IF no clear Relation THEN 　For EACH Data-Point (#N) 　　Read Data Points (Op6) 　　Calculate Measures (Op4)
The complexity of Fig. 4		Best Case: Op5 Worst Case: Op5 + #N*(Op6 + Op4)

with capturing the shape of the data series, because the scatter plot is already structured according to the two variables of interest. If a very clear picture emerges, the task is achieved. Otherwise, again calculation is necessary. In Section 2 in the paragraph about the complexity, we mentioned that complexity analysis in computer science often provides both *upper* and *lower bounds*. In the above example for the complexity measure of Figure 4, best case and worst case are these bounds.

In order to facilitate comparisons of the complexity measures of Figure 3 and Figure 4, we further simplify their measures. As there are several operations, each with constant length, we can introduce scalars (a, b, c, etc.) to express one as a multiple of the other (compare Section 3.3). Let Op1 take X seconds, then Op2 takes a*X seconds, Op3 takes b*X seconds, etc. Transforming the complexity functions from Table 3, we get the following new measures

The complexity of Fig. 3.	#N*(#O + a + b + c)*X
The complexity of Fig. 4.	Best Case: d*X
	Worst Case: (d + #N*(e + c))*X

With further simplifications we can compare the complexity of Figure 3 with the worst case complexity of Figure 4: $\dfrac{\#O+a+b+c}{\dfrac{d}{\#N}+e+c}$. With large #N we arrive at

$\dfrac{\#O+a+b+c}{e+c}$ which shows that the visualization in Figure 3 is much worse as long as Op6 is not more complex than (#O+a+b)*X.

The comparison of the complexities still includes the element c (coming from Op4 – calculating measures). This indicates that – if there is no clear shape – computation is still necessary. For the example given, the data point at 53° strongly shapes the overall impression. If this point would be considered to be an outlier then the picture would be less clear and correspondingly computation necessary. It might be the case that classical statistics is better and more informative to apply then. Still, the visualization of Figure 4 would allow for an extremely quick check whether there is a very clear relationship or whether calculation is necessary.

5 Discussion and Conclusions

In the paper presented, we have argued for *ViCo*, a metric for the complexity of various diagrams or more general approaches dealing with Information Visualization. For this purpose, we defined several conceptual fundamentals: Tasks and Users, Reading and Writing, Comparing and Calculation, and Complexity. Our approach is mainly influenced and guided by two scientific fields, on the one side, the algorithmic thinking and complexity theory in computer science [13] and, on the other side, the study of cognition and perception in psychology [7].

Our goals were to utilize concepts from perception and cognition to arrive at measures to judge the readability and the complexity of visualizations. We are definitely aware that perception and cognition work differently than algorithmic thinking (for example, we did not address, how we are dealing with know-how or any kind of learning effects to ease and facilitate the understanding of diagrams). We have knowingly simplified some cognitive aspects (e.g., memorization of information, know-how, learning) because we argue that in spite of such simplifications meaningful comparisons can be made. Similar considerations hold true for temporal aspects. Sometimes it may be necessary to actually measure times (e.g., with eye tracking). However, in many cases *ViCo* can work without such measurement.

We are not aiming to explain intuitive understanding of diagrams or any kind of visualizations. Additionally, we do not compare oranges with apples or scatter plots with danger signs. *ViCo* goes for a smaller but still reasonable aim. We analyze the readability of diagrams with respect to particular users and tasks. This means we are comparing oranges of kind A with oranges of kind B.

Finally, we aim 'only' for *relative complexity* of visualizations and not for *absolute complexity*. I.e., we do not attempt to develop a metric that covers all possible visualizations, for all possible tasks for all possible user groups. We go for smaller objective: We want to be able to compute the complexity for any given set of visualizations with given tasks and given assumptions regarding the user group allowing for incremental enlargement for any specific field.

ViCo does (to some degree) analogous things in the field of Information Visualization as GOMS does in the field of user interface design. GOMS (Goals, Operators, Methods, and Selection rules) [9] is an analytical analysis technique. The goal of GOMS is to radically reduce the time and cost of designing usable systems through developing analytic engineering models for usability tests based on validated computational models of human cognition and performance. The GOMS family provides various methods to count and measure how long a user needs to accomplish a task using a particular tool. Many variants of GOMS rely on measuring and calculating actual times, which limits the field of application and makes it more difficult to apply it for new types of information processing. However, this is just what Information Visualization aims for. Furthermore, some limitations of GOMS are inherited in our approach *ViCo* too, like differences between users, learning process, mistakes in executing the basic operations and inside the interpretation step.

We are well aware of the fact that the procedures described above touch a high number of questions that cannot be solved in general (e.g., comparison of algorithms, accelerate possible algorithm, definition of minimal operations). However, these questions can be tackled to an acceptable degree in most practical situations. Correspondingly, the procedure can contribute to better-informed decision making on which visualization to use when in a way that is not possible with direct observation or by measuring only the time that is needed to accomplish tasks as a whole.

Acknowledgements. We thank Monika Lanzenberger, Peter Purgathofer, Robert Spence, Jessica Kindred, and the anonymous referees for their useful and constructive comments. The Asgaard Project is supported by „Fonds zur Förderung der wissenschaftlichen Forschung" (Austrian Science Fund), grant P12797-INF.

References

1. Baase, S. & Van Gelder, A.: *Computer Algorithms: Introduction to Design and Analysis* 3rd ed.). Addison Wesley, Reading, Mass. [u.a.], (2000).
2. Berg, M.: Accumulating and Coordinating: Occasions for Information Technologies in Medical Work. *Computer Supported Cooperative Work: The Journal of Collaborative Computing,* 8 (1999) 373-401.

3. Card, S. K., Mackinlay, J. & Shneiderman, B. (Eds.): *Readings in Information Visualization: Using Vision to Think*. Morgan Kaufmann, San Francisco (1999).
4. Chambers, J., Cleveland, W., Kleiner, B. & Tukey, P.: *Graphical Methods for Data Analysis*, Wadsworth, (1983).
5. Gärtner, J.: *Software in Consulting*. Habilitationsschrift Thesis, Technische Universität Wien, (2001).
6. Gärtner, J. & Wahl, S.: The Significance of Rota Representation in the Design of Rotas. *Scandinavian Journal of Work, Environment & Health*, 24(3) (1998) 96-102.
7. Goldstein, E. B.: *Sensation and Perception* 5th ed.). Brooks/Cole Publishing Company, (1998).
8. Grinstein, G. G., Hoffman, P. E. & Pickett, R. M.: Benchmark Development for the Evaluation of Visualization for Data Mining. In Fayyad, U., Grinstein, G. G., et al. (eds.), *Information Visualization in Data Mining and Knowledge Discovery*, Morgan Kaufmann, San Francisco, (2002) 129-176.
9. John, B. E. & Kieras, D. E.: The GOMS Family of User Interface Analysis Techniques: Comparison and Contrast. *ACM Transactions on Computer-Human Interaction*, 3 (1996) 320-351.
10. Kosara, R. & Miksch, S.: Metaphors of Movement: A Visualization and User Interface for Time-Oriented, Skeletal Plans. *Artificial Intelligence in Medicine, Special Issue*, 22(2) (2001) 111-131.
11. MacEachren, A. M.: *How Maps Work* The Guilford Press, New York, (1995).
12. Miksch, S., Horn, W., Popow, C. & Paky, F.: Utilizing Temporal Data Abstraction for Data Validation and Therapy Planning for Artificially Ventilated Newborn Infants. *Artificial Intelligence in Medicine*, 8(6) (1996) 543-576.
13. Papadimitriou, C. H.: *Computational Complexity* Addison Wesley, Reading, Mass. [u.a.], (1994).
14. Schumann, H. & Müller, W.: *Visualisierung: Grundlagen und allgemeine Methoden* Springer, Berlin, (2000).
15. Spence, R.: *Information Visualization* ACM Press, New York, (2001).
16. Tufte, E. R.: *The Visual Display of Quantitative Information* Graphics Press, Cheshire, CT, (1983).
17. Tufte, E. R.: *Envisioning Information* Graphics Press, Cheshire, CT, (1990).
18. Tufte, E. R.: *Visual Explanation* Graphics Press, Cheshire, CT, (1997).

Opening the Information Bottleneck in Complex Scheduling Problems with a Novel Representation: STARK Diagrams

Peter C-H. Cheng[1], Rossano Barone[1,2], Peter I. Cowling[2], and Samad Ahmadi[1,2]

[1]ESRC Centre for Research in Development Instruction and Training,
School of Psychology
[2]ASAP Group, School of Computer Science and Information Technology,
University of Nottingham, Nottingham, NG7 2RD, UK.
peter.cheng@nottingham.ac.uk   rb@psychology.nottingham.ac.uk
pic@cs.nott.ac.uk   s.ahmadi@cs.nott.ac.uk

Abstract. This paper addresses the design of representational systems for complex knowledge rich problems, focussing on scheduling in particular. Multiple tables are ubiquitous in representations of schedule information, but they impose large cognitive demands and inhibit the comprehension of high-level patterns. The application and evaluation of representational design principles in the development of STARK diagrams, a novel system for scheduling problems, is reported. STARK diagrams integrate conceptual dimensions, principal relations and individual cases into a single diagrammatic structure. An experiment compared performance on STARK diagrams and a conventional representation with features typical of current commercial scheduling software interfaces. Subjects using the STARK diagram performed better at improving an examination schedule by minimising constraint violations. This provides support for the validity and utility of the design principles.

1 Introduction

The critical role of problem representations has been well established by research in cognitive science [1, 10, 15]. A problem can be more than an order of magnitude more difficult to solve with a poor representation than a good representation [11]. There are potentially substantial benefits of diagrammatic representations in contrast to sentential representations [12]. Mental effort and computation can be off-loaded onto external representations [14]. By matching the informational dimensions with appropriate visual dimensions it is possible to make representations that support the discovery of patterns and efficient inferences [18].

Although there has been research on how these and other findings on the nature of representations can be applied to the design of effective representations [5, 7, 16], the science of representational design is in its infancy. Given a particular domain, how should a representation be designed to support the different tasks that need to be performed and to support users with different levels of knowledge and experience of the domain? One of the grand challenges to this growing area is the design of effective representational systems for information intensive and conceptually

M. Hegarty, B. Meyer, and N. Hari Narayanan (Eds.): Diagrams 2002, LNAI 2317, pp. 264–278, 2002.
© Springer-Verlag Berlin Heidelberg 2002

demanding domains.   In previous studies with knowledge rich domains, including mechanics, electricity and probability theory, we have designed novel representational systems to support problem solving and learning in instructional contexts [6, 7].   By comparing the new representations with the existing conventional notations, including the empirical evaluation of the representations, principles for the design of representations for complex conceptual domains have been discovered [8].

This paper describes work on a project that is applying the principles to the analysis and design of novel representations for real-world informationally intensive scheduling problems.   Automated systems are typically used to solve such problems, but the way they operate tends to be inflexible and difficult for users to understand. The solutions they produce are only as good as the model of the problem they are given, so rare circumstances or idiosyncratic requirements are not handled.   The context of this project is to attempt to humanise such automated systems by designing new representations that will allow the flexibility and creativity of humans to be integrated with the computational power of automated systems. The aim is to get the best of both worlds by using the particular advantages of the approaches to overcome each others' specific limitations.

Examination timetabling is the first scheduling problem that has been addressed by the project.   A novel representation, STARK diagrams (Semantically Transparent Approach to Representing Knowledge), has been designed for this class of scheduling problem.   This paper describes the principles and how they were used to design STARK diagrams.   It also reports on the design of a conventional representation that served as the basis for comparison in an experiment that has been conducted to compare how well users can manually improve an examination timetable using the two representations.   The outcomes of the experiment are discussed and the implications for the validity and utility of the principles considered.   The nature of the examination scheduling domain will first be outlined.

2 Examination Scheduling

Examination scheduling is a complex organisational problem which occurs in many educational institutions across the world.   It is often solved by modelling as an NP-hard combinatorial optimisation problem that demands the allocation of exams to rooms and time periods under a high density of constraints [4].   These problems are large, typically involving tens of rooms and periods, hundreds of exams, and thousands of students.   Solutions to the problems are typically generated by automated software systems, with the user defining a fixed set of rooms, days and periods within each day.   Together with data for students, exams and constraints this information is used to generate solutions that the user may then edit.

Several different types of constraints exist in examination scheduling problems. Individual universities may differ in the kinds of constraints they employ but common to most are two categories that we term *resource* and *intersection* constraints.   For a description of other constraints used in examination scheduling problems see [3]. Resource constraints concern the availability of space and time for the allocation of an exam to a given room and period.   Room capacity violations occur when there is insufficient space to hold the number of students in the exams allocated to a room. Examination schedules are organised into daily periods of unequal size.   The inequality of period size demands the specification of the second type of resource constraint, which maintains that an exam should not be allocated to a period that is shorter than the duration of the exam. A violation of this constraint is termed a period excess violation.

Intersection constraints are concerned with the temporal proximity of allocated exams that involve the same students. They are termed intersection constraints because they occur as a consequence of the intersection, or sharing, of students between two exams. Clash exam violations occur when intersecting exams are allocated to the same period because students cannot take more than one exam simultaneously. Consecutive exam violations result when intersecting exams are allocated to adjacent periods. In this case students would have to take two exams in succession. Clash and room capacity constraints are often treated as inviolable or *hard* in optimisation approaches, with solution quality evaluated with respect to the degree of violation of *soft* constraints such as consecutive exams. However, this distinction is somewhat artificial in practice, and we treat all constraints as being soft, giving appropriate weights depending upon the relative importance of each one.

Despite the raw computational power of automated systems the simplified models of the problem they employ leaves much scope for human intervention. The solutions produced by automated systems will often satisfy the clash and room constraints but leave many soft constraints still violated, in particular those which were not present in the original model. The user must try to find new allocations for the violated exams that satisfy all the constraints. Resolving these violations is not straightforward otherwise the automated software would satisfy them. To effectively improve a solution the user needs to access and integrate many sources of different kinds of information. The manner in which this information is represented can substantially determine the users capacity to manually improve the solution.

Conventional ways of representing examination scheduling problems typically involve organising the problem data into numerous lists and displaying these lists in different windows or frames of reference. Advanced interfaces may provide *a la carte* tables that the user can construct to show particular perspectives of interest. Whilst conventional scheduling interfaces may be both sophisticated and flexible, the presence of multiple tabular representations tends to inhibit, rather than facilitate, user intervention. There are three common limitations of traditional scheduling interfaces: (a) they impose large demands on cognitive resources; (b) they support local inspection of information but do not provide global overviews; (c) they play little or no role in constraining user behaviour due to their profound lack of conceptual structure.

The principles of representational design have been used to design a novel representation that attempt to overcome these limitations.

3 Principles of Representational Design

The principles were derived in our previous work on the design of novel representations to enhance problem solving and learning by re-codifying knowledge in conceptually demanding educational domains including probability theory and electricity [6, 7, 8]. Six principles have been formulated to date and have been classified according to (1) whether they prescribe that the conceptual structure of the domain should be made apparent in the structure of the representation, *semantic transparency*, or (2) whether they concern efficient problem solving operators and procedures, *syntactic plasticity*.

The semantic transparency principles consider how the meaning of a domain can be made clearly apparent in three ways:

(1) **Integrate levels of abstraction.** Different levels of abstraction should be integrated to reduce the conceptual gulf between (a) overarching laws or general relations that govern a domain and (b) specific cases or instance at a concrete level. In a representation that integrates levels of abstraction extreme cases will help interpret the general nature of the laws and the laws will explain the form of typical cases.

(2) **Integrate alternative perspectives.** Alternative perspectives, including alternative ontologies [9], can be used to describe a domain. Perspectives at the same level of granularity or abstraction should be integrated in an effective representation, to allow the alternative perspectives to act as mutual contexts for each others' interpretation.

(3) **Combine a globally homogenous with a locally heterogeneous representation of concepts.** An interpretative framework should be provided that simultaneously combines (a) a globally coherent interpretative scheme based on the principal conceptual dimensions of the domain with (b) local representational features that make specific conceptual distinctions clear. A principle conceptual dimension is a property or aspect that is universal to the domain. Time and space are such dimensions for many domains. Under a given global dimension, one way a specific conceptual distinction may be identified is by the use of alternative scales of measurement for the different things under the global dimension (i.e., ratio, interval, ordinal or categorical scales). A representation with such an interpretative framework should support the making of valid generalisations whilst reducing the chance of over generalising specific concepts.

There are three ways in which the syntactic plasticity principles can make problem solving with a representation easier:

(1) **Malleable expressions.** The expressions of a representation should not be too rigid nor too fluid, they should be malleable. A rigid representation lacks the procedures to allow all the meaningful expressions needed in problem solving to be generated. This will cause dead ends to occur during problem solving. A fluid representation allows many arbitrary expressions to be generated at each potential solution step, so making the space of possible expression impracticably large to consider. A malleable representation sails a middle course between these extremes.

(2) **Compact procedures.** The procedures for solving problems in a representation should be compact, in the sense that the typical number of operations needed to find a solution from the initial state should be relatively small. The fewer the operators the less computation that is needed and the less the chance of making errors.

(3) **Uniform set of operations.** A representation should have a small variety of consistent and uniform operators making up its problem solving procedures. The fewer the types of operators the simpler the overall approach to problem solving is likely to be.

4 Applying the Principles to Scheduling

The application of the principles to examination scheduling requires the conceptual structure of the domain and the nature of the problem solving activities to be specified. In doing this, the term *slot* is taken to mean the conjunction of a particular room in a specific period to which an exam may be assigned.

In examination timetabling there are two aspects over which levels of abstraction should be integrated. First, there are the implicit relations due to the underlying physical nature of the domain. At the abstract level this means that arithmetic operations apply to the sizes of slots and to non-intersecting exams (e.g., free room capacity = total room capacity - Σ sizes of exams allocated to the room). Also at the abstract level set theoretic notions apply to exams as "members" of a slot and to students as "members" of exams (e.g., students taking both exam-A and exam-B = (exam-A ∩ exam-B)). At the concrete level is information about the size and duration of actual exams and slots. The second aspect concerns the quality of solutions. At the abstract level the quality of a solution is assessed by a mathematical model called the evaluation function. At the concrete level are actual distributions of allocations and constraints over which the evaluation function computes the quality of a solution.

The different perspectives in the domain that should be integrated include: space and time resources that together define available slots; demands in the form of exams to be allocated; constraints that are to be satisfied and violations to be eliminated; intersections amongst exams at the level of students, in particular the *intersection set* for a target exam that comprises the group of all other exams sharing students with the target exam.

The global conceptual dimensions of the domain to be incorporated into an overarching interpretative scheme include: time, space, constraints, and types of entities. Time is obviously uni-dimensional. For timetabling, space is also essentially one dimensional, because it is sufficient to distinguish different rooms without specifying their location in space and because size and capacity, number of students, are one-dimensional properties. On the time dimension local conceptual distinctions to be made include: days, periods, duration of periods, duration of exams. On the space dimension rooms, room capacity, exam location, and exam size are to be distinguished. Constraints and violations can be distinguished on the basis of the things to which they apply. With respect to students there may be exam clashes or consecutive exams. For exams themselves there may be preferences for certain orders of exams or for holding some exams simultaneously. For periods and rooms there may be preferred times or locations, such as a laboratory for a practical exam. Finally, exams and slots are different entities to be distinguished.

What are the problem solving procedures that must be supported by an effective representation for exam scheduling? The class of examination scheduling problems that was chosen for the experiment was the manual improvement of an exam schedule solution by the reallocation of exams. In this context, a specification of a solution constitutes an expression and a new expression is generated whenever an exam is moved to a new slot. The nature of the problem and the kinds of problem solving procedures that human problem solvers typically use were considered. Users' choices in solution improvement should be guided by the information present in the current solution. It would be ideal for a user to be able to: (a) clearly identify the constraint violations in the solution; (b) prioritise those exams that are causing the most problems; (c) estimate with some degree of precision how easy it is to satisfy the problem exams; (d) make an informed decision about which exams to reallocate based on some estimation of the cost of moving the exam relative to the gain expected in solution improvement; (e) clearly evaluate the improvement or lack of it following each reallocation.

In examination scheduling much of the cognitive work relies on the capacity to access, compare and integrate detailed information. An examination scheduling representation that is a bottleneck to such information is likely to result in greater errors, inhibit exploration and the generation of novel strategies, and may ultimately frustrate the user. Our interviews of professional examination schedulers suggest that this is precisely the case. These schedulers expressed little confidence in their capacity to manually improve a solution with current systems and appeared to depend entirely on automated solution generation. For effective examination scheduling ease of access to relevant information appears to be an important factor which is missing from the current generation of decision support software for the problem.

The next two sections consider how the structure of the new STARK diagram representation and the conventional representation were designed according to the principles discussed above.

5 STARK Diagram Representation

Figure 1 shows a screen span shot of the STARK diagram interface designed with the principles and as used in the experiment. For ease of inspection a magnified view covering three days and nine rooms is shown, but a full size schedule can be reasonably viewed on a standard computer monitor. STARK diagrams combine globally homogeneous and locally heterogeneous representations of concepts. For global homogeneity, a general interpretive scheme is provided that includes space, time, constraints and entities. Time is represented along the horizontally axis and space is represented along the vertical axis. Basic timetable entities are represented by rectangular icons. Most constraints are represented by lines connecting different graphic components and others by specific configurations of icons. The specific conceptual distinction between exams and slots (a room for a given period) is made using light yellow icons and dark blue icons, respectively. On the time dimension periods are shown by columns of slot icons. Days consist of three periods and are represented by a group of three columns. The duration of a period or an exam is given by the width of its icon. Under the space dimension each row of slot icons represents a room (available over time) and the exams allocated to a particular slot are shown by exam icons contained within its slot icon. The capacity of a room or the size of an exam is shown by the height of the respective icon. This representational structure provides an interpretative scheme for identifying exam and room quantities and also provides a frame of reference for locating and making inferences about the temporal proximity of specific exams, rooms, periods and days.

The specific conceptual distinctions under the global constraints dimension also exploit this scheme, as shown in Figure 2. The schemes allows three types of information to be encoded: (1) the type of constraint violation; (2) the exams involved and their present allocation; (3) the number of students involved. Resource constraints such as room capacity and period excess are represented as exam icons overflowing the slot icons containing them, Figure 2a. Such constraints are distinguishable from intersection constraints, which are represented by lines connecting offending exams, Figure 2b. Types of intersection exam are themselves differentiated by the colour and orientation of the connecting lines.

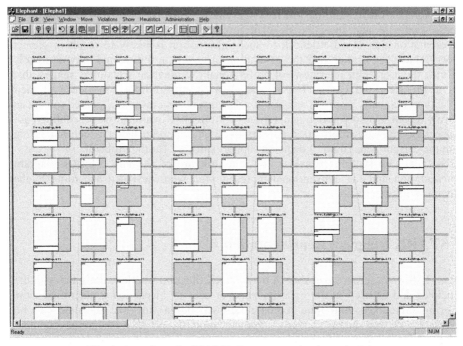

Fig. 1. A section of a STARK diagram examination schedule

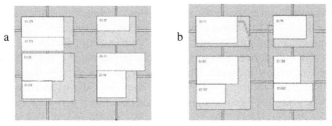

Fig. 2. Some constraints in the STARK diagrams: (a) top left – over capacity, bottom right – period exceeded; (b) vertical line – exam clash, diagonal connecting line – consecutive exam

With respect to the underlying physical regularities of the domain, levels of abstraction are integrated in STARK diagrams by: (1) using geometric configurations to capture arithmetic relations and spatial arrangements to capture set theoretic relations; (2) encoding quantities as sizes of icons. The assignment of a number of students to an exam and the allocation of an exam to a slot are represented by viewing exam and slot icons as containers. For example, the exam icons for two exams that are allocated to the same slot are not permitted to overlap. The total time a period is exceeded is given by the amount that the offending exam icon horizontally extends beyond the boundary of the slot icon. Connections between the underlying physical regularities of the domain and actual cases are thus directly visualised.

Another important aspect of abstraction is the evaluation function. The STARK diagram allows the user to view the solution and evaluate its quality, at some level of precision, using his/her own model of the evaluation function rather than relying solely on a given mathematical model. This is feasible because STARK diagrams not only support the recognition of individual instances but allow distributions of allocations and constraints to be judged. For instance in Figure 1 the distribution of exams is fairly uniform, but some slots are empty .

It is not surprising that the perspectives of resources, demands and constraints are integrated in STARK diagrams, because the global interpretive scheme combines space, time, entities and constraints. The resource perspective is represented by the space-time matrix of slots. The demands perspective is represented by the layer of exam icons distributed over that matrix. The constraints perspective is shown by lines connecting exam icons and by the overflow of exam icons over slot icon boundaries. The intersection set perspective is also integrated with the others under the interpretative framework, but information from this perspective is only made available when a target exam is selected. Figure 3 shows how when a target exam is selected all exams that share students with that exam are highlighted in dark red. The black region at the top of each of these intersecting exams denotes the number of students that are also taking the selected exam. The integration of the intersection set perspective with the other perspectives has a substantial impact on the nature of the problem solving procedures used with STARK diagrams.

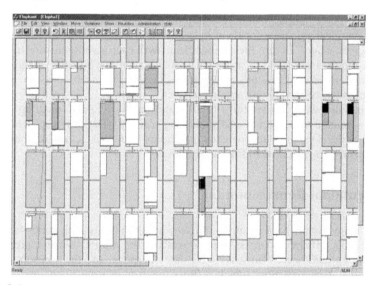

Fig. 3. Intersection set for a selected target exam coloured white (row 3, column 5)

STARK diagrams constitute a malleable representation for examination scheduling, because they make problem relevant information salient that greatly constrains the choice of manual changes that are meaningful to make. Consider two examples of this. First, a user can also readily judge whether the target exam will violate resource constraints in a period by comparing the size of the target exam icon with the size of the rooms considered for potential reallocation. Second, being able to identify those exams that share the same students with a target exam, intersection set information, is particularly important. Suppose the selected exam in Figure 3 is to be reallocated. If it is reallocated to the same period as any of its intersecting exams (in red), such as the immediately preceding or following period, a clash violation would occur. If reallocation is to any period immediately before or after a period with intersecting exams, such as column 3 in Figure 3., then consecutive violations would again occur. The only periods free from such potential constraint violations are the 10th and 11th columns. By supporting such meaningful judgements the representation is malleable because it restricts the user's options to ones that are good.

The accessing of information is critical to problems in exam scheduling. If the procedures for finding relevant information are not simple the overall problem solving procedures are likely to be complex and involved, so not compact. In this respect STARK diagrams are compact in three ways. First, there is a single frame of reference for all the major classes of information, so information needed for solution improvement can be read directly from the diagram. Second, many useful relations that are not usually stated explicitly in examination schedules appear as emergent features in STARK diagrams. Third, the representation exploits the built-in discriminatory power of the human visual system to make relevant information "pop out"; for example the red exam icons for the intersection set, as in Figure 3. These representational devices mean that laborious searches for information, in the form of indistinct textual labels spread across multiple lists, are avoided.

This ease of information access allows sophisticated problem solving strategies to be adopted. For example, in many solutions it is impossible to reallocate a given exam without clearing space for them by reallocating other exams in the intersection set of the given exam. In turn these other exams will each have their own intersection set. This requires a recursive approach that users of the STARK diagram in the experiment were seen to execute successfully. This would not be possible unless the representation has relatively uniform procedures.

The principles were successfully used to design the STARK diagram so that it possessed semantic transparency and syntactic plasticity. The design process initially focussed on the semantic transparency principles and satisfaction of the syntactic plasticity principles appeared to follow naturally.

6 Conventional Representation

There are many commercial examination scheduling systems with sophisticated user interfaces. The principles can be used to analyse the merits of the representations behind those interfaces. However, for the purpose of the present study a "conventional" representation, which reflects the current state of the art of existing systems, was designed. It is informationally equivalent to STARK diagrams, largely diagrammatic, but does not satisfy the principles well, so as to give a good basis of comparison for STARK diagrams.

Figure 4 shows the conventional representation. It uses a tabular format that is common to traditional timetables, with time represented on both axes of the 2-D plane. Days are represented along the x-axis and for each day there are three periods shown by three rectangular areas down the y-axis. Within each rectangle rooms are shown by the dark blue icons on the left and the exams allocated to the rooms shown by the light yellow icons to the right.

The spatial and temporal information for each exam and room are shown by the numbers on the icons. The identity of the room is written above its icon along with the time that the room is available. On the room icons the room capacity is given and along with the current free space in parentheses (negative values denote how many students exceed the room capacity). For exams the number of students taking the exam and the duration of the exam are shown.

a   b

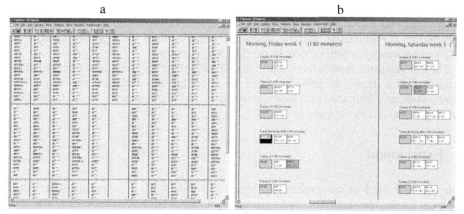

Fig. 4. Conventional schedule representation: (a) overview, (b) detailed view

Constraint violations are shown by changing the icon colour. Black room icons indicate some resource constraint is violated. The type and reason for the violation must be inferred from the values, or alternatively users can display the details of the violation as listed in a separate window. A red exam icon indicates an intersection constraint violation. To see if it is a clash or consecutive exam violation the user must inspect a list in a separate window that gives relevant details.

The conventional representation does a poor job of combining a globally homogenous with a locally heterogeneous representation of concepts. It does not use the principal visual dimensions in a coherent way to encode space, time, constraint and entity dimensions. The y-axis encodes both time (period) and spatial (room) information. The specific conceptual distinction between periods and days is made using principal spatial dimensions, but other specific temporal distinctions are associated with numbers attached to icons. The same is true for information about sizes of entities.

Levels of abstraction are not well integrated in the conventional representation. Set theoretic relations and arithmetic relations are not directly encoded in the representation. Judgements on these bases require the user to make mental inferences and computations rather than perceptual ones. In relation to the evaluation function, the distribution of the exam allocations is apparent but information about spare capacity and the distribution of types of constraints requires deliberate search across separate lists.

The conventional representation does not integrate alternative perspectives well. Although resources and demands are shown together, detailed information about violations and intersection sets for particular exams are not available in the main diagrammatic window. Combining the information from different perspectives is a task left to the user. For example, in the simple scenario where there are two consecutive periods that do not have intersecting exams, the user must infer this by working through the lists for intersection constraints for all the exams in the periods.

The need to search for information across separate lists means the representation is not compact and tends to be rigid (not malleable). To make a meaningful move many separate operations are needed to find relevant information about intersection sets and suitable slots. This information bottleneck in turn makes it difficult to take a solution in one state and transform it into some better state. The conventional representation does not have particularly uniform operators, because there are a greater number of information lookup functions than in the STARK diagram.

The conventional representation and the STARK diagram have been considered in detail to show how the principles can be used for the design and analysis of representations. This also provides a good sense of the complexity of the task faced by the participants in the experiment.

7 Experiment

The problem used in the experiment was a real full scale examination scheduling problem for the University of Nottingham, which involved: 800 exams, 20 rooms and 32 periods (i.e., 640 slots), and 10113 instances of exam intersections. Two schedule solutions were generated by automated software of comparable power to commercial systems. One was used for a practice session and one was used for the test session.

Participants were six research students of the Automated optimisation and planning group (ASAP) at the University of Nottingham. Two were conducting research on automated approaches to examination timetabling and were allocated to different conditions (STARK vs Conventional representation). The others had little knowledge of the domain so were randomly allocated. Participants were paid for doing the experiment, which lasted between three and four hours.

The experiment involved an independent subjects design comprising of the two interface conditions. The training and test sessions were almost identical for both groups and involved three sections: (1) learning about the examination scheduling problem; (2) learning to use the scheduling software; (3) a practice session. In the 20 minute practice session participants attempted to improve a timetable solution by minimising the number of constraint violations in a simplified problem with approximately 50% of the exam intersections randomly removed. Before the test session began a brief screening interview established that subjects' level of understanding was sufficient to proceed. The test problem was a full Nottingham University data set without simplification and so was substantially harder than the practice problem. The test session lasted an hour and participants were told to improve the solution to the best of their ability. They were also told a simple weighting scheme for the importance of different violated constraints and instructed to use the weightings to prioritise their work. The weightings were: 5 points for clashes and room capacity violations; 2 points for period excess violations; 1 point for consecutive exam violations. The software automatically logged all the actions performed by the subjects.

Table 1. Frequencies of operation types for each subject in the test session

Participants	STARK			CON			STARK	CON
	1	2	3	1	2	3	mean	mean
Show Intersections	110	171	58	15	59	70	113	48
Reallocate exam	67	105	41	8	28	52	71	29.3
Swap exam	0	1	0	0	0	0		
Exam to clipboard	0	0	0	0	20	0		
Allocate from clipboard	0	0	0	0	16	0		
Undo	24	17	13	1	33	36	18	23.3
TOTAL	294	201	112	158	24	156	202.3	112.6

Six types of operations performed by the subjects on the systems are of interest here: (1) showing the intersection set of an exam; (2) reallocating an exam; (3) swapping exam allocations; (4) placing an exam on the clipboard; (5) moving an exam from the clipboard to the schedule; (6) undoing operations 2 to 5. Table 1 shows their frequencies for all six subjects. There is considerable variability between subjects in the number of operations performed, but for the STARK group the mean was nearly twice that of the conventional group. The mean number of exam reallocations and show intersection operations were also substantially greater for the STARK group suggesting that in general these participants were more productive or at least more adventurous. Only one subject performed clipboard operations (Con 3) and one subject swapped a single pair of exams (STARK 1). On average the STARK group needed to do fewer undo operations relative to the number of reallocations made.

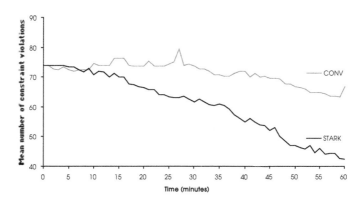

Fig. 5. Mean number of constraint violations over time for each group

The number of constraint violations present in the solution were recorded at the end of each minute. Figure 5 shows how the number of violations changed over time for the two groups as a whole. On average the maximum number of constraint violations resolved by STARK interface group was nearly three times greater than Conventional interface group (STARK=32.4, Con=11.7). Figure 6 shows how individuals in each group performed. All the STARK interface subjects removed more violations than any of the conventional interface subjects. It is noteworthy that two of the STARK group succeeded in less than half the time to eliminate as many violations as the best Conventional group participants.

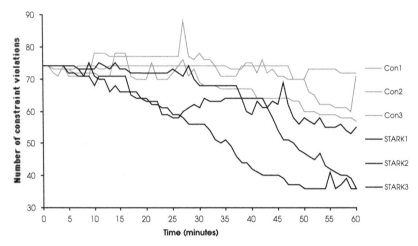

Fig. 6. Change in the number of constraint violations over time for each subject

8 Discussion

The pilot experiment reported here provides further evidence for the validity and utility of the representational design principles and suggests that the STARK diagram gives superior support for manual exam schedule improvement compared to the conventional interface. In general the STARK group was more productive than the Conventional group. Analyses are currently underway to reveal detailed differences in cognitive behaviour that result from the differences in the representations.

The success of the STARK diagram design and the outcomes of the experiment are notable because they extend previous studies addressing the principles in at least two ways. First, the domains previously studied were conceptually demanding topics in science and mathematics education that had substantial abstract components [8]. The examination scheduling problem is more concrete, with no high level laws referring to intangible properties. However, the design of a new representation using the principles still conferred a substantial advantage to its users compared to the conventional representation. Second, exam scheduling is a very information intensive domain compared to the previous domains studied. There are many more types of data and the sheer quantity of data is great. Nevertheless a new successful representation was designed using the principles. Together these points provide support for the claim that the principles are valid and have utility. The benefits of satisfying the semantic transparency principles are directly apparent from the comparison of the structure of STARK diagrams and the conventional representation.

The syntactic plasticity principles must be considered in the light of the experimental outcomes. The substantially greater productivity and extent of constraint violations removed by the STARK diagram group, relative to the Conventional group, supports the hypothesis that the STARK diagram interface has more compact procedures comprising fewer operations. The basic analysis reported here does not provide direct evidence that the STARK diagram representation is more malleable than the conventional representation. To address this it will be necessary to examine how well the representations enhance the ability of users to sample information resulting in the selection of effective operations. Further analysis will attempt to asses how effective the different users' choices were in the context of the information available.

The development and evaluation of the principles may be viewed as an extension of previous work on the cognitive implications of the nature of the relation between represented domains and representing symbolic systems [2, 13, 17]. Such work has addressed relatively information lean and knowledge sparse domains compared with the richness and complexity of the real-world domains to which we are attempting to apply the principles considered here.

The application of the design principles to the information rich domain of scheduling has revealed some interesting new aspects about the design of effective representations. First, one important difference between STARK diagrams and the conventional representation is the major reduction in the number of discrete expressions required to represent the problem. This appears to be a consequence of the demands of semantic transparency principles that information be organised and integrated in a conceptually coherent manner. This confers obvious benefits to users in computer based tasks. In Figure 1 a portion of the full schedule is shown, but a user can view all the data for a full scale examination schedule on a standard computer monitor, which has a positive impact on memory, cognitive loads and the complexity of operations. Second, STARK diagrams appear to allow users to view examination schedules in a way that is abstracted from the domain. It is often unnecessary to think about the actual details of the domain itself. Problem solving can occur at the level of relations that underpin the domain by considering the containment and spatial arrangement of exam icons and room icons along with the constraint violation links that connect them. Such an approach is arguably less demanding and depends upon how well the representational structure reflects the conceptual structure of the problem. It is possible that this might allow schedule generation to be undertaken by individuals with less experience and knowledge of the domain. Third, the scheduling domain involves many different levels of granularity, ranging from the individual student through to complete schedules. Such levels are distinct from levels of abstraction and alternative perspectives, because they focus on concrete aspects of solutions and alternative levels of granularity can be identified under each perspective. STARK diagrams support the integration of these levels of granularity in a manner that does not occur with the conventional representation. In the previous instructional domains where the principles were applied, levels of granularity were not as important in the present domain. This suggests that an additional semantic transparency principle – integrate levels of granularity – should be proposed. Work on other information intensive scheduling domains in the current project will consider whether such a principle is necessary.

Acknowledgements. The research was supported by an ESRC/EPSRC research grant (L328253012) under the PACCIT programme and by the ESRC through the Centre for Research in Development, Instruction and Training CREDIT).

References

1. Anderson, M., Cheng, P C-H., & Haarslev, V. (eds.): Theory and Application of Diagrams: First International Conference, Diagrams 2000. Berlin: Springer (2000)
2. Barwise, J., & Etchemendy, J.: Heterogenous Logic. In: Glasgow, J., Narayanan, N. H., & B. Chandrasekaran, B., (eds.): Diagrammatic Reasoning: Cognitive and Computational Perspectives. AAAI Press Menlo Park, CA (1995) 211-234
3. Burke, E.K., Elliman, D., Ford, P., Weare, R.: Examination Timetabling in British Universities: A Survey, Practice and Theory of Automated Timetabling I, Vol. 1153. Springer LNCS (1996) 76-90
4. Carter, M. W., Laporte, G.: Recent Developments in Practical Examination Timetabling, Practice and Theory of Automated Timetabling I, Vol. 1153. Springer LNCS (1996) 3-21
5. Casner, S. M.: A task-analytic approach to the automated design of graphic presentations. ACM Trans. on Graphics, 10(2), (1991) 111-151
6. Cheng, P. C-H.: Law encoding diagrams for instructional systems. Journal of Artificial Intelligence in Education, 7(1), (1996) 33-74
7. Cheng, P. C-H.: Interactive law encoding diagrams for learning and instruction. Learning and Instruction, 9(4), (1999) 309-326
8. Cheng, P. C-H.: Unlocking conceptual learning in mathematics and science with effective representational systems. Computers in Education, 33(2-3), (1999) 109-130
9. Chi, M. T. H.: Conceptual change within and across ontological categories: examples from learning and discovery in science. In: Giere, R. N. (ed.): Cognitive models of science. Minneapolis: University of Minnesota Press (1992) 129-186
10. Glasgow, J., Narayanan, N. H., & Chandrasekaran, B. (eds.): Diagrammatic Reasoning: Cognitive and Computational Perspectives. Menlo Park, CA: AAAI Press (1995)
11. Kotovsky, K., Hayes, J. R., & Simon, H. A.: Why are some problems hard? Cognitive Psychology, 17, (1985) 248-294
12. Larkin, J. H., & Simon, H. A: Why a diagram is (sometimes) worth ten thousand words. Cognitive Science, 11, (1987) 65-99
13. Palmer, S. E.: Fundamental aspects of cognitive representation. In: Rosch, E., & B. B. Lloyd, B. B. (eds.): Cognition and Catergorization. Hillsdale, N.J.: Lawrence Erlbaum (1978) 259-303
14. Scaife, M., & Rogers, Y.: External cognition: how do graphical representations work? International Journal of Human-Computer Studies, 45, (1996) 185-213
15. Simon, H. A.: Models of Discovery and other topics in the methods of Science. Dordrect: Reidel (1977)
16. Vincente, K. J.: Improving dynamic decision making in complex systems through ecological interface design: A research overview. Systems Dynamics Review, 12(4), (1996) 251-279
17. Zhang, J.: A representational analysis of relational information displays. International Journal of Human Computer Studies, 45, (1996) 59-74
18. Zhang, J.: The nature of external representations in problem solving. Cognitive Science, 21(2), (1997) 179-217

Using Brightness and Saturation to Visualize Belief and Uncertainty

Joseph J. Pfeiffer, Jr.

Department of Computer Science
New Mexico State University
Las Cruces, NM 88003 USA
pfeiffer@cs.nmsu.edu

Abstract. In developing a visual language for mobile robots, it is necessary to represent the uncertainty present in robot position, obstacle position, and even obstacle presence. In developing a visualization of the robot's model of its environment, this uncertainty should be presented to the experimenter, in order to be able to evaluate the extent to which the robot's sensors and sensor fusion rules are providing consistent and reliable information.

In Isaac, a project developing a rule-based visual language for mobile robots, a time-varying diagram is used to represent the robot's current world model. Hue is used to represent object classes, and brightness is used to represent the degree of belief of an object's presence. A region in which there is confidence that no object is present is shown as white, while a region with high confidence in the presence of an object is represented with color. Saturation is used to represent confidence in the assessment of object presence (or absence): a totally unsaturated (*i.e.* grey) area represents an area in which there is no belief at all either in favor of or against the presence of any object; a fully saturated area represents an area in which there is high confidence in the region's classification. The combination of hue to distinguish between object classes with brightness and saturation for belief and confidence results in a three-dimensional color space for model visualization.

Sensor characteristics are encoded in belief functions; upon receiving sensor information, both belief functions and confidence levels can be modified. Belief functions in the presence and absence of obstacles in the model are maintained through Dempster-Shafer evidential reasoning.

1 Introduction

An important consideration in programming a mobile robot is having the capability of visualizing the robot's model of its environment. The visualization includes aspects such as the robot's location and orientation relative to environment features known *a priori*, new features and obstacles which have been discovered by the robot in the course of exploring and interacting with its environment, and pseudo-objects added to the environment model such as markers identifying paths which have been explored and do not need to be re-examined.

M. Hegarty, B. Meyer, and N. Hari Narayanan (Eds.): Diagrams 2002, LNAI 2317, pp. 279–289, 2002.

A well-known characteristic of a robot's environment model is that information is not perfect. Sensors return spurious values (indicating the presence of nonexistent objects), return inexact values (giving rise to incorrect estimates of either object or robot location), and fail to return values at all (failing to recognize the presence of objects). Modifications to the model must take these uncertainties into account, and a visualization should also display it. Also, since objects in the model may overlap, the visualization must display that as well. In this paper, we discuss the use of saturation to represent uncertainty in the Isaac mobile robotic environment. Following this Introduction, Section 2 will briefly describe the Isaac geometric reasoning language. Section 3 will review Dempster-Shafer theory and describe its application and visualization in Isaac. Section 4 describes related work, and Section 5 will present some preliminary conclusions and future work.

2   Isaac

Isaac is a rule-based visual language for geometric reasoning, intended for the control of mobile robots[1]. As usual in rule-based languages, a rule has a left-hand side containing preconditions (facts which must be present to enable the rule), and a right-hand side with postconditions (changes which will be made to the environment as a result of executing the rule). A typical Isaac rule, implementing obstacle avoidance, is shown in Figure 1.

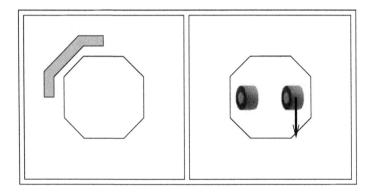

Fig. 1. Obstacle Avoidance Rule in Isaac

In this figure, the robot is shown as an octagon. The robot's direction of travel is upward in the figure. The region forward and to the left of the robot in the precondition is an avoidance region (in the actual environment and in the on-line version of this paper this region is colored red); if this region has a non-zero intersection with a similarly colored region in the robot's current environment model then the rule is enabled. The postcondition shows the result of activating the rule: two new objects (icons representing wheels, with the left wheel stopped and the right wheel turning in reverse) will be inserted into the environment. These objects represent actions to be taken by the robot; these are specialized output objects which will actuate the motors as specified[2].

For purposes of this introduction, objects in the world model are represented as fully-saturated geometric objects. This will be generalized in Section 3.2; in addition to hue, objects will also have variable saturation and brightness.

2.1   Rule Enabling and Application

Figure 2 shows a typical application of the rule shown in Figure 1.

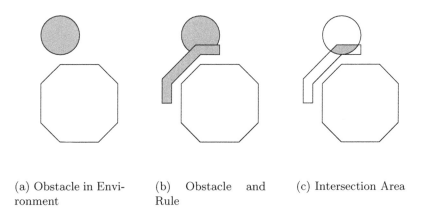

(a) Obstacle in Environment   (b)   Obstacle   and   Rule   (c) Intersection Area

Fig. 2. Rule Enabling in Isaac

In Figure 2(a), the robot is shown approaching an obstacle, and Figure 2(b) shows the obstacle avoidance rule precondition in combination with the environment. Finally, Figure 2(c) shows the intersection of the precondition with the obstacle. As the intersection is non-empty, the rule is enabled. Color is significant for rule enabling; a rule precondition is intersected only with objects in the environment with the same hue as the precondition (in the actual system, and in the on-line version of this paper, the object and the precondition are both red).

2.2   Rule Combination

Situations frequently arise in which more than one rule is enabled simultaneously. In these circumstances, conflicts between the rules must be resolved in order to select a course of action.

Isaac uses a weighted average to combine the rules. Rules have weights associated with them; when several rules are activated simultaneously and have incompatible right-hand sides, the result is the weighted average of the right-hand sides of the enabled rules.

As an example, consider a situation in which a robot is following a planned path, but must avoid an obstacle. The situation is shown in Figure 3. The straight line is the planned path; it is in the model but not in the actual environment. The area with both path and obstacle is shown in a striped pattern.

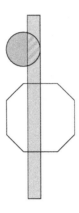

Fig. 3. Robot Following Path in Presence of Obstacle

In this situation, the robot has veered slightly to the right of the planned path, and the rule shown in Figure 4 will be activated to bring the robot back onto the path. This rule's postcondition stops the left motor while driving the right motor forward (in the actual implementation, and the on-line version of this paper, the path and corresponding object in the rule are shown in green).

The obstacle is also shown, as described above; consequently the object avoidance rule from Figure 1 is also active. These two active rules both call for the left motor to be stopped; the course correction rule calls for the right motor to go forward, while the obstacle avoidance rule calls for the right motor to go back.

In order to resolve the competing rule postconditions, weights are assigned to the rules.[1] A reasonable weighting for the two rules described here might be 1 for the line following and 10 for the obstacle avoidance; in this case, the net

---

[1] This is a change from previous descriptions of Isaac. In the original conception, rule weighting was defined in terms of the area of the matching rule precondition, as described in [1]. Our intent in using this definition was to avoid the necessity of defining rule weights; unfortunately, the effect was to make anticipating rule interaction nearly impossible.

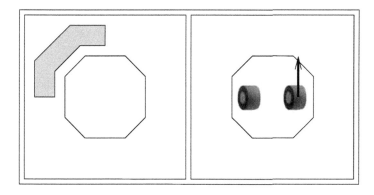

Fig. 4. Path-Following Rule

result would be that the left motor would stop (as both rules call for this), while the right motor would be given a value of $\frac{(1.0)(-1)+(.1)(1)}{1.0+.1} = -.82$

3 Uncertainty

The example in the previous section assumes perfect knowledge: knowledge of the location of the robot, and knowledge of both the presence and location of the obstacle. Neither of these is typically known in an actual environment. Instead, sensors provide data that is both inaccurate and unreliable. Sonar, in particular, is prone to false returns, failure to reliably generate echos on some substances, and spreading. Consequently, we can only regard sensor inputs as evidence (rather than certain knowledge) of possible features, and lack of input as evidence of lack of features. The solution to this will be to respond to sensor returns by putting objects in the model with a size and uncertainty derived from the characteristics of the sensors themselves. Multiple sensor returns will be fused to form a coherent picture of the environment.

3.1 Dempster-Shafer Belief Functions

Dempster-Shafer theory, and particularly Dempster's rule of combination, provides a means of explicitly maintaining uncertainty and combining evidence from multiple sources[3]. In Dempster-Shafer theory, propositions are represented as subsets from a set Θ of mutually exclusive alternatives for a parameter (referred to as a *frame of discernment*). In our case, the only two alternatives are the presence or absence of an object of a given color at a location in the world model, and the subsets are $\{\phi, T, F, \Theta\}$ where ϕ is the empty set, T and F are the presence or absence of an object, and Θ is $T \cup F$. A belief function $Bel(\theta)$ is used to represent the degree of belief in each of the subsets, where $Bel(\theta)$ must satisfy the following conditions:

1. $Bel(\phi) = 0$
2. $0 \leq Bel(T), Bel(F) \leq 1$
3. $Bel(\Theta) = 1$

A point which is implicit in this definition are that the sum of the belief in the two alternatives can be no greater than 1 (as $Bel(\Theta) = 1$), but can be less than 1, allowing some portion of the total belief to be "unallocated."

Dempster's rule of combination provides a means of updating these belief functions in the presence of new evidence. The interval $[0,1]$ is divided into three parts according to the belief functions:

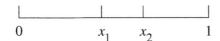

where $x_1 = Bel(T)$ and $x_2 - x_1 = Bel(F)$. The remainder of the interval is the unallocated portion of the belief. To combine two belief functions, the line segments representing their belief functions are combined as in Figure 5. The relative areas of the nine regions of the figure provide the updated belief function.

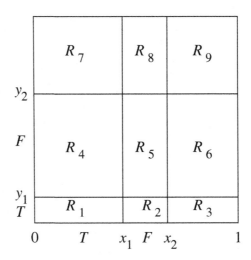

Fig. 5. Orthogonal Combination of Belief Functions

Upon combining the two functions, we have a total of nine regions. In the result, region R_9 reflects that part of the belief function which remains unallocated. The combined areas of $R_1 + R_3 + R_7$ represents the new belief assigned to T; the combined areas of $R_5 + R_6 + R_8$ represents the new belief assigned to F. Regions R_2 and R_4 are incompatible assignments (they represent a belief that both of the mutually exclusive alternatives are true), so the new values are normalized by dividing them by the total area of the compatible assignments, $1 - (R_2 + R_4)$.

As an example, consider a situation in which we presently believe an area to be clear of obstacles with certainty 0.7. For compactness, we will represent a belief assignment with a tuple $(Bel(T), Bel(F), 1-(Bel(T)+Bel(F)))$. The third component in the tuple is redundant, however we prefer to show the unallocated belief explicitly. The belief assignment in this case is $(0, 0.7, 0.3)$. Now assume a sensor input is received indicating that there is an obstacle in the region; we have a confidence of 0.8 in this sensor, so its belief assignment is $(0.8, 0, 0.2)$. Combining the sensor input with our previous belief function, the new belief assignment is $(0.14, 0.54, 0.32)$.

3.2   Visualizing Uncertainty

In visualizing the belief function, we map $Bel(T)$ to saturation and $Bel(F)$ to brightness. The $Bel(T)$ mapping uses the full range of saturation from 0 to 1; the $Bel(F)$ mapping only uses brightnesses from 0.5 to 1, in order that darker lines can be used for emphasis. Combined with hue for object classification, this results in a three-dimensional (H, S, B) color space. The mapping of belief and uncertainty to a color is shown in Figure 6.

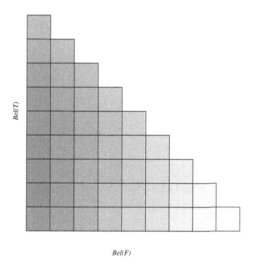

Bel(T)

Bel(F)

Fig. 6. Visualization of Belief and Certainty for Object

In this figure, the three corners have the following interpretation: the lower left corner is the visualization of a region whose contents are completely unknown. The saturation is 0, and the brightness is 0.5. The top left corner, with $Bel(T) = 1$ and $Bel(F) = 0$, has a saturation of 1 and brightness of 0.5. Finally, the lower right corner, with $Bel(T) = 0$ and $Bel(F) = 1$, has a brightness of

1. The figure is triangular, reflecting the fact that $Bel(T) + Bel(F) \leq 1$. In the on-line version of this paper, the upper corner is red.

The saturation of an object with a brightness of 1 is always 0, and its hue is undefined. This does not cause an inconsistency in Isaac, as Isaac rule preconditions can only be conditioned on the presence of an object and never on its absence.

3.3   Sensors and Sensor Rules

The rules exhibited in Section 2 are one of the three types of rules available in Isaac: actuator rules. Sensor rules also exist for adding objects to the environment model as a result of sensor input, and deduction rules are able to use objects currently in the environment to perform operations such as path planning.

Isaac's response to a sensor input is the invocation of an "input rule" to place an object in the robot's world model at a location determined by the robot's position and the sensor's parameters[2][4]. Due to the uncertainty in the sensor return, the object's location and extent are also uncertain. Consider a rangefinding sensor such as sonar. A reading from such a sensor indicates two things: that there is an object at the indicated distance and location, and that there is no object closer (in that direction). A rule representing this for a sensor with a 30° spread, and with a belief assignment of $(0, .5, .5)$ for the close region and $(.5, 0, .5)$ for the obstacle region would appear as in Figure 7.

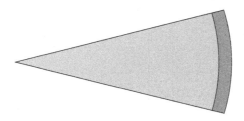

Fig. 7. Example Sensor Rule

A simulation has been developed to show the results of a series of sonar readings taken as a robot traverses a model room. The room, diagrammed in Figure 8, is a simulated 5 meters on a side with a single 1.25 meter square obstacle in its center. In this simulation, the robot proceeds across the left-hand wall while taking readings to the right. The robot moves one centimeter forward after taking each reading.

For purposes of this simulation, we use a belief assignment of $(0, .1, .9)$ for the close region and $(.1, 0, .9)$ for the obstacle region. For a first example, the sensor return is assumed to be perfect; the range returned is exactly the range to the nearest object within the sensor's spread. The results of the simulation are displayed in figure 9.

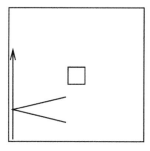

Fig. 8. Room to be Explored

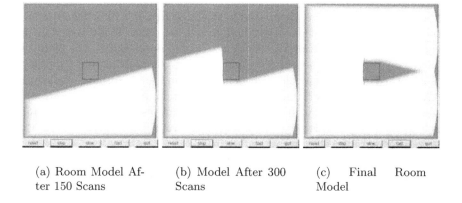

(a) Room Model Af-   (b) Model After 300   (c)   Final   Room
ter 150 Scans   Scans   Model

Fig. 9. Simulation of robot exploring room

Initially, we assume no information regarding the room contents, so the environment model is a uniform grey representing a belief assignment of $(0, 0, 1)$. After each sonar reading, the environment model is modified according to Dempster's Rule, and the robot proceeds forward. Following 150 readings, the appearance of the room model is as shown in Figure 9(a). In the interest of clarity, the outlines of the room and the obstacle are included in this Figure (even though it is not actually part of the model). In the simulation, and in the on-line version of this paper, a red arc is visible at the far extent of the white area, coincident with the far wall.

After another 150 readings, the room model will have changed as shown in Figure 9(b). At this point, the obstacle has also been located. More importantly from the perspective of this work, the extent to which the robot has belief in the presence or absence of obstacles, and the areas which have not yet been explored, are clearly visible.

The process continues as the robot makes its way across the room; the final model is shown in Figure 9(c). As the overlapping sensor returns have been fused, the system shows higher confidence in the lack of obstacles in those regions which have returned "empty" several times, and likewise shows greater confidence in the presence of obstacles in the regions that have had multiple echo returns.

3.4 Uncertainty and Rule Activation

In Section 2.1, rule enabling is presented as the intersection between the robot's environment model and the preconditions of a rule. In the presence of uncertainty, this is modified: a rule is enabled to the extent that an object is believed to be present in the area defined by a rule's precondition. Objects in which the robot has only a slight belief have only a slight effect on the robot's behavior.

4 Related Work

The previous work most closely related to this is Anderson's Inter-Diagrammatic Reasoning[5][6]. As with this work, Anderson's diagrams divide the plain into polygons with common properties (he refers to this as tesselating the plain); the properties of these tesserae are represented using color. In his most developed version of the theory, these colors are selected from a cyan-magenta-yellow subtractive color space. Operations are defined for manipulating diagrams; these are gemerally similar in nature to fuzzy logic operations (for instance, the intersection of two diagrams is defined by taking the minimum of each of the three color coordinates at each point in the plain).

Inter-diagrammatic reasoning is more purely a diagrammatic reasoning system than Isaac's uncertainty visualization. Anderson defines a semantics of operations on diagrams based on color, while this work's use of color is only as a visualization tool. It would be possible to redefine Dempster's Rule in terms of operations on the saturation and brightness of the polygons, however, this would not be fruitful.

Dempster-Shafer theory has been used in robot localization (with a much more detailed sonal model than described here) by [7]. This paper contains figures representing belief functions from sonar returns; the authors use a series of figures (one figure for presence, one for absence, and one for unallocated belief) for a single belief assignment, with intensity representing a component of the belief assignment.

5 Conclusions

We have described a visualization of belief functions for uncertain geometric reasoning. This visualization is well-suited for use with Isaac's rule processing mechanism, as it extends the previously one-dimensional color space used by Isaac (hue) into a three-dimensional space better making use of the capabilities of the display device.

Our present work is focused on creating more detailed simulations of Isaac's behavior on mobile robots, developing models of our sensors in order to define rules using them, and merging models produced by several robots in a single, global model.

References

1. Pfeiffer, Jr., J.J.: A language for geometric reasoning in mobile robots. In: Proceedings of the IEEE Symposium on Visual Languages, Tokyo, Japan (1999) 164–171
2. Pfeiffer, Jr., J.J., Vinyard, Jr., R.L., Margolis, B.: A common framework for input, processing, and output in a rule-based visual language. In: Proceedings of the IEEE Symposium on Visual Languages, Seattle, Washington, USA (2000) 217–224
3. Shafer, G.: A Mathematical Theory of Evidence. Princeton University Press (1976)
4. Vinyard, Jr., R.L., Pfeiffer, Jr., J.J., Margolis, B.: Hardware abstraction in a visual programming environment. In: Proceedings of the International Multiconference on Systemics, Cybernetics, and Informatics, Orlando, Florida, USA (2000)
5. Anderson, M., McCartney, R.: Inter-diagrammatic reasoning. In: Proceedings of the International Joint Conference on Artificial Intelligence, Montreal, Canada (1995)
6. Anderson, M., Armen, C.: Diagrammatic reasoning and color. In: Proceedings of the 1998 AAAI Fall Symposium on Formalization of Reasoning with Visual and Diagrammatic Representations, Orlando, Florida (1998)
7. Hughes, K., Murphy, R.: Ultrasonic robot localization using dempster-shafer theory. In: SPIE Neural and Stochastic Methods in Image and Signal Processing, San Diego, CA, Society of Photo-Optical Instrumentation Engineers, Society of Photo-Optical Instrumentation Engineers (1992) 2–11

Structure, Abstraction, and Direct Manipulation in Diagram Editors

Oliver Köth and Mark Minas

Lehrstuhl für Programmiersprachen
Universität Erlangen-Nürnberg
Martensstr. 3, 91058 Erlangen, Germany
minas@informatik.uni-erlangen.de

Abstract. Editors for visual languages should be as simple and convenient to use as possible; at the same time, programmers should be able to create such editors without prohibitive effort. We discuss the benefits that can be gained from combining the following aspects in an editor-generator approach:

- direct-manipulation editing (as in drawing programs)
- structure-based editing (as in common diagram tools)
- structural analysis and a common formal model

As a major practical example, we present an editor for UML class diagrams. We show that direct-manipulation editing capabilities can enhance the usability of such an editor in comparison to standard tools. A further improvement is obtained by including selective abstraction features similar to the well-known "fisheye-viewing" and "semantic zooming" paradigms. We show that the proposed generator architecture provides an excellent base for implementing such features. The resulting technique can be applied to a wide range of different diagram languages; in contrast to other general solutions, it takes into account the abstract structure and specific abstraction features of the individual languages.

1 Introduction and Overview

Working with visual languages requires appropriate tools, particularly editors, that are specially tailored to the respective language. It is possible to use a general-purpose drawing program to create, for example, UML diagrams. But this support immediately turns out to be insufficient when it comes to *working* with a language, i. e. modifying and extending those diagrams iteratively and extracting their "meaning" for use with other tools, e. g. a code generator. Consequently, for visual languages, language-specific editors are a must.

In this paper we present DiaGen, an editor-generator toolkit, which allows to create editors for specific diagram languages[1] with comparatively little effort. In the next section we contrast two different editing modes and corresponding

---

[1] In this paper, we use the terms "visual language" and "diagram language" as synonyms.

M. Hegarty, B. Meyer, and N. Hari Narayanan (Eds.): Diagrams 2002, LNAI 2317, pp. 290–304, 2002.
© Springer-Verlag Berlin Heidelberg 2002

architectures, syntax-directed editing and direct manipulation editing, and show how they have been combined in a general concept (Section 3). Section 4 gives an example how this approach can be applied to a complex VL by the example of UML class diagrams, and why direct manipulation can be an improvement over conventional tool behavior. Section 5 explains why support for diagram abstraction is important for working with large diagrams and Section 6 shows how language-specific abstraction views can be supported in a diagram editor through the use of diagram transformations and diagram parsing information. We also consider how these features affect the automatic layout correction. Section 7 discusses the advantages and disadvantages of the presented abstraction technique, and Section 8 shows how it relates to other work. Finally, Section 9 repeats the main results of this work and indicates directions for future research.

2 Structure-Based Editing and Direct Manipulation

Programming specialized VL editors requires a lot of effort. A conventional architecture for such editors is built around an abstract internal representation of the diagram; in the case of UML diagrams that would be an object model based on the UML metamodel [18]. Most diagram modifications are accomplished through structure-based editing operations that modify this internal model, e. g. to add an attribute to a UML class. The actual diagram is a visualization of this internal model according to the rules of the diagram language as they are stated, for example, in the UML Notation Guide.

The whole editor architecture is highly dependent on the internal model and, since a VL typically has its own internal model, an editor for a new language must be implemented from scratch. That means that the effort to implement new experimental languages is quite high.

Editor toolkits or generators like DIAGEN [14] aim to improve this situation and support the programmer in creating a specialized editor for a given diagram language. The programmer only needs to specify the graphical symbols of the language and the rules that describe how these symbols can be arranged to form correct diagrams. The toolkit then generates a complete visual editor that can typically output some sort of "abstract description" to be used by other tools.

As the specification of a new VL should be as easy and concise as possible, the generator should not force the programmer to define a complete set of language-specific editing operations that allows convenient creation and manipulation of all language features. Instead, generated editors often allow the user to directly manipulate the visual symbols of the language and the abstract internal model is generated by analyzing the visual representation of the diagram.

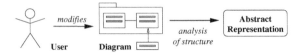

An important characteristic of this editing style is that it is possible for the user to create "diagrams" that violate the rules of the language. While this may seem to be a disadvantage, we have found that it often greatly simplifies the editing process. For example, it is possible to duplicate parts of a diagram using "copy & paste" features and fix up the arising errors (unconnected links etc.) later.

In contrast to a simple drawing program, such a diagram editor knows about the rules of the diagram language and the abstract "meaning" of the diagram. It can thus indicate, which parts of the diagram conform to the rules, and it can execute automatic layout corrections and beautifications; as a very simple example, when a class box in a UML diagram is moved, all attached links should stay connected to it.

3 DiaGen

The DIAGEN toolkit combines the principles of direct manipulation and syntax-directed editing and allows to modify the visual and abstract representations of a diagram at the same time: A concise specification language is used to define the syntax of the VL; the specification document is then processed by a program generator to create a simple direct manipulation editor customized to that language. (Direct manipulation takes place by dragging "handles" that determine the shape and position of symbols.) If necessary, the specification can be extended with syntax-directed editing operations that manipulate the internal diagram representation [12,16]; those "diagram transformations" combine elementary modifications of the diagram in the way of a powerful macro capability. The toolkit as well as the generated editors are all expressed in the Java programming language.

To achieve this integration of direct manipulation and syntax-directed editing, DIAGEN uses attributed graphs[2] as a common base for the internal representation of different VLs. This allows us to use graph grammars as a common means for the analysis of diagrams; graph transformations can be used to model syntax-directed editing operations in a language-independent manner. It should be noted that this internal graph representation does by no means restrict the application of the DIAGEN toolkit to graph-like VLs; the representational form is suited to almost any kind of diagrammatic language. It has been applied to UML class diagrams, petri nets, finite automata, control flow diagrams, Nassi-Shneiderman diagrams, message sequence charts, visual expression diagrams, sequential function charts, ladder diagrams etc.

---

[2] Actually, DIAGEN uses hypergraphs whose edges are allowed to visit more than just two nodes.

Figure 1 shows the architecture of a diagram editor based on DiaGen (for details see [15,16]): Direct manipulation is supported by the *drawing tool*. The resulting diagram is transformed into a (hyper-)graph model by the *modeler* after each diagram manipulation. *Reducer* and *parser* analyze this model; this is similar to traditional compiler techniques for textual languages: Reducer and parser correspond to lexical and syntactic analysis, resp., however based on graph grammars instead of string grammars. Syntax-directed diagram modifications are carried out by the *hypergraph transformer* which modifies the hypergraph model by graph transformations. Please note that results of any "diagram transformation" are analyzed by reducer and parser, i. e. syntax-directed editing operations cannot change the syntax of the VL [15,16].

Figure 1 also shows how semantic analysis *attribute evaluation* and *layout correction* are integrated into the generated editor: Structural information which is gathered during diagram analysis is used by the *layouter* which adapts the diagram layout after each diagram modification (cf. Section 6).

Generated editors can be easily extended to support more convenient editing operations and new language constructs. The tight integration with the Java language allows to freely extend the user interface as well as other aspects of the editors and to use them as modules (Java-Beans) in larger programs. The toolkit as well as some documentation and publications is available at http://www2.cs.fau.de/DiaGen.

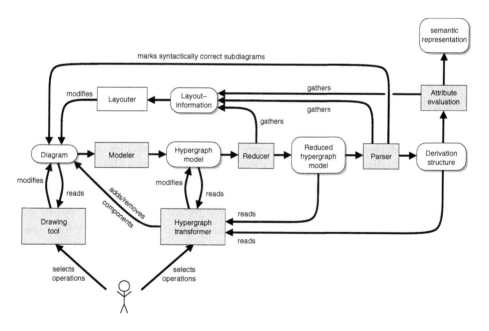

Fig. 1. Architecture of a diagram editor based on DiaGen.

4   Diagram Editing Using Direct Manipulation

Former publications on DIAGEN have centered on the theoretical concepts and underlying mechanism for diagram analysis and transformation. In this paper we want to focus on the application of the toolkit and show the advantages that the approach presents in contrast to other methods.

As a continuous example we will use (a subset of) UML class diagrams; we assume that the reader is familiar with the most frequently used parts of this notation (classes, the different link types, attributes and packages). We have implemented a DIAGEN editor for this type of diagrams and we believe that it strongly supports our claim that the techniques outlined above can indeed be used for complex diagram languages with practical relevance. With relatively little effort, it was possible to include support for a lot of notation variants that even commercial tools do not generally support (graphically nested packages, association classes, associations between more than two classes etc.). The internal graph representation of this editor is modeled after the standard UML meta-model and thus provides an adequate base for interfacing the editor with other CASE tools. To demonstrate this, we have implemented an XMI export feature, so we are able to exchange the created diagrams with other XMI-conforming tools. e. g. Together[3].

When we compare such a direct manipulation editor to standard (commercial) editing tools for UML diagrams, we find in fact that it often provides more convenient and direct ways to achieve a desired diagram modification.

As a simple example, consider the task of moving an attribute from one class to another. If a syntax-directed editor does not provide a special operation "move attribute", the only possible solution is to delete the attribute and re-create it in the destination class. In a direct manipulation editor, the user can simply drag the graphic symbol for the attribute from one class to the other. The diagram analysis recognizes the modified structure and updates the abstract representation; the layout correction takes care of resizing the class frames and aligning the attribute properly at its new place. Other examples one could think of could be "redirecting" an association that has some associated information (name and multiplicities) from one class to another or the duplication of arbitrary diagram parts using "copy & paste", when the resulting diagram is not structurally correct, but needs to be completed by additional editing operations.

From a more general point of view, a syntax-directed editor always forces the user to think in terms of "deep" modifications of the abstract model represented by the diagram. Direct manipulation features, in contrast, permit the user to additionally consider "shallow" modifications of the visual representation, which may sometimes be more direct and convenient and fit better into the immediate visual perception.

Figure 2 shows a screenshot of the UML editor. Several features relating to direct manipulation are visible here:

---

[3] http://www.togethersoft.com/

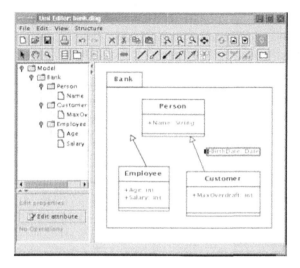

Fig. 2. Screenshot of a UML editor supporting direct manipulation

– The toolbar at the top of the window allows to create all kinds of elements that may occur in the supported UML class diagrams; there is no need to search through any menus, e. g. to add a stereotype to a class or a link.
– The selected diagram element (the "BirthDate" attribute) has been dragged out of the "Customer" class and is going to be moved to the "Person" class. Currently it is, of course, not placed correctly and therefore not recognized by the diagram analysis.
– The generalization arrow originating from the "Employee" class does not end at a class and is therefore incorrect. This is indicated by giving it a different color, which cannot be seen in the figure due to the grayscale display. To correct this error, the user would have to select the arrow and drag the endpoint into the "Person" class.
– Whenever parts of the diagram are manipulated, the automatic layout correction tries to adapt the whole layout. For example, the generalization arrow between the "Person" and "Customer" classes ends exactly at the boundaries of those elements, and the "Bank" package frame is shaped as the enclosing rectangle of its parts plus some border width.

The lower left window pane allows to execute syntax-directed editing operations; for the selected element (an attribute), there are no such operations available. Like in many other CASE tools, the tree display on the upper left window pane shows a different view on (parts of) the abstract internal diagram representation; in our case, it is generated from the results of the diagram analysis

Of course, similar features can also be included in standard (syntax directed) editors; for example, it is possible to implement drag & drop support for attributes and trigger operations to modify the respective classes accordingly. The point is, that all these manipulations would have to be programmed explicitly, whereas the combination of direct manipulation, diagram analysis and automatic layout correction provides them almost "for free"; the programmer only needs to specify the structure of the language (and the layout), but does not need to care about every possible way to modify it. In fact, direct manipulation may permit manipulations of the diagram that the programmer has not even thought of.

For more complex modifications that are frequently needed, our architecture still allows to define powerful syntax-directed editing operations. As an additional feature, because all manipulations are ultimately broken down into a small set of elementary steps (insertion/deletion of graph parts, modification of attributes), it is comparatively easy to implement a multi-level undo/redo capability that works across all editing operations.

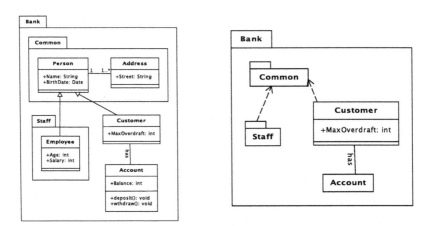

Fig. 3. A sample UML diagram and a "Focus and Context" view

5   Working with Abstraction

We have found that the combination of graph parsing and graph transformation that DIAGEN uses to analyze and modify diagrams is particularly well suited to express the concept of *abstraction* in diagram editing. Editor support for abstraction is necessary, because, with most VLs, when you scale them from toy applications to practical problems, the visualizations often get uncomfortably large and it is not possible to display all the information readily on a single computer screen. To permit the user to keep an overview of the whole system, it is necessary to suppress detailed information and display it in a simplified form (a principle that has been termed "semantic zooming" [2]).

To permit convenient working with large diagrams, an additional key feature is *selective abstraction*: Normally the user is working on a restricted region of the diagram (the "focus" area) and needs to see that in detail. The surrounding regions (the "context"), however, should still be visible so that it is clear how the focus area fits into the context and what other diagram parts may be affected. Therefore it is desirable to "shrink" the context visually (by using abstraction) so that a larger context area can be displayed. This can be achieved by selectively abstracting the context while keeping the focus area unchanged.

Figure 3 demonstrates how this idea can applied to UML class diagrams. The left side shows a (very simplistic) model of a banking system that contains several nested packages. The diagram on the right side has been modified to facilitate working with the "Customer" element (the focus): all the details of this element remain visible, while the surrounding diagram parts only show their name and their interconnections and hide all details that are assumed to be irrelevant for the current task. The abstraction reduces the total size of the diagram, therefore it can be displayed at a larger scale (using standard optical zoom) so that the interesting details are better visible. Evidently, this technique yields most benefits, when it is applied to large models, where even a screen full of information can only show part of the whole system at an acceptable detail (text) size.

This *focus and context* viewing principle has been introduced by Furnas in 1981 [9] and since been used in a variety of applications; the best known implementations are probably optical nonlinear magnification systems, also known as "fisheye lenses" [10,23]. In contrast to such systems, DIAGEN offers the possibility to use structure-based abstraction instead of optical distortion, because the generated editors know about the abstract structure of the diagram and the graph-based internal models offer a standardized way of dealing with it.

Most visual languages already provide explicit means for abstraction. Two examples from the UML that we have integrated into our editor example are:

- the abstraction of classes by "hiding" their attributes and operations (cf. the "Account" class in Figure 3), and
- the abstraction of packages by "hiding" all contained elements (cf. the "Common" and "Staff" packages in Figure 3).

As can be seen in the example, the abstract forms for both classes and packages have visualizations that differ slightly from the standard forms, to distinguish them from elements that are merely empty and do not contain any hidden information. This implies that, like many other parts of the editor, the handling of abstraction must be tailored to the specific VL.

In fact there may be even more language-specific adaption required; in the case of UML package abstraction we need to redirect links that connect hidden package contents to the "outside" so that they end at the package. (In Figure 3, this is the case for the two generalization arrows ending at the "Person" class). However, according to the UML semantics, this way of redirecting generalization or association arrows changes the meaning of the diagram and may not even lead to a correct diagram. We can deal with this problem by turning those arrows into dependency arrows, which simply imply that there is some connection between the outside element and the package (which, in our case, cannot be shown in detail unless the package contents are expanded again). Here we have an example for an adaption that obviously needs to be coded explicitly with specific knowledge about the VL semantics.

6   Abstraction by Means of Diagram Transformations

Now the architecture of the DIAGEN system, which we have sketched in Section 2, allows to treat this kind of abstraction as just another form of diagram transformation. That means that the specification of a diagram language is augmented with syntax-directed editing operations that

- remove details from the diagram (and store them, so they can be re-inserted later),
- mark abstract elements visually or (optionally) replace them with different abstraction-representations,
- and perform other necessary adaptations like the arrow modifications mentioned above.

Of course, "reverse" transformations are also required to refine the diagram and re-insert the abstracted details. Note that this concept does not require specific geometric relationships between the "detail" and "higher-level" elements; in most cases, the details are visually contained in the higher-level elements, but, to give a different example, we have also used the same technique to implement an editor for rooted trees that allows to selectively hide all subtrees rooted at designated nodes. In this case, the "details" (lower-level nodes) are indicated not by containment but by the interconnection of the nodes; still the results of the diagram analysis allow to find all subordinate elements for a certain node easily.

Like the UML example shows, such abstraction transformations are usually based on the structural hierarchy of the diagram (although the concept also allows for other applications like the abstraction of selected parts of "flat" petri nets). This hierarchical structure is also expressed in the grammar for a diagram language. In our example, all the contents of a UML package are derived from a "Package" nonterminal symbol at some level of the grammar. DIAGEN uses a sort of extended context-free graph parser for the analysis of diagrams, which creates a derivation structure (cf. Figure 1) in the form of a directed acyclic graph (DAG). The graph transformations that describe abstractions can make use of this DAG structure to select the diagram parts that are to be transformed. (If necessary, they may additionally include selection patterns that refer to the geometric relations between diagram parts, like "arrow a connects to class c" or "package p_1 is contained in package p_2".)

Figure 4 gives an idea of how the derivation structure can be used for selecting the details of a certain entity in the diagram, i. e. all the diagram elements that should be hidden by an abstraction transformation. It shows part of the (simplified) derivation structure that is created by analyzing the UML diagram on the left side of Figure 3. The lowest level contains the visual elements of the

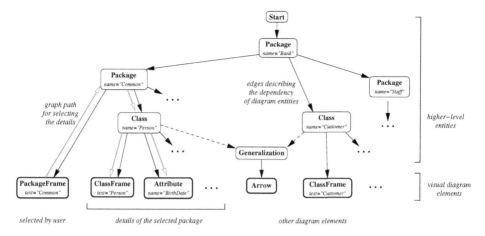

Fig. 4. Retrieving hierarchy information from the derivation structure of a diagram

drawing, which serve as terminal symbols for the grammar.[4] The other elements of the graph structure are higher-level entities (nonterminal symbols) found by the analysis module; the two types of black arrows represent how they depend on each other. Selected attributes of the symbols are also shown to make clear how the derivation structure relates to the actual diagram.

We have found that, in most cases, abstraction transformations can be defined by a simple standard pattern: The selection starts from a diagram element selected by the user (e. g. a package frame), then it walks "up" in the derivation structure to the first nonterminal symbol of a certain type (e. g. "Package") representing the desired higher-level entity; finally it follows the edges "down" again to include everything derived from that nonterminal. The gray arrows in the figure illustrate this selection process; it can be described by *path expressions* on the derivation DAG quite similar to *XPath* expressions [6] for XML documents. Matching entities are then handled by graph transformation rules, i. e. by the hypergraph transformer (cf. Figure 1). Thus the graph-grammar approach is very well suited to describe diagram abstraction.

The basic abstraction transformations for individual diagram units (like a single class or package) can be combined by higher-level transformations to create different views of the whole diagram. For example, the principles described by Furnas [9] can be used to compute a "degree of interest" for every diagram element, which can then be used to determine how far it should be abstracted, resulting in a view with a selected focus element shown in detail and a simplified context.

---

[4] In fact, these terminal symbols are found by the *reducer* (cf. Figure 1) which performs a preprocessing step that also takes into account the geometric relations (containment etc.) between the elements; a more detailed description of the analysis process can be found in [14,16].

This desired focus and context effect stems from the fact that abstraction visualizations take up less screen space than the diagram part they represent. Obviously, if they just remained in their previous positions, the result would be a very "sparse" diagram of almost the same size with lots of free space between the small abstractions. To achieve the desired result, the layout correction mechanism must be able to handle these size changes properly.

Layout corrections for a DIAGEN editor (cf. Figure 1) must, of course, be specified individually, along with the diagram language. DIAGEN supports a constraint-based layout mechanism[5] as well as directly programmed layout algorithms. Since most diagram languages do not permit overlapping elements, the case of "growing" diagram parts (after refinement) can typically be handled by the standard layout procedure that pushes overlapping elements apart.

The case of "shrinking" diagram parts (after abstraction), on the other hand, must be addressed explicitly, because a "spread-out" layout does generally not violate the rules of the diagram language and should therefore not always be corrected automatically. Therefore the abstraction operation must explicitly request a "layout compaction" to use the free space. In the case of a constraint-based layout this might be done by inserting temporary "weak" constraints, which strive to reduce the distance between certain diagram elements as long as this does not violate other "hard" correctness constraints.

In the UML example, we have used a programmed layout based on *force-directed* layout algorithms (cf. [4]) to handle the class and package placement. The definition of adequate attracting or repelling forces between layout elements (which take into account their size and shape) allows to achieve the desired expansion or contraction of the diagram while preserving the relative element positions. We have found that the force-directed layout principle is a very flexible and powerful tool for expressing such interactive layout corrections; even if multiple elements change their size and shape simultaneously, we obtain a visually pleasing result.

In contrast to most applications of force-directed layout algorithms, we are not dealing with the *creation* of a complete layout from an abstract graph representation only, but instead, we are merely using them to *adapt* an existing layout to relatively minor changes. As Misue et al. [17] have already stated, this situation is ideally suited to heuristic optimization techniques like those employed by force-directed layout: We do not have to look for a globally optimal layout from a randomly chosen initial setup, but instead only need to do a few iterations to find the nearest local optimum from a relatively good start position. Complex optimizations like the minimization of link crossings can be left to the user; automatic corrections of this sort would appear to be confusing.

---

[5] based on the QOCA constraint solver [13] from Monash University, http://www.csse.monash.edu.au/projects/qoca/

7 Discussion

When we compare our approach of treating abstraction as a special case of diagram transformation with other focus and context techniques, we find that it provides some important advantages:

- The proposed technique can be applied across a wide range of diagram types, but can always be customized to the requirements of the specific diagram language. The resulting displays conform to the language rules and are easy to understand.
- Abstraction operations are well integrated with the normal editing behavior, which means that, instead of adding an extra layer of visualization control, the display can be manually adjusted and fine-tuned using standard editing functions.
- Features of "normal" diagram transformations directly extend to abstraction operations. In particular, transformations can be undone and redone to an arbitrary depth and the transitions can be animated smoothly; both capabilities improve the usability of the editor (cf. [23]).

Of course, there are also some drawbacks:

- Abstraction operations must be implemented separately for every diagram language. Fortunately, we found that they usually employ a simple common scheme (cf. Section 6), thus their specification is usually short and can be partly automated for general cases.
- Layout support for the changing size of diagram parts is expensive to program. We hope to remedy this by providing a library of customizable layout algorithms for common diagram classes (e. g. graph-like diagrams).
- Abstraction operations modify the internal representation of a diagram instead of merely providing a different view on them. This implies that it is up to the programmer to ensure that the "meaning" of the diagram remains consistent; it also does not allow to present simultaneous views on the same diagram with different abstraction levels in separate windows.

While the proposed technique is certainly not a panacea for dealing with complex VL descriptions, we think that it may serve as a valuable tool for easily creating VL editors that can be conveniently used for large, "real-world" problems. Some preliminary experiments have been conducted which support this hypothesis: Some subsystems of the DIAGEN editor framework have been re-engineered to UML class diagrams. Students have been asked to compare a commercial UML tool with the UML class diagram editor with abstraction support. They confirmed that those abstractions offered valuable help for faster software understanding.

8 Related Work

There are several other approaches for generating visual language editors from a formal specification. However, we know about no approach whose editors support abstraction with varying level of detail as it has been described in Section 5. All of the existing approaches fall into one of two categories: the ones *with* a graph-like model and the ones *without* such a model. DIAGEN is a member of the first category. Other approaches and tools which are based on a graph-like model are GENGED [1], KOGGE [8], VISPRO [24], and the approach by Rekers and Schürr [20]. With the exception of DIAGEN and Reker's and Schürr's approach, all of them support syntax-directed editing only. Editing operations primarily modify the graph-like model, and the visual representation of the model is adapted accordingly. Direct manipulation – as described in Section 2 – which requires graph parsing is only supported by DIAGEN and Reker's and Schürr's approach.[6] Unfortunately, their approach has never been implemented whereas DIAGEN has already been applied to many different diagram languages.

VLCC [7] and PENGUINS [5] are examples of the approaches of the approaches which do not depend on an internal graph-like model, but parse the diagram directly. Whereas VLCC makes use of *Positional Grammars*, PENGUINS uses *Constraint Multiset Grammars* for specifying diagram syntax. These approaches, therefore, also support direct manipulation editing, but – unlike DIAGEN and Reker's and Schürr's approach – they do not support syntax-directed editing which requires the use of an abstract internal model.

Our approach to diagram abstraction builds on a lot of work about information visualization. It fills a gap between optical fisheye techniques as described, for example in [10,23], and solutions for specific diagram types, like hierarchical graphs [21,17]. The latter do not provide a general basis for other diagram types, while the former can deal with arbitrary information but cannot be customized to follow the rules of a specific visual language and result in unnaturally distorted views. In [22] an empirical study is presented with the result that focus and context techniques indeed lead to improved usability compared simple zooming. "Semantic zooming" toolkits [2,3,19] provide a general base for applications using abstraction views, but they require that the shape and size of objects remain constant at different detail levels and they cannot support selective abstraction to create focus and context views.

The layout techniques that we use for the UML editor are very similar to those described by Misue et al. [17], who especially advocate the use of force-directed layout algorithms for layout adjustment as opposed to layout creation. Brandenburg et al. [4] give a broad overview over different variants of force-directed graph layout algorithms. A complete discussion of the background and implementation of our abstraction technique, the UML example and the layout algorithms used for it can be found in [11].

9   Conclusion

We have shown that the combination of direct manipulation and parsing techniques can be used to create editors for complex diagram languages and may

---

[6] VISPRO does not support direct manipulation editing although it depends on graph parsing and could, therefore, support direct manipulation.

provide more intuitive manipulation capabilities than structure-based editing alone. The use of graphs as a common framework for specifying abstract diagram representations makes it possible to augment these editors with powerful syntax-oriented editing operations.

Such editing operations can also be used as a means to support language-specific abstraction of diagram parts. They can be easily defined based on information about the hierarchical structure of a diagram that is inherently contained the grammar describing the VL and the parsing structure for an actual diagram. By combining individual abstraction operations, it is possible to generate focus and context views of complex diagrams. These present the user's area of interest in sufficient detail and still show how it fits into the whole diagram, and thus make working with large VL constructs lot easier. To create such focus and context views, the diagram layout correction needs to provide special support for the changing sizes of diagram parts.

The principle of direct manipulation can still be taken a step further by supporting pen-based editing (e.g., [5]): When creating a diagram, editors force the user to select among several diagram component types and thus force the user into a editor-specific way of editing. Pen-based editing, i.e. sketching a diagram with a pen and diagram recognition by the editor, would simplify the editing process. For instance, such an editor could be used as an easy means for sketching ideas in group discussions, e.g., when analyzing and designing software with UML diagrams. Future work will examine how the fault-tolerant parser of DiaGen can be used for recognizing diagrams which have been sketched with a pen. Pen-based editing will profit from current DiaGen features, in particular from abstract views: Instead of sketching diagram components with full detail level (e.g., left side of Figure 3), the user may sketch a simpler abstract diagram (e.g., right side of Figure 3) first and add more details later.

References

[1] R. Bardohl. GenGEd: A generic graphical editor for visual languages based on algebraic graph grammars. In *Proc. 1998 Symp. on Visual Languages (VL'98)*, pages 48–55, 1998.

[2] B. Bederson and J. Hollan. Pad++: a zooming graphical interface for exploring alternate interface physics. In *Proc. Symposium on User Interface Software and Technology 1994 (UIST'94)*, pages 17–26, 1994.

[3] B. Bederson and B. McAlister. Jazz: an extensible 2d+zooming graphics toolkit in Java. HICL Technical Report 99–07, University of Maryland, 1999.

[4] F. Brandenburg, M. Himsolt, and C. Rohrer. An experimental comparison of force-directed and randomized graph drawing algorithms. In *Proc. Graph Drawing 1995 (GD'95)*, LNCS 1027, pages 76–87, 1995.

[5] S. S. Chok and K. Marriott. Automatic construction of user interfaces pen-based computers. In *Proc. 3rd Int. Workshop on Advanced Visual Interfaces, Gubbio, Italy*, 1996.

[6] J. Clark and S. DeRose. XML path language (XPath). W3C recommendation 16 November 1999, W3C, 1999. http://www.w3.org/TR/xpath.

[7] G. Costagliola, A. De Lucia, S. Orefice, and G. Tortora. A parsing methodology for the implementation of visual systems. *IEEE Transactions on Software Engineering*, 23(12):777–799, Dec. 1997.

[8] J. Ebert, R. Süttenbach, and I. Uhe. Meta-CASE in practice: a case for KOGGE. In *Advanced Information Systems Engineering, Proc. 9th Int. Conf. (CAiSE'97)*, LNCS 1250, pages 203–216. Springer, 1997.

[9] G. W. Furnas. The fisheye view: a new look at structured files. Technical Memorandum #81-11221-9, Bell Laboratories, 1981.

[10] T. A. Keahey and E. L. Robertson. Techniques for non-linear magnification transformations. In *Proc. IEEE Symp. on Information Visualization, IEEE Visualization*, pages 38–45, 1996.

[11] O. Köth. Semantisches Zoomen in Diagrammeditoren am Beispiel von UML. Master thesis, University of Erlangen-Nuremberg, 2001. (in German).

[12] O. Köth and M. Minas. Generating diagram editors providing free-hand editing as well as syntax-directed editing. In *Proc. International Workshop on Graph Transformation (GRATRA 2000), Berlin*, March 2000.

[13] K. Marriott, S. S. Chok, and A. Finlay. A tableau based constraint solving toolkit for interactive graphical application. In *4th International Conference on Principles and Practice of Constraint Programming (CP'98), Pisa, Italy*, pages 340–354, Oct. 1998.

[14] M. Minas. Creating semantic representations of diagrams. In M. Nagl and A. Schürr, editors, *Int. Workshop on Applications of Graph Transformations with Industrial Relevance (AGTIVE'99), Selected Papers*, LNCS 1779, pages 209–224. Springer, Mar. 2000.

[15] M. Minas. *Specifying and Generating Graphical Diagram Editors [in German: Spezifikation und Generierung graphischer Diagrammeditoren]*. Shaker-Verlag, Aachen, 2001.

[16] M. Minas. Concepts and realization of a diagram editor generator based on hypergraph transformation. Appears in *Journal of Science of Computer Programming (SCP)*, 2002.

[17] K. Misue, P. Eades, W. Lai, and K. Sugiyama. Layout adjustment and the mental map. *Journal of Visual Languages and Computing*, 6:183–210, 1995.

[18] Object Management Group. *Unified Modelling Langugage Specification*. http://www.omg.org/uml/.

[19] S. Pook, E. Lecolinet, G. Vaysseix, and E. Barillot. Context and interaction in zoomable user interfaces. In *Proc. Advanced Visual Interfaces 2000 (AVI'2000)*, pages 227–231, 2000.

[20] J. Rekers and A. Schürr. A graph based framework for the implementation of visual environments. In *Proc. 1996 Symp. on Visual Languages (VL'96)*, pages 148–155, 1996.

[21] M. Sarkar and M. H. Brown. Graphical fisheye views of graphs. In *Proc. Conference on Human Factors in Computing Systems 1992 (CHI'92)*, pages 83–91, 1992.

[22] D. Schaffer, Z. Zuo, S. Greenberg, L. Bartram, J. Dill, S. Dubs, and M. Roseman. Navigating hierarchically clustered networks through fisheye and full-zoom methods. *ACM Transactions on CHI*, 3(2):162–188, 1996.

[23] M. Sheelagh, T. Carpendale, D. Coperthwaite, and F. Fracchia. Making distortions comprehensible. In *Proc. 1997 Symp. on Visual Languages (VL'97)*, pages 36–45, 1997.

[24] D.-Q. Zhang and K. Zhang. VisPro: A visual language generation toolset. In *Proc. 1998 Symp. on Visual Languages (VL'98)*, pages 195–201, 1998.

On the Definition of Visual Languages and Their Editors

Paolo Bottoni[1] and Gennaro Costagliola[2]

[1] Dipartimento di Scienze dell'Informazione, Università di Roma, La Sapienza
[2] Dipartimento di Matematica ed Informatica, Università di Salerno

Abstract. Different diagrammatic languages are concrete variants of a core metamodel which specifies the way in which to express relations, and which is the basis for a semantic interpretation. In this paper, we identify families of diagrammatic languages exploiting the notion of metamodel as introduced in UML, i.e. through an abstract syntax, given as a class diagram, and a set of constraints in a logical language. The abstract syntax constrains the types of expressable relations and the types and multiplicities of the participating entities. The constraints express contextual and global properties of the relations and their participants. We propose a set of metamodels describing common types of diagrammatic languages. The advantages of this proposal are manifold: the analysis of constraints in the metamodel can be used to assess the adequacy of a type of language to a domain semantics and it is possible to check whether a concrete notation or syntax complies with the metamodel or introduces unforeseen constraints. Finally, we discuss how this characterisation allows the definition of flexible editors for concrete diagrammatic languages, where a specific editor results from the specialisation of some high-level construction primitives for the relevant family of languages.

1 Introduction

Diagrammatic languages are increasingly used to specify configurations, behaviours or protocols of complex, virtual or concrete, systems. For such languages to be useful, sufficiently precise definitions of the well-formed sentences in the language and of their semantics must exist. The intended domain semantics is often expressed through some formal abstract syntax, to which additional constraints and translation mechanisms can be attached. On the contrary, the set of well-formed diagrams is usually defined through some concrete syntax, often introducing irrelevant details and obscuring the relation to the semantics.

We explore the possibility of defining visual languages with a metalevel approach, which defines well-formed diagrams languages by an abstract syntax and a set of constraints for the notation. From this approach, it becomes apparent that several languages share a common abstract structure, which can be exploited in the construction of diagram editors or parsers, where a tool is obtained by progressively specialising a core of functionalities common to all classes. Specialisations progressively refine families of languages up to the characteristics of a specific language.

M. Hegarty, B. Meyer, and N. Hari Narayanan (Eds.): Diagrams 2002, LNAI 2317, pp. 305–319, 2002.
© Springer-Verlag Berlin Heidelberg 2002

As a working definition, a *diagrammatic language* is a language of images associated with an interpretation which considers as significant some well-defined spatial relations among graphical tokens. Hence, we do not consider data visualisation languages where metric properties of graphical tokens have semantical relevance. A diagrammatic sentence defines the (types of) entities one is interested in and the types of relations among these entities. Hence, types must be distinguishable from one another and no ambiguity may arise as to their interpretation. Moreover, the set of spatial relations to be considered must be clearly defined, and the holding of such relations among entities must be decidable.

The specific types of visual terms used in a diagrammatic language either derive from conventional notations employed in a user community or are elements specially designed to convey some meaning. In general, such types are used to denote entities in a given domain and domain relations are expressed via spatial relations, such as *touch, containment, adjacency, left_of*, etc. In this case we talk of *implicit representations* of (domain) relations. On the other hand, specific types of visual terms can be used to denote relations among elements. In this latter case, we talk of an *explicit representation* of a (domain) relation. An explicit representation is in any case founded on the existence of some spatial relations. For example, drawing an edge between two nodes requires that a relation of type *touch* exists between an edge end and a node border. However, when reasoning about explicit connections, one takes these relations for granted.

We propose to identify families of diagrammatic languages by exploiting the notion of metamodel as introduced in UML, i.e. through an abstract syntax, given as a class diagram, and a set of constraints in a logical language. The abstract syntax constrains the types of expressable relations and the types and multiplicities of the participating entities. The constraints express contextual and global properties of the relations and their participants. We propose a set of metamodels describing common types of diagrammatic languages, which could be used to assess the adequacy of a type of language to a domain semantics. Moreover, it is possible to check whether a concrete notation or syntax complies with the metamodel or introduces unforeseen constraints. Finally, we show how this characterisation allows the definition of flexible editors or parsers for concrete diagrammatic languages, where a specific tool results from the specialisation of some high-level construction primitives for the relevant family.

2 Related Work

Several authors have studied abstract definitions of visual languages as a basis to define formal semantics on them, or for the construction of visual tools.

Erwig introduced the notion of abstract syntax for visual languages, and adopted a graph-based approach to it [10], in contrast with the tree-based approach typical of abstract syntaxes for textual programming languages. While he is interested in defining abstract syntax graphs for specific visual languages, we focus here on the abstract description of the characteristics of the adopted diagrammatic languages.

On the other hand, approaches based on graph transformations usually exploit a distinction between a low-level and a high-level interpretation, possibly occurring on distinct graphs [16]. In the approach proposed here, we assume the existence of suitable realisations of geometry so that we can omit considering the low-level interpretation, and we concentrate on the definition of constraints relating the different components of a visual language. A recent proposal refers to *category theory* to characterise families of connection-based languages in terms of morphisms among elements [5]. Classes of languages were defined according to an Entity-Relationship approach in [9].

Metamodel approaches are gaining interest in the visual language community, following their success in defining the UML different visual languages. In particular, the UML approach combines diagrammatic sentences (usually class diagrams) and textual constraints for the definition of the semantics of visual languages. A metamodel approach is implicit in most generators of diagrammatic editors, in which at least an abstract notion of graphical object has to be defined. Most such generators, however, are based on the translation of some formal syntax into a set of procedures controlling user interaction, and on constructing the semantic interpretation of the the the diagram in a parsing process. Examples of such tools where the formal syntax is some variant of graph rewriting are Diagen [15] (based on hypergraph rewriting) and GenGEd [2] (in which constraints are used to define lexical elements and two separate graph rewriting systems are used, one to guide diagram construction, and one to define the parsing process).

In MetaBuilder [6], a designer defines a new visual language by drawing its metamodel class diagram, from which an editor is automatically generated. MetaBuilder does not provide support for the management of textual constraints, nor does it define basic metamodels for families of diagrammatic languages.

We are interested here not only in the definition of editors, but also of parsers. In particular, we want to define visual rewriting systems which are the basis for editing or parsing. We will discuss how the definition of a visual language as an element of a family of visual languages described at the metalevel, can be systematically translated into a pair of rewriting systems, an incremental one which provides the specification for an editor, and a grammar-like one, defining the parser. Moreover, as metalevel definitions are progressively refined, editing or parsing primitives can be derived from each level of refinement.

Such a perspective unifies the approaches to incremental editing and to parsing as separately developed by the authors in previous experiences. The GenIAL tool [4] allows the definition of connection-based languages starting from two basic classes *Entity* and *Connection*. A designer specialises *Connection*s by defining their arities and orientations. Moreover, attributes taking values in a set of basic primitive types can be associated with each token, thus defining classes of visual terms. Rules of a visual rewriting systems are interactively defined in accordance with the class of each term; conditions and a global check are expressed in textual form. An incremental editor is automatically generated from the specification. The Visual Language Compiler Compiler project [7], [8] allows the definition of the logical and graphical components of visual terms and of the attaching zones associated with them. The designer defines a positional grammar on the alphabet of visual terms. An editor is generated from the definition of the alphabet, taking care that placement of visual tokens is performed in accordance with the definition of the attaching zones, while the parser checks the acceptability of a generated sentence.

3 Visual Sentences

In order to proceed with our discussion, we introduce here the notion of visual sentence. A visual language is a set of visual sentences. In general, we will assume that membership in such a set is decidable. Informally, we consider a visual sentence as the combination of an image and its description as a diagram, i.e. one in which structures are identified and uniquely described. Structures are here considered as arrangements of pixels which satisfy some predicate. To be more formal, we start from an algebraic definition of the set of images.

Let $[\cdot] : \mathbf{N} \to \mathcal{P}(\mathbf{N})$ be the function defined by $[n] = \{1, \ldots, n\}$ if $n \geq 1$, $\{0\}$ otherwise. Given an alphabet of *colours* C, a *picture* π on C is a function $\pi : [\overline{r}] \times [\overline{c}] \to C$ where $\overline{r}$ and $\overline{c}$ are two integers. An element $(r, c, \pi(r, c))$ is called a *pixel* of π. A set of pixels in π is called a *structure*. A *pictorial language PL* on C is a subset of $\Pi(C)$, the set of all pictures on C. Two operations *shift* and *sup*, allow the generation of all pictures in any PL by arranging and superimposing elementary images from a finite set[1].

Let $\Sigma = \{t_1, \ldots, t_n\}$ be a finite set of *type symbols*, $\Gamma = \{\gamma_1, \ldots, \gamma_r\}$ a finite set of *generators* of *structure classes*, $A = \{\alpha_1, \ldots, \alpha_m\}$ a finite set of *attribute names* and $\Delta = \{\delta_1, \ldots, \delta_k\}$ a finite set of *domains*. In the following γ indicates both a generator and the associated class. For each class γ_i in Γ, we assume the existence of some recursive procedure p_i which decides the membership in the class of any subset of pixels in a picture. Moreover, for each γ_i, a set $\Theta_i = \{\theta_{i_1}, \ldots, \theta_{i_{n_i}}\}$ is defined where each θ_{i_j} is an operation closed with respect to γ_i. We assume that the domains in Δ are algebraic domains, well-behaved with respect to some set of operations.

A mapping $\nu : A \to \Delta$ associates an attribute name α_i with its domain δ_j. A mapping $\lambda : \Sigma \to \bigcup_{i=0}^{|A|} A^i$ assigns to each symbol t_k a tuple $(\alpha_{k_1}, \ldots, \alpha_{k_{n_k}})$ of attribute names such that all the names in it are distinct. The attributes in $\lambda(t_k)$ are said to be the *characterising attributes* of t_k and n_k is called the *arity* of t_k. A mapping $\sigma : \Sigma \to \Gamma$ assigns to each symbol t_k a graphical class. Finally, let X be a possibly infinite set of distinct *variable names*.

A *visual term* is a construct of the form $t_i(s, d)$, where s is either the generator $\sigma(t_i)$ or a structure such that $p_{\sigma(t_i)}(s)$ terminates with success, and $d = (u_1, \ldots, u_{n_i})$ is a tuple of arity n_i where each element u_j is either a variable name from X or a value from $\nu(a_j)$. s is called a *characteristic structure* and d a *description*. Whenever u_j is a value, its value is compatible with the concrete

---

[1] To be precise, this requires the use of a *transparent* symbol, which is not of interest here, see [3].

geometry of s and the context in which s is embedded. We will also use *terms* of the form $t_i(d)$ when the structure s can be left understood. We assume the existence of an attribute *identifier*, in each tuple $\lambda(t_k)$, whose value, from a domain ID, uniquely designates the visual term. An attribute taking values in $\mathcal{P}(ID)$, or in tuples where elements of ID can appear, is called a *relational attribute*.

A finite set of visual terms such that no two terms share exactly the same structure constitutes a *visual sentence*. In this paper we are interested not in images as abstract mathematical models, but in images which are produced in an interactive environment. Hence, we relax the requirement of unicity for structures, by using *layers*, so that we can also deal with structures hidden by others. Given an image i, the sets Σ, A, Δ, Θ, the mappings ν, σ, λ, and the procedures p_i defined as before, there always exists at least one visual sentence vs defined on it. Given a visual sentence vs, there always exists at least one image i which contains all the structures in vs.

We indicate with $H(\Sigma, A, \Delta)$ the set of visual terms over (Σ, A, Δ) and with $VS(\Sigma, A, \Delta)$ the set of visual sentences constructed using visual terms in $H(\Sigma, A, \Delta)$.

As an example, consider matrices of characters in a Cartesian plane. Let $\Sigma = \{a, b, c\}$, $A = \{id, x\_coord, y\_coord\}$ and $\Delta = \{N\}$ where elements in Σ represent the types of character to be arranged, and their attributes indicate the position in which they appear in an image, i.e., $\lambda(X) = (id, x\_coord, y\_coord)$ $\forall X \in \Sigma$. Moreover, $\nu(x\_coord) = \nu(y\_coord) = N$. A typographical character $\mathbf{a}$ in the plane is defined as the visual term $a(\mathbf{a}, (a_1, x\_coord, y\_coord))$ where a_1 is the identifier. The set $\{a(\mathbf{a}, (a_1, 0, 0)), a(\mathbf{a}, (a_2, 0, 1)), b(\mathbf{b}, (b_1, 0, 2))\}$ is a visual sentence and $\mathbf{aab}$ is the corresponding image.

As an other example, consider the labeled flowchart in Figure 1. Let $\Sigma = \{begin, pred, end, conn\}$, $A = \{AP_1, AP_2, AP_3, APSet\}$ and $\Delta = \{ID, \mathcal{P}(ID)\}$ where each attribute indicates the set of elements connecting to the i-th attaching point of a symbol. More formally, $\lambda(begin) = (id, AP_1)$, $\lambda(pred) = (id, AP_1, AP_2, AP_3)$, $\lambda(end) = (id, AP_1)$, $\lambda(conn) = (id, APSet)$, and $\nu(AP_i) = ID$ for $1 \leq i \leq 3$ and $\nu(APSet) = \mathcal{P}(ID)$. Then, the flowchart image[2] in Figure 1 corresponds to the following visual sentence (we only reproduce terms; the corresponding graphical elements can be derived from the picture):

$\{begin(b_1, c_1), pred(p_1, c_1, c_4, c_2), pred(p_2, c_2, c_4, c_3), pred(p_3, c_3, c_4, c_1), end(e_1, c_4), conn(c_1, \{b_1, p_1, p_3\}), conn(c_2, \{p_1, p_2\}), conn(c_3, \{p_2, p_3\}), conn(c_4, \{p_1, p_2, p_3, e_1\})\}$.

4   Diagrammatic Syntaxes

When defining a diagrammatic language, one is first interested in defining which types of visual constructs are to be considered important, and then in how to realise such constructs. For instance, one wants to express the fact that an edge is connected to nodes, and only then does one need to specify which spatial

---

[2] For sake of clarity the connections have been labelled with their identifiers.

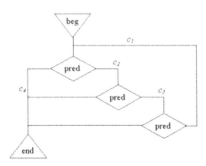

Fig. 1. A flowchart

relation is construed as representing such a connection. Hence, we distinguish an abstract level and a concrete syntax in the definition of visual languages.

In this paper we explore the metamodel approach to the definition of diagrammatic languages, to allow an abstract specification of the nature of a diagrammatic language, of which concrete syntactic definitions constitute instantiations. We adopt the UML definition of class diagrams for the specification of what we regard as the *abstract syntax* behind a diagrammatic language. A set of constraints, expressable in some logical language, complete the metamodel. Due to lack of space, we will express these constraints using natural language.

In UML, metamodels are used to define the semantics of the different visual languages composing it. Here, we rely on the existence of such semantics for the domain, but we use the metamodel to describe common characteristics of different visual notations, independently of the conveyed semantics. The separate definition of the concrete notation allows then the check of the desired properties against the adopted notations. For example, one could observe that the use of the *adjacency* relation to indicate a transitive relation is not adequate. Such studies have been undertaken by several authors, e.g. [17], [12], [14]. We propose here to arrive at a metalevel definition of families of visual languages. In this way, it becomes possible to abstract from the specific notations used for members of a same family, to obtain an abstract characterisation of such families. The adequacy of a representation can thus be checked by comparing metamodels for classes of domain problems and for classes of visual languages.

Class diagrams and textual constraints differ in their representational power. The abstract syntax imposes a set of constraints which may be deemed as *local*. For example, it can establish which types of elements can be put in which type of relations (and according to which cardinality), thus implicitly excluding the possibility of creating an instance of a relation in which some required element is missing. The use of a logical language with quantifiers allows instead the expression of *global* or *contextual* constraints. Checking such constraints on a concrete diagram might require the inspection of sub-diagrams of arbitrary size.

For example, the requirement that cycles be not produced in an inheritance relation cannot be expressed by local constraints alone.

We illustrate our approach by presenting a set of metamodels for some common families of diagrammatic languages. Such metamodels are all refinements of a basic one for defining spatial relations among identifiable elements.

4.1   Semantics Expressed through Spatial Relations

We first present the general metamodel for diagrammatic languages, where a set of relations among identifiable elements is used to express some domain semantics. The abstract syntax is given in Figure 2.

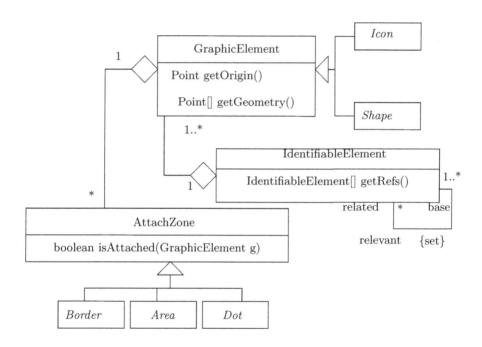

Fig. 2. The basic abstract syntax for diagrammatic languages.

Basically, *IdentifiableElement*s define the classes of visual terms in the language and can be put in relation through a set of **relevant** associations. For each *IdentifiableElement* there is an aggregation of *GraphicElement*s which takes care of its geometry and define a structure in a visual term. For each association, we can establish the **base** of the association and the set of its **related** elements.

An *IdentifiableElement* can be queried about the elements which are associated with it. All associations are navigable to obtain the collection of associated elements. For each *GraphicElement* a collection of *AttachZones* can be defined, which can be of type *Area, Border,* or *Dot.* The method `isAttached()` checks whether a given *GraphicElement* is related to the queried one. Membership in an association must be decidable on the basis of the definition of *AttachZones* of the *GraphicElements.* For each type of relevant association, the textual constraints in the metamodel express the type of *AttachZone* responsible for it.

We do not assume any specific structural feature, while basic implementations can be replaced in specific classes for the concrete notations. Hence, all classifiers in the diagrams are intended as interfaces or abstract classes (only basic implementations for some of the interfaces). For reasons of space we avoid repeating the stereotype `interface` and leave names of interfaces in upright characters and of abstract classes in italic. The abstract classes realise *Attach-Zone* and the operations in Θ_i for a *GraphicElement.* In particular, *Icon* defines elements which are not modifiable except for direct or inverted gray levels, and *Shape* defines elements which are deformable, scalable, and rotatable.

Specific families result from the definition of peculiar types of spatial relations and the consequent specialisation of constraints, and concrete languages will require the definition of geometries.

An important family of languages based on spatial relations is that of *containment* languages. Due to its relevance, rather than identifying the relative set of constraints, it is useful to specialise its abstract syntax so that specific types of *IdentifiableElement* are considered as *Containers,* which can be queried about the *IdentifiableElements* they contain. The containment relation can be based only on the *Area* of the *GraphicElement* associated with the *Container.* The constraints stipulate that *Containers* are associated with *Shapes,* while a generic *IdentifiableElement* can also be an *Icon.*

4.2   Explicit Representation of Domain Relations

A very important specialisation is that in which some designated *IdentifiableElement* explicitly represents a domain relation. We thus distinguish between *ReferrableElements* and *Connections.*

The diagram of Figure 4 provides the abstract syntax for several types of diagrammatic languages usually considered as possessing different syntaxes. In particular, it can accommodate languages which are variants of graphs (directed or non directed, multigraphs), as well as hypergraphs and plex languages, where restrictions exist on the way in which connections can be attached to elements. It also offers the possibility of having *Connections* as *ReferrableElements,* as for example is needed in UML class diagrams to define association classes or to attach notes to an association. In this family, we introduce a new type of *GraphicElement* namely a *Path.* A *Path* is formed by *Segments* and a *Segment* is a *Path* in turn. Typically, a *Path* can be queried for its end points. The constraints stipulate that the *GraphicElement* corresponding to a *Connection* must be a *Path,* one corresponding to an *Entity* must be an *Icon* or a *Shape.*

The set of constraints usually imposes that at least one *Entity* must exist and that no two *GraphicElements* relative to some *Entity* have overlapping

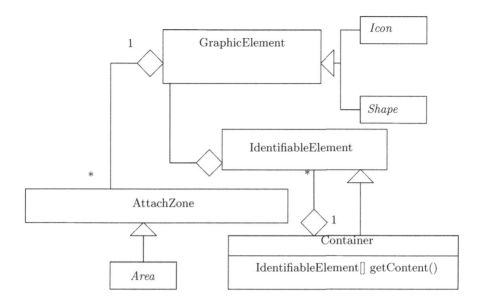

Fig. 3. The abstract syntax for diagrammatic languages based on containment.

geometries. In case a *Connection* is directed, it must have non empty collections at the `source` and `target` association ends and an empty collection at the `members` end; vice versa for non directed *Connections*. As before, the relation between a *Connection* and its *ReferrableElement*s must be expressed through the `isAttached` operation. More specific constraints allow us to distinguish between, say graph languages where the multiplicity of the association must be exactly one, and hypergraphs, where it can be arbitrary or restricted to fixed arities. Languages which support different types of connections can be defined. In this case, `isAttached` is specialised to the different types of *Connection*.

A specific language can be obtained by specialising the metamodel to the different types of *GraphicElement* to be used, possibly restricting which types of *ReferrableElement* can be referred by which type of *Connection*, and providing specific implementations for the considered operations. Such a specialisation can occur through the addition of constraints, or through the creation of a model abstract syntax, which must comply with the metamodel. For example, in Petri Nets, two types of *Entity*, namely *Place* and *Transition*, exist and an *Arrow*, which is a type of *Connection*, can only link two *Entity*s of different type.

The common kernel in all these metamodels favors the possibility of defining hybrid languages where, for instance, explicit connections coexist with containment, as is the case of the Higraph family of languages [13], of which the several variants of Statecharts are instances.

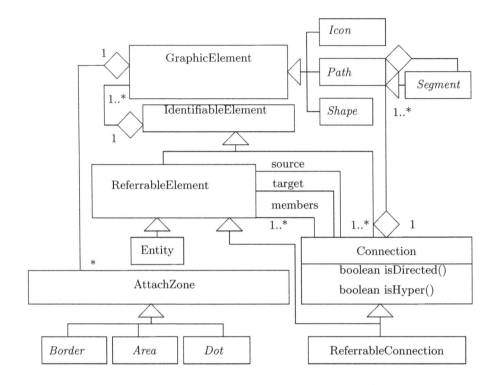

Fig. 4. The abstract syntax for languages with explicit connections.

5   Levels of Definition in Diagrammatic Languages

In view of (partially) automating the construction of editors or parsers from specifications, it is important to understand how the constraints expressed in the metalevel, either textually or through abstract syntax class diagrams, can be satisfied by defining suitable visual rewriting systems.

An incremental visual rewriting system can be defined from the abstract syntax, starting from two basic steps: 1) *IdentifiableElements* may be created freely only if they do not appear at the **related** association end of a **relevant** relation; 2) *IdentifiableElements* at a **related** association end may be created only if the identifiable elements at the **base** end exist.

For connection-based diagrammatic languages, these can be redefined as follows: 1) *Entitys* may be created freely; 2) *Connections* may be created only if the elements to be connected exist.

Multiplicity constraints can be managed by defining attributes counting the number of elements in a given relation to an *IdentifiableElement* and checking their compliance with the multiplicity limits.

We will therefore have four levels of specialisation in the definition of editors:

Free addition of elements. In this case, any visual term can be inserted in the current visual sentence through a sequence of operations: 1) selection of a symbol $t \in \Sigma$; 2) selection of the generator $\sigma(t) = \gamma_i \in \Gamma$; 3) application of operations from Θ_i to obtain a *dummy* instance s'; 4) application of *shift* and *sup* to s' so that an actual instance s is positioned in the current image i; 5) evaluation of attributes in $\lambda(t)$ in the context of the visual sentence vs so far defined on i, to obtain a tuple d; 6) insertion of the visual term $t(s, d)$ in vs. The only constraint which applies to the free addition is the requirement that the resulting set of visual terms must constitute a visual sentence, hence it is not possible to generate two identical structures which lie in the same layer.

Addition constrained by abstract syntax. The insertion of a visual term is such that if spatial relations are established with existing visual terms, these are defined in accordance with the abstract syntax. For instance, if the abstract syntax dictates that an association of a given type can exist only among some types of visual terms, then the generation of such an association must obey this constraint. This rules out the possibility of generating dangling edges. Actually, the abstract syntax defines a set of *local* constraints, in the sense that the predicates implicitly defined by fragments of the abstract syntax can be evaluated by checking a fixed set of terms.

Constrained placement of elements. Relevant spatial relations, or the establishing of a connection, are defined through attaching zones. For instance, containment is intended as complete containment and a connection may be considered to touch a referrable element if it has a common point with its border. The insertion of a new visual term in relation with another must hence obey the constraints implicitly imposed by the `isAttached` operation. In this approach, so called *emerging relations* are not considered. That is to say that the casual insurgence of a spatial relation between graphical tokens is not relevant if the new visual term was not created to be explicitly put in relation with the other terms involved in such a relation.

Global and contextual constraints. Some logical constraints cannot be expressed in the abstract syntax through class diagrams, but involve navigation of associations, assertions of existence or unicity. Intermediate steps in the construction of a diagram may not satisfy the whole set of constraints, but insertion of a new element can occur as long as it does not prevent these constraints to be satisfied in the visual sentences which can develop from the modification of the current one. For example, if it is required that no cycle can be created in the representation of a given relation, then no visual term which closes a cycle should be added to the current visual sentence. This requires the dynamic check of complex conditions on the whole visual sentence. A final validating step is needed to check if the produced sentence obeys the whole set of constraints, or the diagram must be completed.

In the case of free-hand editing each arrangement drawn must be assigned to a class of structures, hence decided by a p_i. The adoption of an incremental perspective for free-hand editing imposes that no ambiguity may arise. This is a quite restrictive requirement, so that usually it is demanded to a parsing process to take the final decision on the actual type of a structure. However, some basic distinction between basic shapes, closed curves, and paths may be employed to assist a user in drawing a diagram, for example taking care of beautifying curves, or positioning path ends on the correct attachment zones.

In the parsing perspective, one has to reduce a set of visual terms to some well-defined visual sentence, i.e. to an axiom. In principle, parsing should be able to reconstruct the abstract structure underlying a visual sentence, i.e., an instance diagram complying with the class diagram and the constraints giving the language model. In turn, such a model is an instance of the metamodel for one family of visual languages.

The parser recognises *IdentifiableElements* and their relations, and checks the constraints on these relations. If the input to the parser is produced by an editor which takes care of guaranteeing the correct management of spatial relations through attaching zones, then parsing need deal only with checking that only legal types of identifiable elements are put in relation and that contextual constraints are satisfied.

Parsing can in turn be seen as a process in which visual sentences are rewritten into visual sentences. Independently from the approach used for parsing, whether using textual (Chomsky-style or constraint or logical based) grammars or visual rewrite rules (graph-transformation or set rewriting), we can model the parsing process as a sequence of transformation of visual sentences, in which some visual terms are considered to be non-terminal. In particular, such non-terminals designate aggregations of *IdentifiableElements*. The rules for their definition follow a common pattern:

1. A set of *IdentifiableElements* is chosen as a seed for the aggregation and redefined as a nonterminal *Aggr*.
2. Further *IdentifiableElements* in the aggregation are progressively embodied in *Aggr*. The embedding of *Aggr* in the rest of the diagram derives from the relations of the aggregated elements with elements outside the aggregation.
3. Further non-terminals are used as place-holders for *IdentifiableElements* as long as there are relations between these elements and elements outside the aggregation.
4. When all relations have been resolved, the place-holders are consumed and embodied in *Aggr*.

Non-terminal visual terms can possess structures (as defined through *GraphicElements*) which result from the composition of the structures of the visual terms which were reduced to the non terminal. In this case, the process ends when the whole set of *GraphicElements* in the original visual sentence is associated with the unique instance of the grammar start symbol. Alternatively, non-terminals can have specific representations and the structures in the reduced visual sentence are removed and substituted with non terminal structures. The process ends when a specific visual sentence, either empty or containing at most a well defined set of non terminal visual sentences, is produced.

Constraints requiring the existence of some element (or even its uniqueness) can be satisfied by introducing the required elements in the axiom.

As stated before, constraints involving arbitrary navigation, for instance exploiting the transitive closure of some relation, require in principle, the inspection of the whole visual sentence. In the parsing perspective, this requires that information on the already parsed sub-sentence be maintained to perform such checks. Hence, relational attributes must be defined to assign to a non-terminal the set of (terminal) terms involved in a relation, and conditional rules must be used to check such attributes. A less refined strategy could be to delegate such checks to a post-parsing contextual analyzer, much as a parser for a programming language delegates type checking to the semantic analysis phase.

6   An Example

We provide an example by considering a simple flowchart language with sequential, selection and loop constructs. The language is based on the plex structure as defined by Feder [11]. This flowchart language can be easily described by creating a model instance of the abstract syntax described in Figure 4.

The instantiation is developed at different levels of abstractions. At a first level we identify all the types of *Entity*, *Connection*, and *ReferrableElement* in the language and a first set of constraints characterising the plex structure of the language, in which all *ReferrableElements* are represented by *Icons* and have *AttachZones* of type *Dot*. As subtypes of *Entity*, we have: *Begin* and *End* of which only an instance can occur and which have only one *AttachZone*. The other subtypes of *Entity* are always components of aggregations of *ReferrableElements*. These components are of type *Block* (with two *AttachZones*) and *Predicate* (with three *AttachZones*). The aggregations are in turn *ReferrableElements* of type *Statement*, with two *AttachZones*, which can be subtyped to *Block* and to *Alternative*. An *Alternative* is an aggregation of one *Predicate* and two *Statements*, and statements can be recursively aggregated to form new *Statements*.

Connections are directed hyperlinks, having arbitrary multiplicities at both ends (except those ending in *End* or starting from *Begin*, which have multiplicity 1 at the corresponding end). They connect *Statements*. Some *Connections* may originate from *Predicates*, in which case we distinguish a left and a right *Connection*, while a *Connection* originating from a *Block* is referred to as an out *Connection*. Figure 5 illustrates the abstract syntax for this language.

Some constraints are common to the family of plex languages: a *ReferrableElement* cannot connect to itself through a *Connection*; every *AttachZone* of every *ReferrableElement* must "touch" a *Connection*; no dangling *Connections* are allowed; only one *Connection* can "touch" an *AttachZone* of a *ReferrableElement*.

This level of formalization provides the basis for the construction of a graphical structure editor that forces the users to draw well-formed plex sentences towards the creation of flowcharts. Such an editor takes care of all the local constraints deriving from the type and aspect of the language symbols, connec-

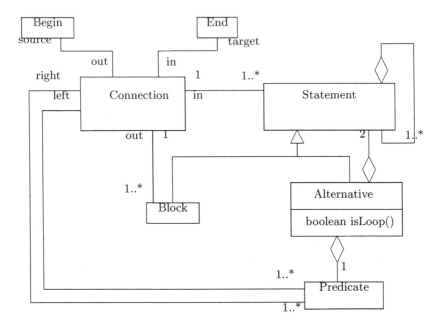

Fig. 5. The abstract syntax for a simple language of flowcharts.

tions and local interaction among them. In a second phase, the abstract syntax and the constraints can be further instantiated to produce a complete definition of the language under examination. Once a full characterization – down to the concrete syntax – is provided, the abstract syntax and constraints may be translated into a rewriting rule system to produce an interactive visual editor that not only takes care of the basic structure of the language but also of the global and contextual constraints that precisely define the language syntax. In this way the user can only draw "legal" sentences. Due to lack of space, we cannot show the whole model for the flowchart language, nor the visual rewriting rule system derived from the model along the lines described in the previous section.

7   Conclusions

The paper has discussed a metamodel approach to the definition of diagrammatic languages, in which a language can be defined through a class diagram specifying its abstract syntax, and a set of constraints. Some constraints, such as multiplicity ones, can be directly expressed in the class diagram, while others, of a more contextual nature, must be expressed in some textual notation. Concrete notations must conform with the metamodel. A strategy for the systematic derivation of editors and parsers has been given. An advantage of this method is that the metamodel can be put in relation to a metamodel description of the intended domain semantics of the language, so as to assess its adequacy. The exploration of this possibility is the subject of future work.

Acknowledgements. Work supported by the Italian Ministry of University and Research under the Research Project of National Interest "Specification, Design and Development of Visual Interactive Systems" and Esprit Working Group APPLIGRAPH.

References

1. M. Anderson, P. Cheng, V. Haarslev eds. *Theory and Application of Diagrams.* Springer, 2000.
2. R. Bardohl, T. Schultzke, G. Taentzer. Visual Language Parsing in GenGEd. *Proc. 2nd International Workshop on Graph Transformation and Visual Modeling Techniques GT-VMT'01*, 2001.
3. P. Bottoni, M. F. Costabile, S. Levialdi, P. Mussio. Defining visual languages for interactive computing, *IEEE Transactions on Systems, Man, and Cybernetics – A*, **27**(6):773–783, 1997.
4. P. Bottoni, M.F. Costabile, P. Mussio, "Specification and Dialogue Control of Visual Interaction through Visual Rewriting Systems", *ACM TOPLAS*, **21**(6):1077-1136, 1999.
5. Z. Diskin, B. Kadish, F. Piessens, M. Johnson. Universal Arrow Foundations for Visual Modeling. In [1], 345–360.
6. R. I. Ferguson, A. Hunter, C. Hardy. MetaBuilder: The Diagrammer's Diagrammer. In [1], 407–421.
7. G. Costagliola, G. Polese. Extended Positional Grammars. In *Proceedings of 2000 IEEE Symposium on Visual Languages.* 103–110, 2000.
8. G. Costagliola, A. De Lucia, S. Orefice, G. Tortora. A Framework of Syntactic Models for the Implementation of Visual Languages. In *Proceedings of 1997 IEEE Symposium on Visual Languages*, pp.8–67, 1997.
9. J. Ebert, A. Winter, P. Dahm, A. Franzke, R. Süttenbach. Graph Based Modeling and Implementation with EER/GRAK. In *Proc. ER'96*. B. Thalheim, ed., 163–178, Springe, 1996.
10. M. Erwig. Abstract Syntax and Semantics of Visual Languages. *Journal of Visual Languages and Computing.* **9**(5):461–483, 1998.
11. J. Feder. Plex Languages. *Information Science*, **3**:225–241, 1971.
12. C. Gurr. On the Isomorphism, or Lack of It, of Representations. In *Visual Language Theory.* K. Marriott, B. Meyer eds., 293–305, Springer, 1998.
13. D. Harel. On Visual Formalisms. *Comm. of the ACM.* **31**(5):514-530, 1988.
14. A. von Klopp Lemon, O. von Klopp Lemon. Constraint Matching for Diagram Design: Qualitative Visual Languages. In [1], 74–88.
15. M. Minas. Concepts and Realization of a Diagram Editor Generator Based on Hypergraph Transformation. *Journal of Computer Programming*, to appear.
16. J. Reekers, A. Schuerr. Defining and Parsing Visual Languages with Layered Graph Grammars. *Journal of Visual Languages and Computing.* **8**(1):27–55, 1998.
17. D. Wang, H. Zeevat. A Syntax Directed Approach to Picture Semantics. In *Visual Language Theory.* K. Marriott, B. Meyer eds., 307–323, Springer, 1998.

Describing the Syntax and Semantics of UML Statecharts in a Heterogeneous Modelling Environment

Yan Jin[1], Robert Esser[1], and Jörn W. Janneck[2]

[1] Department of Computer Science, Adelaide University, SA 5005, Australia
yan@cs.adelaide.edu.au, esser@computer.org
[2] EECS Department, University of California at Berkeley, CA 94720-1770, USA
jwj@acm.org

Abstract. In this paper UML statechart diagrams are used as an example of a generic approach to integrating a visual language in a heterogeneous modelling and simulation environment. A system represented in a visual language is syntactically defined as an attributed graph, with well-formedness rules specified by a set of first-order predicates over the abstract syntax of the graph. The language semantics are specified by an *Abstract State Machine* (ASM) parameterized with syntactically-correct attributed graphs. In this paper the key issues in the definition of UML statechart semantics are highlighted.

1 Introduction

As visual languages become widely accepted for system design and analysis, the precise specification of their syntax and semantics is becoming an important issue. Various techniques have been suggested in the literature to define visual languages. These techniques differ significantly in the way they conceptualise visual notations as well as in the way they formulate the semantics.

On the other hand, as systems are becoming more and more heterogeneous (i.e. they are often composed of components with very different behavioural characteristics), various visual languages are being used to model different parts of a system. This approach makes full use of the advantages of visual languages in helping to generate more concise, accurate and understandable system models. However, it also presents new challenges such as conformance analysis between system components and system correctness checking. These are more complicated when the visual languages are defined in different and incompatible ways.

Taking UML statechart diagrams as an example, this paper will present a generic approach to the definition and integration of visual languages, specifically for heterogeneous modelling and simulation. With their syntax and semantics defined in a uniform way it can be shown that the total design is consistent and that further analysis of these systems is possible. The definition of a visual language is shown in figure 1 and consists of three parts:

- First of all, the diagrams are defined as attributed graphs using the Graph-type Definition Language (GTDL)[9], while their syntax well-formedness rules are specified by a set of first-order predicates over the abstract syntax of the graphs.

M. Hegarty, B. Meyer, and N. Hari Narayanan (Eds.): Diagrams 2002, LNAI 2317, pp. 320–334, 2002.

- Secondly, language semantics are specified by an *abstract state machine* that represents the operational semantics of syntactically-correct attributed graphs. This also includes a definition of scheduling mechanisms.
- Thirdly, based on the specified semantics, runnable components of these diagrams are generated for execution and simulation.

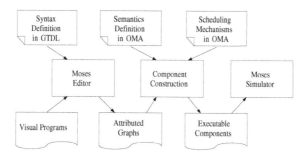

Fig. 1. Visual language definition

UML statechart diagrams is an important part of the standard UML language [16]. It is a highly expressive hierarchical modelling language with well defined syntax [16]. Unfortunately, its precise semantics are not well formalized. This paper presents the definition of the syntax and semantics of UML statechart diagrams and also facilities for supporting the modelling of heterogeneous systems.

Our work builds on the Moses tool suite [5]. Moses addresses the definition of a certain class of visual languages, viz. graph-like notation (bubble-and-arc diagrams) and supports the modelling and simulation of discrete-event heterogeneous systems. These systems consist of communicating components, each modelled using an appropriate visual language. The collaboration between components is independent of the exploited formalisms. The Moses graph editor and simulator are generic and parameterized by visual language definitions. This facilitates the modelling and simulation of not only components written in various languages but also the complete composite system.

In this paper we present a definition method of visual languages that is generic and independent of the particular visual language being defined. The interfaces (GTDL definition, scheduling mechanisms, etc.) to the editing, execution and analysis platform are designed to be language independent. This paper also discusses issues covering the whole process of the language definition of the UML statechart diagram from syntax and semantics to diagram execution or simulation. In particular, we will show how ASMs are employed to formulate the semantics, taking advantage of the high abstraction level of ASMs to be able to easily adapt a definition should the need arise.

2   Related Work

Statechart diagrams were originally introduced by David Harel [8,7] in the mid 80's. The notation and semantics of UML statecharts were adapted from the Harel's original version with the addition of object-oriented features [16]. UML Statecharts extend ordinary state transition diagrams with notions of hierarchy and concurrency [16]. They are a highly expressive visual language and are typically used to model the dynamic behaviour of a class of UML objects. This language has proved useful in modelling complex control aspects of many software systems.

The syntax and semantics of UML statecharts are described in [16]. Although the UML meta-models do capture the language relatively precisely, they often fail to give an exact interpretation of nontrivial statecharts structures. This has been the motivation for the precise definition of its formal semantics which, due to the richness of the language, has proved extremely challenging. A variety of proposals have been made in the literature, e.g. [13,14,17,4,3,7].

Latella *et al.* [13] exploited hierarchical automata as an intermediate language for formulating the semantics in terms of Kripke structures. However, the approach does not directly support transitions that cross the border of composite states, and the hierarchical structure must be flattened for formal analysis.

Our semantics definition is inspired by Lilius and Paltor [14] which formalized the structure and operational semantics of UML statechart diagrams separately. They developed an operational semantics on the basis of a term-rewriting system and a predetermined static priority relation on transitions. This then serves as the theoretical basis of their vUML tool. In contrast our approach is more flexible as it solves transition conflicts by using dynamic computation of transition priority.

Schäfer *et al.* [17] also proposed a dynamic computation algorithm. They model states as individual processes communicating through channels. In order to determine firable transitions, the control is transferred from the top state process down to leaf state processes. However, this communication mechanism results in a high overhead dependent upon the number of states and transitions in a statechart diagram and makes it difficult to ensure the correctness of the algorithm. Instead, we exploit a simpler method by modelling a statechart diagram as an ASM that executes two sequential steps – selecting firable transitions followed by concurrently executing them. The firable transitions are selected directly based on the state hierarchy.

Both [14] and [17] translate UML models into the SPIN input language Promela [15] for verification in SPIN. However, Promela is not a formal language suitable for defining the semantic model of a language [4]. In our approach an ASM language called *Object Mapping Automata* (OMA) is employed. Using OMA, the semantics of statecharts are defined at a high abstraction level, making it much easier to ensure the correctness of our algorithms with the flexibility to easily modify them should the need arise.

ASM semantic models of UML statecharts were also studied by [4,3]. In [4] an extended statechart diagram is built by adding a transition for every affected state of a firing transition. This makes explicit the exiting and entering of states of the original transition. This can also simplify the ASM transition rules for all kinds of states. However, this approach only describes single transition execution issues and it is not clear how to control the order of execution of the original transition and the newly introduced ones. Moreover, the transition selection strategy was not mentioned.

Börger *et al.* [3] proposed an ASM model and an approach to dealing with all the features that break the thread-of-control. The model covers the event handling and the run-to-completion step, and formalizes object interaction by combing control and data flow features. However, the authors did not give a complete solution to solving transition conflicts and it is not clear how firable transitions are selected.

3   UML Statechart Diagrams

A statechart diagram specifies the states that a object can reside in and describes the event triggered flow of control, as the result of the object's reactions to events from the environment. A state depicts a situation where the object satisfies some condition, performs some activity, or waits for some event. Reactions include changing states, executing internal actions and sending events to other objects. The event-driven nature of statecharts is especially useful in modelling reactive systems. See figure 2 for an example.

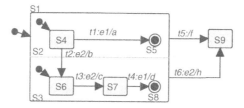

Fig. 2. A startchart diagram

According to [16], a statechart diagram consists of states, pseudostates and transitions between them. States can be simple, composite (sequential or orthogonal), or final. For example, $S4$, $S2$, $S1$, and $S5$ are examples of the above types respectively. Simple states and final states are the basic constructs, with final states indicating the completion of their containing states. Composite states are constructs containing a group of states. Transitions to/from a composite state affect all the states inside it. An orthogonal composite state contains several concurrent regions (sequential composite states), graphically separated by dashed lines. All of its regions are active while the state remains active. In addition to the above primitive constructs, pseudostates such as initial (a black dot •) and history ((H) or (H*)) are used to extend the notation. An initial state is the source of a transition which points to the default substate of the composite state containing the initial state, while a history state records the most recent active state information of its containing state. The state hierarchy of a UML statechart is graphically depicted by the physical containment relationship between (pseudo) states. For instance, $S2$ and $S3$ are the substates of $S1$; $S6$, $S7$ and $S8$ are of $S3$.

Transitions are directed arcs between (pseudo) states representing the complete response of the diagram to a particular event, e.g. $t1$ to $t6$. Transitions may be guarded by conditions and specify actions to be executed or events to be sent when the transition is fired. A transition without a trigger is called as a *completion* transition and may be fired when its source has completed all its internal activities or reached a final state(s), e.g. $t5$.

The semantics of statecharts is based on the *run-to-completion* assumption [16]. Incoming events are processed one at a time and an event can only be processed if the processing of the previous event has been completed – when the statechart component is in a stable state configuration. This will be discussed in detail in section 5.

4   Syntax Definition

In this section we present our approach to the syntax definition of UML statechart diagrams using the Graph-type Definition Language (GTDL). As GTDL itself is similar to common mathematical notations and can be easily understood, we shall omit a description of it. For the language specification, cf. [9].

The syntax definition is based on a notion of an attributed graph. An attributed graph is a formal structure that represents the pictures we would like to draw and interpret. It consists of a set of vertices V, a set of edges E, a function src from edges to their source vertices, a function dst from edges to their target vertices, and an attribute function $attr$ from graph elements and their attribute names to attribute values. A graph element may be the graph itself, a vertex or an edge. For the formal definition of attributed graphs, cf. [11]. In this paper we consider a statechart diagram to be an attributed graph that represents a class of components that can be instantiated from the diagram.

The syntax of UML statechart diagrams is defined in two parts: (1) a definition of language-specific *attributes* and *appearance* of graph elements (such as states and transitions), and (2) a specification of the *well-formedness* rules of statecharts. While the former is (mostly) a context-free property, the latter can be very complex and based on the attributes of states and transitions or the inter-dependency between states, transitions and the connection structure.

The syntax of a statechart is defined by V (the set of states and pseudostates), E (the set of transitions), src and dst functions that map transitions into their source states and their target states respectively and the $attr$ function where, as an example, given a transition and an attribute name "Guard", an $attr$ function returns the transition's guard expression. On the other hand, the state hierarchy need not be defined here but can be deduced from the physical containment between vertices by calling an attribute function $vertex\_id("Container")$.

Figure 3 illustrates a part of the GTDL syntax specification for UML statecharts. Three kinds of vertices (initial states, simple states and composite states) and one kind of edge (transitions) are defined. Following a vertex or arc type name, the element's language-specific attributes are defined. Simple states may have an entry action, an exit action, an activity and a set of deferrable events. Similarly, composite states may have an additional attribute *NumOfRegions* to specify the number of their subregions. If the number is greater than 1, the state is orthogonal; otherwise, it is a sequential state. The attributes of a graph element can be declared as expressions, strings or other basic data types. Expressions can be assignments, boolean formula and message sending primitives. These data types are supported by a small expression language ELAN embedded in the GTDL (see [2] for a specification of ELAN).

In addition attributes representing the graphical appearance (e.g. shape, colour and default size) of graph elements are specified in the *graphics* section. For instance, the

```
1   graph type UMLstatechart {
2       attribute expression declarations.
3       attribute expression associations.
4       attribute expression parameters.
5
7       vertex type InitialState()
8           graphics (Shape = "Oval", Width = 8, Height = 8).
9       vertex type SimpleState
10          (expression EntryAction, ExitAction, Activity, DefEvents)
11          graphics (Shape = "RoundRectangle").
12      vertex type CompositeState(..., integer NumOfRegions)
13          graphics (Shape = "RoundRectangle").
14      edge type Transition(expression Trigger, Guard, Action)
15          graphics (Head = "ClosedTriangle").
16
18      predicate "Transitions exist only between states
19          and do not end at initial states or start from final states."
20          forall e ∈ Transition :
21          src(e) ∈ (InitialState + SimpleState + CompositeState)
22          && dst(e) ∈ (SimpleState + CompositeState + FinalState) end
23      optional predicate "Some Transitions may not have triggers."
24          forall e ∈ Transition : src(e) ∈ (InitialState + HistoryState)
25          ⇒ e("Trigger") = null end
26  }
```

Fig. 3. The syntax definition of UML statechart (abridged)

appearance of an initial state is oval, 8 points wide, and 8 points high. Also the graph itself may also have attributes, e.g. the attribute *"declarations"* is used to declare diagram-specific variables, *"associations"* defines the references to other system components, and *"parameters"* is used during component instantiation of this diagram.

The definition of graph elements is followed by a list of predicates that describe syntactic constraints on statecharts. For instance, it is required in [16] that transitions exist only between states while an initial state has no incoming transitions and a final state has no outgoing transitions. The first predicate formulates this requirement. Following the keyword *predicate*, a diagnostic string and a predicate formula are given. In this case, the formula contains a universal quantifier e over the set *Transition* that contains all graph elements of this type. The second predicate is declared to be optional since an outgoing transition from the initial state at the top level may have a trigger with the stereotype «create».

There are some built-in functions available in predicate formulae. For instance, as mentioned previously, $src(e)$ and $dst(e)$ return the source and the target (pseudo) state of the transition e respectively. The attribute function $e("Trigger")$ in the second predicate returns the trigger expression of the transition e, where e is applied to its attribute name *"Trigger"*. The absence of an attribute value can be tested using the *null* keyword in GTDL.

Having defined the syntax of statecharts, we can make use of the Moses graph editor to create and edit a statechart diagram while its syntax correctness is checked in the background. This is possible because the editor is parameterized by a syntax (GTDL) definition. In addition a statechart diagram can be embedded into other visual formalisms such as Timed Petri Nets and Process Networks to support heterogeneous modelling. However, in order to execute or simulate the diagram, we need to define statechart semantics and the interface with the Moses run-time system.

5   Semantics Definition

In this section, the semantics of the UML statecharts, as described in [16], will be formulated using Object Mapping Automata (OMA) [12] – an extension of Gurevich's Abstract State Machines [6]. The definition consists of two parts: the structure formalisation and the operational semantics specification. The structure is directly mapped into the vocabulary of an OMA which is a finite collection of functions. Each valuation of these functions forms an abstract state of the (OMA) automata. This structure serves as the basis of the specification of the operational semantics. The operational semantics includes mechanisms for event queueing, dispatching and processing of statecharts. It is implemented as a set of rules which update the vocabulary of the OMA thereby changing the abstract states of the automata.

In order to specify the semantics in a general way, we intend to keep it as independent as possible of diagram-specific features. For example, we delegate the evaluation of the transition guard and the action execution to the run-time environment. This is achieved by calling the run-time library offered by the Moses tool suite. As the interface to the library is well-defined it is possible to reuse the semantics in any other platform that supports an equivalent interface.

5.1   Object Mapping Automata

Gurevich's Abstract State Machines (ASMs) [6] are operational formalisms that model algorithms at various abstraction levels, from abstract views down to implementation details. ASMs are very expressive formal languages with simple formal semantics. These formalisms have proven their applicability and usefulness to system specification, design and formal analysis through numerous case studies, cf. [1].

Although ASMs are easily used to describe complex hardware and software (sub) systems, they lack support for combining individual subsystems. As a result one subsystem may influence another subsystem in an unexpected way. Object Mapping Automata [12] makes the object oriented flavour of ASMs more explicit, having both state and behaviour attached to objects. In OMA, only local updates of the state are allowed and communication between objects is restricted to message passing. Consequently OMA can ensure the locality and compositionality of objects in the sense that algorithms based on the local state of an object are not influenced by other system components. On the other hand, OMA generally allows a full range of models to be specified from those that are tightly or loosely coupled, distributed systems, as well as timed systems by supporting timing and scheduling facilities. For detailed descriptions about the syntax and semantics of OMA, cf. [12,10].

5.2 Formalising the Structure

We have shown that the structure of syntactically-correct statecharts is given by attributed graphs. Since the graph data can not easily be processed, we transform the graph data into the algebraic structure required by OMA.

Similar to [14][1], we define the OMA algebraic structure of UML statecharts as a tuple $< \Sigma, \mathcal{S}, \Phi, \mathcal{T}, \mathcal{C} >$ where,

- Σ is the union of two sets: states Σ_s and pseudostates Σ_p. Σ_s is composed of four disjoint sets: simple states Σ_{ss}, sequential composite states Σ_{cs}, orthogonal composite states Σ_{ocs}, and final states Σ_{fs}, while Σ_p consists of three disjoint sets: initial states Σ_{is}, shallow history states Σ_{sh}, and deep history states Σ_{dh}.
- $\mathcal{S}$ is a group of state attribute functions ($\Sigma_s \rightarrow \mathbb{E}^*$): $\mathcal{S}^{ent}$, $\mathcal{S}^{ex}$, $\mathcal{S}^{act}$, and $\mathcal{S}^{def}$, which map a state into its attributes (entry action, exit action, activity, and deferrable events respectively). $\mathbb{E}^*$ is a set of ELAN expressions plus the special value *undef* representing an absence of an attribute value in OMA.
- Φ is the set of transitions such that $\Phi \subseteq \mathcal{P}(\Sigma - \Sigma_{fs}) \times \mathcal{P}(\Sigma - \Sigma_{is})$, where $\mathcal{P}(\Sigma)$ is the power set of Σ.
- $\mathcal{T}$ is a group of transition functions: $\mathcal{T}^{src}$ ($\Phi \rightarrow \mathcal{P}(\Sigma - \Sigma_{fs})$), $\mathcal{T}^{dst}$ ($\Phi \rightarrow \mathcal{P}(\Sigma - \Sigma_{is})$), $\mathcal{T}^{trg}$ ($\Phi \rightarrow \mathbb{E}^*$), $\mathcal{T}^{grd}$ ($\Phi \rightarrow \mathbb{E}^*$), and $\mathcal{T}^{act}$ ($\Phi \rightarrow \mathbb{E}^*$). Given a transition these functions return its sources, destinations, trigger, guard, and action respectively.
- The function $\mathcal{C}$ is a partial function $\Sigma \hookrightarrow (\Sigma_{cs} + \Sigma_{ocs})$ which maps a (pseudo) state into its direct containing state. With no containers, the states or pseudostates at the top level are mapped into *undef*.

The compilation from attributed graphs to the above structure for statecharts is straightforward except for compound transitions. Basically attributed graphs and their elements are functions from attribute names to attribute values. As we have seen in the previous section the attribute values of a graph element can be accessed in form of $element\_id("AttributeName")$, e.g. $vertex\_id("EntryAction")$. The values may be expressions, sets and strings. Their types are specified by the syntax definition in GTDL. Moreover, there are some built-in attributes whose values are given by attribute graphs. For instance, the formula $g("Vertices")$ returns the set of vertices of the graph g, while $v("Container")$ returns a graph element of the vertex v's direct container. From this information the algebraic structure can be uniquely determined.

The derivation of compound transitions is more complex. Compound transitions represent "semantically complete" paths consisting of transitions connected by pseudostates such as fork, join, junction and choice from a set of states to a set of states. We employ the translation method proposed by Lilius and Paltor in [14] to derive compound transitions. For more information see section 2.1 of [14].

In addition to the signature defined above, we also define two derived functions. The set of transitively nested substates of a composite state $s \in \Sigma_{cs} \cup \Sigma_{ocs}$ is defined as $\Omega(s) \equiv \{v \in \Sigma | \exists v_1, ..., v_k \in \Sigma_{cs} \cup \Sigma_{ocs}, \mathcal{C}(v) = v_1, \mathcal{C}(v_1) = v_2, ..., \mathcal{C}(v_k) = s\}$, where $k \leq |\Sigma_{cs} \cup \Sigma_{ocs}|$. The transitive containment relationship is defined as a boolean

---

[1] In contrast to [14] we make the structure of statecharts more complete by including the state container function $\mathcal{C}$ from which the state hierarchy can be deduced.

function $cover(s, s\prime) \equiv s \in (\Sigma_{cs} \cup \Sigma_{ocs}) \land s\prime \in \Omega(s)$. In addition, we assume that in every statechart there is an inherent composite state called *the top state* which covers all the (pseudo) states and is the container of the states with $C(s) = undef, s \in \Sigma$.

5.3 Operational Semantics

As defined in the UML standard [16], a state machine semantically represents a statechart diagram and consists of an event queue for storing incoming events, an event dispatching processor and an event processor. As the nature of the event queue and the order of event dequeuing is not defined in the standard, without loss of generality we implement the event queue as a set and an event is dequeued randomly. When the state machine is ready to execute a step, the event dispatching processor nondeterministically selects an event from the set. The machine then goes on processing this event during a *run-to-completion (RTC)* step, bridging two stable state configurations. The step is executed under the assumption that an event can only be dequeued and dispatched if the processing of the previous event is fully completed [16].

Run-to-completion step. When dispatched, an event e, called the *current event*, may enable one or more transitions. According to [16], we define that a transition is *enabled* by e if and only if (1) all of its source states are in the current state configuration, (2) its trigger is matched by e, and (3) all its guards are evaluated true where, if no guard is specified for a transition, true is returned. If we define the current state configuration as a set Δ of current active states, the set of enabled transitions e is defined as $enabled(e) \equiv \{t \in \Phi | \mathcal{T}^{trg}(t) = e \land evaluate(\mathcal{T}^{grd}(t)) \land \mathcal{T}^{src} \subseteq \Delta\}$.

When a number of transitions are enabled, a subset, called a *maximal consistent set* of enabled transitions, is selected for firing. If the dispatched event does not enable any transition, it is consumed unless it is contained in the union of the sets of deferrable events of all active states. If deferred, the event will be returned to the event queue after the machine has changed its state configuration.

The step written in OMA is shown in figure 4, where *activeStates* refers to Δ, *firable(e)* and *execute(t)* represent the maximal consistent set and the transition execution procedure respectively (described later). The keyword *once* indicates that the following block runs only once.

Determining the maximal consistent set. The transition selection algorithm finds a maximal consistent set of transitions satisfying the following conditions: (1) all the transitions in the set are enabled and consistent with each other; (2) no enabled transition is omitted that is consistent with all the transitions in the set; (3) no enabled transition outside the set has *priority* over a transition in the set.

Two enabled transitions are *consistent* if they do not exit a common source state, more precisely, the intersection of the set of states they exit is empty, e.g. in figure 2, $t1$ and $t4$ (if enabled). Otherwise the two enabled transitions are said to be *in conflict*, e.g. $t2$, $t3$ and $t6$.

We use $conflict(t, e)$ to represent the set of enabled transitions conflicting with a transition t with respect to the current event e. Formally, $conflict(t, e) \equiv \{t\prime \in$

```
1  rule RunToCompletion :
2  once
3    choose e ∈ eventQueue :
4      if enabled(e) = ∅ then
5        if exists s ∈ activeStates : e ∈ deferEvents(s) then
6          deferQueue(e) := true
7      else
8        do forall t ∈ firable(e) : execute(t) :
9        if deferQueue ≠ ∅ then
10           moveinto(deferQueue, eventQueue)
```

Fig. 4. The run-to-completion step

$enabled(e)|t \neq t\prime \wedge exitSet(t) \cap exitSet(t\prime) \neq \emptyset\}$, where the set of exiting states $exitSet(t) \equiv \{mainSource(t)\} \cup (\Omega(mainSource(t)) \cap \Delta)$. The computation of the main source state of a transition will be discussed later.

Priority is given in [16] to the inner-most enabled transitions. We define a priority relation $\succ$ between two transitions t and $t\prime$ as $t \succ t\prime \equiv \exists s\prime \in \mathcal{T}^{src}(t\prime) \Rightarrow (\forall s \in \mathcal{T}^{src}(t) : cover(s\prime, s))$, where t has priority over $t\prime$. For instance, $t2 \succ t6$ and $t3 \succ t6$ in figure 2. $\neg(t \succ t\prime) \wedge \neg(t\prime \succ t)$ means that there is no priority relation between t and $t\prime$, e.g. $t2$ and $t3$.

Fireable transitions are selected by defining an intermediate set $nonPriority(e)$ which does not contain any enabled transition with lower priority over others in $enabled(e)$. Formally it is defined as: $nonPriority(e) \equiv \{t \in enabled(e)|\forall t\prime \in enabled(e)$,
$t \neq t\prime \Rightarrow \neg(t\prime \succ t) \vee t \succ t\prime\}$. Thus a maximal consistent set $firable(e)$ of enabled transitions with respect to the current event e is calculated recursively using the OMA rule in figure 5.

```
1  rule TransSelection[e] :
2  then
3    choose t ∈ nonPriority(e)
4      with forall f ∈ firable(e) : not (t ∈ conflict(f, e)) end :
5        firable(e, t) := true
```

Fig. 5. The transition selection algorithm

Figure 5 shows that a transition t that is not in conflict with all the chosen transitions is nondeterministically selected and placed into the set *firable(e)* during a single pass of the *then* block. This block will be executed until a fixed point is reached, i.e. no transition can be chosen. Upon completion of this rule, the set *firable(e)* will contain a set of mutually consistent enabled transitions.

Transition execution. In this stage, due to their mutual consistency, the chosen transitions are fired simultaneously. The execution of a transition consists of a sequence of steps: (1) the main source state is exited, (2) transition actions are executed, and (3) the target states are entered.

The main source of a transition is either the direct active substate of the least common ancestor (LCA) of this transition if the LCA is a sequential composite state, or the LCA itself if the LCA is an orthogonal composite state. Note that the main source may be the source state itself. The LCA of a transition t is defined as $\mathcal{LCA}(t) \in \Sigma_{cs} \cup \Sigma_{ocs}$ such that $(\forall s \in \mathcal{T}^{src}, cover(\mathcal{LCA}(t), s)) \wedge (\forall d \in \mathcal{AT}(t), cover(\mathcal{LCA}(t), d))$, where $\mathcal{AT}(t)$ is the set of the actual target states of t. The function *cover* is defined as before and represents the transitive containment relationship. $\mathcal{AT}(t)$ is defined as: $(\mathcal{T}^{dst}(t) - \Sigma_{sh} - \Sigma_{dh}) \cup \{content(h) | h \in \mathcal{T}^{dst}(t) \cap \Sigma_{sh}\} \cup \left(\bigcup_{\forall h \in \mathcal{T}^{dst}(t) \cap \Sigma_{dh}} content(h)\right)$. Note that $\mathcal{AT}(t)$ contains no pseudostates. The function $content(h)$ returns the content that a history state records. The content of a shallow history is the the most recent active substate of its containing state, while the content of a deep history is a set of the most recent active state configuration of its containing state [16]. For instance, the main source and the LCAs of $t1$, $t2$ and $t5$ are $S4$ and $S2$, $S1$ and $S1$, and $S1$ and *the top state* respectively in figure 2.

Exiting the main source includes exiting its transitively nested or direct substates. Exiting of states must be executed in an inside-out order, i.e. the deepest states in the current state configuration tree are exited first. Figure 6 shows the exiting rule of the main source of a transition t, where the contents of all history states covered by the main source are updated before the exiting occurs.

```
 1  once
 2      do forall h ∈ historyState :
 3          if cover(mainSource(t), h) then setcontent(h) :
 4      do forall s ∈ activeStates :
 5          if s ∈ simpleState + finalState and cover(mainSource(t), s)
 6          then exitingStates(s) := true
 7  then
 8      do forall xs ∈ exitingStates with
 9          not ( xs ∈ OrthogonalCompositeState and
10              subregions(xs) − activeStates ≠ ∅ ) :
11          exitingStates(xs) := false :
12          activeStates(xs) := false :
13          stopActivity(xs) :
14          doExitAction(xs) :
15          if xs ≠ mainSource(t) and not container(xs) ∈ exitingStates
16          then exitingStates(container(xs)) := true :
17  end
```

Fig. 6. Exiting the main source in an inside-out order

In our approach we exit states starting from the deepest states. As the deepest states must be simple or final states, this rule starts from these leaf states and exits them concurrently. If a state in the set of exiting states is an orthogonal state, it does nothing until all of its subregions have been exited. When exited, a state deactivates itself, stops its activity and executes its exit action. After each run of the *then* block in figure 6, containers are set as the exiting states for the next iteration. This iteration completes when the main source state is exited.

In the second stage of transition execution, transition actions are executed in a fashion similar to the execution of state exit actions. The third stage is shown in figure 7. The main target and its (transitively nested or direct) substates are entered in an outside-in order. If the $\mathcal{LCA}(t)$ is a sequential composite state, the main target is the direct substate of $\mathcal{LCA}(t)$ that contains all the states in $\mathcal{AT}(t)$; Otherwise, it is the $\mathcal{LCA}(t)$ itself, e.g. the main target of $t2$ is $S1$ in figure 2. Note that a main target must not be a pseudostate. Upon entering a state it activates itself, executes its entry action and starts the execution of its activity. If this state is an orthogonal composite state, all its subregions are also entered. If it is a sequential composite state, there are two cases: when the entered state contains any actual target state, the target state is entered, otherwise the $initState(s)$ is entered. The function $initState(s)$ returns the direct substate of s pointed to by the transition from the initial state in s. In this situation, $initState(s)$ must be defined – this is ensured by syntax predicates.

```
 1   once
 2      enteringStates(mainTarget(t)) := true
 3   then
 4      do forall es ∈ enteringStates :
 5         enteringStates(es) := false :
 6         activeStates(es) := true :
 7         doEntryAction(es) :
 8         startActivity(es) :
 9         if es ∈ OrthogonalCompositeState then
10            do forall r ∈ subRegions(es) : enteringStates(r) = true end
11         else
12            if es ∈ SequentialCompositeState then
13               choose ts ∈ actualTargets(t) with cover(es, ts) :
14                  choose ns ∈ States
15                  with container(ns) = es and (cover(ns, ts) or ns = ts) :
16                     enteringStates(ns) = true :
17               else
18                  let ns = initState(es) : enteringStates(ns) := true
19   end
```

Fig. 7. Entering the main target in an outside-in order

6   Construction of Runnable Components

As mentioned previously, we employ UML statechart diagrams to represent the dynamic behaviour of individual components. In order to execute a statechart diagram, we need define not only the language semantics but also scheduling and inter-component communication mechanisms. The scheduling mechanisms allow individual components to coordinate with the Moses run-time environment.

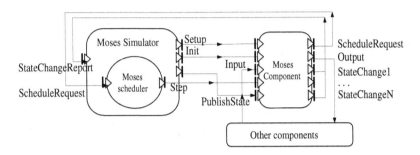

Fig. 8. Scheduling mechanisms

Figure 8 shows how components are integrated with these mechanisms. Five component interfaces are defined: *Setup, Init, Input, Step, PublishState*. These interfaces are defined as OMA rules which are executed when a service request occurs in the corresponding interface.

Setup initiates the construction of the algebraic structure of the semantics from an attributed graph and the creation of communication channels for output, while *Init* initializes the component. These two interfaces are usually called by the scheduler when a component is created from an attributed graph. The *Input* rule handles incoming messages from other components. When a message arrives, this rule causes the message to be stored and a scheduling request to be generated through the predefined channel *ScheduleRequest* to the scheduler. When granting this request according to its scheduling policy, the scheduler sends the acknowledge through the *Step* Interface. The *Step* rule executes a step defined in the semantics of the language. Finally *PublishState* reports the current state of components and is mainly used in animation, statistics and debugging.

Figure 9 shows an abridged version of the component specification in OMA for UML statechart diagrams. The macro $CompileGraph(g)$ transforms the attributed graph of a statechart into OMA algebraic structure defined in the section 5.2, while $enterState(topState)$ is similar to the rule in figure 7 except that the top state is the main target and there are no actual target states. The *Input* rule places the incoming event into the event queue and schedules the component, while the *Step* rule starts a run-to-completion step defined in the section 5.3. These rules provide the statechart component the facilities of communication with the Moses scheduler as well as other system components.

With this specification, executable components of statecharts diagrams can be dynamically generated during the execution or simulation of a system. The scheduler en-

```
class UMLStatechart is
  output ScheduleRequest ;
  input Setup[g] : once CompileGraph(g)   end
  input Init    : once EnterState(topState) end
  input Step    : once RunToCompletion()  end
  input Input[e] :
    once eventQueue(e) := true ; [ScheduleRequest] ← now end
endclass
```

Fig. 9. The component specification for UML statecharts

sures that components are coordinated with each other to represent the desired behaviour of the complete system.

7 Conclusion

In this paper we presented a method for defining the syntax and semantics of UML statechart diagrams. First of all, by using an independent graph definition language GTDL not only was the syntax of UML statecharts defined, but also its well-formedness rules. Next, the operational semantics of statecharts was defined by using an Abstract State Machine while simpler methods were exploited for solving transition conflicts and executing transitions in a run-to-completion step. This semantic model was built at a high abstraction level and is much easier to understand and adapt to future changes or enhancements of the language. Finally, the component specification for statecharts was given facilitating the parameterized instantiation and execution or simulation of the diagrams. Once these aspects have been defined, UML statechart diagrams can be employed in heterogeneous modelling and simulation together with any other visual formalisms supported in Moses.

Furthermore, the method proposed in this paper is applicable for the definition of not only UML statecharts but other visual languages. The use of small domain-specific languages like GTDL, OMA and ELAN ensures that new visual languages can be incorporated within relatively short development cycles.

References

[1] ASM Michigan web page. http://www.eecs.umich.edu/gasm/.
[2] Moses project, Computer Engineering and Communications Laboratory, ETH Zurich. http://www.tik.ee.ethz.ch/~moses.
[3] E. Börger, A. Cavarra, and E. Riccobene. Modeling the Dynamics of UML State Machines. In Y. Gurevich and P. Kutter and M. Odersky and L. Thiele, editor, *Abstract State Machines: Theory and Applications*, LNCS 1912, pages 223–241. Springer-Verlag, 2000.
[4] K. Compton, J. Huggins, and W. Shen. A semantic model for the state machine in the Unified Modeling Language. In *Proceeding of Dynamic Behavior in UML Models: Semantic Questions, UML 2000 workshop*, pages 25–31, York, England, October 2000.

[5] R. Esser and J. W. Janneck. Moses - a tool suite for visual modelling of discrete-event systems. In *Symposium on Visual/Multimedia Approaches to Programming and Software Engineering, HCC01*, 2001.

[6] Y. Gurevich. Evolving Algebras. In B. Pehrson and I. Simon, editors, *IFIP 13th World Computer Congress*, volume I: Technology/Foundations, pages 423–427, Elsevier, Amsterdam, Netherlands, 1994.

[7] D. Harel and A. Naamad. The STATEMATE Semantics of Satecharts. *ACM Transactions on Software Engineering and Methodology*, 5(4):293–333, October 1996.

[8] D. Harel, A. Pnueli, J. P. Schmidt, and R. Sherman. On the formal semantics of statecharts. In *Proceedings of the 2nd IEEE Symposium on Logic in Computer Science*, pages 54–64. IEEE Computer Society Press, 1987.

[9] J. W. Janneck. Graph-type Definition Language (GTDL) - Specification. Technical report, Computer Engineering and Networks Laboratory, ETH Zurich, 2000.

[10] J. W. Janneck. *Syntax and semantics of graphs - An approach to the specification of visual notations for discrete-event systems*. PhD thesis, ETH Zurich, June 2000.

[11] J. W. Janneck and R. Esser. A predicate-based approach to defining visual language syntax. In *Symposium on Visual Languages and Formal Methods, HCC01*, Stresa, Italy, September 2001.

[12] J. W. Janneck and P. W. Kutter. Mapping automata - simple abstract state machines. In *Proceedings of ASM 2000*, 2000.

[13] D. Latella, I. Majzik, and M. Massink. Towards a formal operational semantics of UML statechart diagrams. In *Proceedings of IFIP TC6/WG6.1 Third International Conference on Formal Methods for Open Object-Based Distributed Systems (FMOODS'99)*, Florence, Italy, February 1999. Kluwer.

[14] J. Lilius and I. P. Paltor. Formalising UML state machines for model checking. In R. France and B. Rumpe, editors, *UML'99 - The Unified Modeling Language. Beyond the Standard.*, LNCS 1723, pages 430–445, Fort Collins, CO, USA, October 1999. Springer.

[15] E. Mikk, Y. Lahnech, M. Siegel, and G. Holzmann. Implementing Statecharts in Promela/SPIN. In *Workshop on Industrial-Strength Formal Specifications Techniques (WIFT'98)*, Boca Raton, FL, USA, 1998. IEEE Computer Society Press.

[16] The Object Management Group. *OMG Unified Modeling Language Specification*, September 2001. Version 1.4, http://www.omg.org.

[17] T. Schäfer, A. Knapp, and S. Merz. Model Checking UML State Machines and Collaborations. In *Electronic Notes in Theoretical Computer Science*, volume 47, pages 1–13. Elsevier Science B. V., 2001.

The Learnability of Diagram Semantics

Pourang Irani

Faculty of Computer Science, University of New Brunswick
Box 4400 Fredericton-NB, Canada E3B 2A3
s71h@unb.ca

Abstract. Conveying information through a diagram depends to some extent on how well it is designed as an input to our visual system. Results of previous work by the author (and collaborators) show that diagrams based on a perceptual syntax (Geon diagrams) can improve the legibility of the semantic content in a diagram. The present work evaluates the learnability of semantic information using a perceptual notation. Results of one experiment are reported. These show that Geon diagrams are easier to learn and remember in comparison to equivalent diagrams using conventional line and box drawings.

1   Introduction

Diagrams are essential in designing complex software systems. They are used as tools for communicating between software architects, developers and more recently end-users. They enhance cognition by mapping problem elements to a display such that solutions become immediately evident [6][7]. Often, their syntax has been derived without special attention to the semantic information they carry. This has led to notations that can be easily understood only by experts in the given field.

This poster presents ongoing work to develop a set of diagramming conventions based on structural object perception. According to this theory, as objects are perceived they are decomposed into a set of 3D primitives called Geons and a skeleton structure connecting them. In previous work the author (with Ware) has shown that diagrams based on these 3D primitives (Geon diagrams) are more easily remembered and analyzed than equivalent 2D UML diagrams [3] and 2D diagrams without any shaded components [4]. Here, the author reports on the use of Geon structural descriptions and their application as an aid for learning diagram semantics.

Biederman and his co-workers [1] elaborated the theories of structural object recognition proposed by Marr [2]. In their scheme an object is recognized by first being decomposed into components that remain invariant across viewpoints [1]. The extraction of the structure that specifies how the Geon components interconnect follows the decomposition stage. This structure, known as the Geon structural description (GSD), consists of Geons, their attributes, and their relations with adjacent Geons. The GSD is described in terms of Structural Descriptive Properties (SDPs). The following set of SDPs is based on the set proposed by Biederman:

M. Hegarty, B. Meyer, and N. Hari Narayanan (Eds.): Diagrams 2002, LNAI 2317, pp. 335–337, 2002.
© Springer-Verlag Berlin Heidelberg 2002

SDP1: Geon *shape* is primary while color and texture are secondary.
SDP2: Geon A can be *ON-TOP-OF*, *BOTTOM-OF* or *BESIDE* Geon B.
SDP3: One Geon is *relatively larger* or *relatively smaller* than another.
SDP4: A Geon can be *contained* within another.

In previous work, the author (with Ware et al.) extended the syntax of structural description to include mapping of semantics to data elements [5]. We focused on semantics found in software modeling diagrams specifically those in UML: 1.generalization ("is-a"), 2.aggregation ("has-a"), 3.dependency, 4.multiplicity ("many-of"), and 5.relationship strength. By applying the structural descriptive properties (SDP1-SDP4) defined above we constructed a set of representations for each of the five semantics. Based on the results of three experiments, the following Geon-based notation was formulated:

SM1: Inheritance (is-a) - Same shape Geons can denote objects of the same kind.
SM2: Dependency - A on-top-of B suggests that A is dependent on B.
SM3: Aggregation - Containment shows that Geon A is aggregated with Geon B.
SM4: Multiplicity - Multiple links between two entities can denote multiplicity.
SM5: Relationship Strength - This can be denoted using thicker connections.

2   Experiment

The purpose of the experiment was to compare the explanatory effectiveness of Geon diagrams with that of equivalent UML diagrams. We measured the error rate for a subject matching a given problem to one correct diagram out of a set of possible diagrams. We hypothesized that it should be possible to learn, recall, and match problem descriptions more accurately to diagrams created with the perceptual notation presented above.

2.1 Method for the Experiment

Diagrams. Five non-fictitious problem descriptions with the semantics of generalization, dependency, multiplicity, and aggregation were constructed. The semantics in the problems were clearly presented using their common terminology. The problem descriptions were created with a comparable number of relationships. For each problem description, a set of four UML diagrams and a set of their four equivalent Geon counterparts were created. Only one of the four diagrams accurately depicted the relationships in the corresponding problem description.

Procedure. Subjects were given an hour-long instruction on the various semantics and their respective notations in UML and in Geon diagrams. A week later, subjects were asked to read each problem description and to match one of the four diagrams created for that problem. The problem descriptions were available to the subjects while reading the diagrams, and so they could occasionally consult the description. They were restricted to two minutes for matching a diagram to a problem description. Half the subjects were matching the UML diagrams first and the other half matched the Geon diagrams first.

Subjects. Twenty-six paid volunteers participated of whom 12 had previous exposure to UML diagrams (experts) while the 14 others had not (novices).

2.2 Results of the Experiment

Results are summarized in Table 1, which reports error rates by level of experience and type of diagram. The results are obtained by averaging each subject's scores. A One-Sample T-Test shows that performance is better with the Geon diagrams (p < 0.0001). A Two-Sample T-Test shows that subjects' level of experience is not a relevant factor in comparing the performance between the two types of diagramming notation (p-values are 0.704 and 0.425 respectively). The most striking result is the performance of expert subjects with respect to the interpretation of Geon diagrams. Combining the results we can say that there were more than twice as many errors in analyzing and matching the UML diagrams than the Geon diagrams.

Table 1. Average error rates of matching Geon and UML diagrams to problem descriptions.

	Geon	UML
Novices	22.3%	44.3%
Experts	2.9%	25.9%
Combined	14.6%	36.2%

3   Conclusion

Mapping of application semantics to Geon connections can be used for making effective diagrams. In comparing the learnability of the perceptual notation to UML notation, we found that subjects, regardless of their experience in software modeling, were capable of learning and applying the perceptual syntax with fewer errors. Further experimentation needs to be conducted in order to determine whether the learning of the more intuitive notations could facilitate the creation of diagrams and the conceptualization of abstract ideas for problem solving.

References

1. Biederman, I., "Recognition-by-Components: A Theory of Human Image Understanding", Psychological Review, 94:2, 115-147, 1987.
2. Marr, D., "Vision: A computational investigation into the human representation and processing of visual information", San Fransisco, CA: Freeman, 1978.
3. Irani, P. and Ware, C., "Diagrams Based on Structural Object Perception", Conference on Advanced Visual Interfaces, Palermo, Italy, 2000.
4. Irani, P. and Ware, C., "Should the Elements of Diagrams Be Rendered in 3D?", Late Breaking Hot Topics, IEEE Information Visualization Proc., Salt Lake City, Utah, 2000.
5. Irani, P., Ware, C. and Tingley, M., "Using Perceptual Syntax to Enhance Semantic Content in Diagrams", IEEE Computer Graphics & Applications, 21:5, 76-85, 2001.
6. Zhang, J., "The nature of external representation in problem solving", Cognitive Science, 21:2, 179-217, 1997.

Understanding Simultaneity and Causality in Static Diagrams versus Animation

Sarah Kriz

Department of Psychology, University of California Santa Barbara
Santa Barbara, CA 93106 USA
kriz@psych.ucsb.edu

Abstract. This study assesses how the mode of presentation affects the way in which people structure their mental models of a mechanical system, namely a flushing cistern. Subjects were assigned to one of three learning conditions: diagram only, 3-phases diagram, or animation. After learning the material, subjects generated written descriptions of the workings of the cistern. An analysis of temporal conjunctions used and the number of causal events mentioned indicates that for understanding simultaneity and causality, animation does not provide any benefit over seeing the same information in static diagrams.

1 Introduction

Understanding complex mechanical systems requires that a person obtains information about the movement of the system's parts and integrates that information into a mental model. Furthermore, the relationship between the parts in time and space must be understood in order to correctly infer the chain of events.

The mode in which information is presented may have an impact on how mental models are constructed. Since the advent of computers, researchers have been interested in the possible benefits of presenting material via an animation. Both benefits and deficits of using an animation for instruction have been found [1]. Some studies indicate that animation can facilitate learning better than static diagrams or text [2], other studies have shown that viewing an animation does not provide significant learning benefits over seeing the same material in static diagrams [3,4,5]. Because animations are programmed to play at a certain speed, a person does not have control over the input, and this could cause a cognitive overload. One benefit of viewing static diagrams is that there is not a rigorous time constraint.

This study addresses how subjects create mental models of a mechanical system when learning about the system through varying degrees of visual (non-linguistic) information. Because this experimental design does not include language input, it can assess how the amount of visual information given affects learning outcomes. Additionally, subjects' linguistic descriptions of the machine's functioning can be evaluated without the possibility that the description is a regurgitation of the language instruction.

M. Hegarty, B. Meyer, and N. Hari Narayanan (Eds.): Diagrams 2002, LNAI 2317, pp. 338–340, 2002.
© Springer-Verlag Berlin Heidelberg 2002

2 Methodology

Sixty undergraduates from the University of California, Santa Barbara participated in the study.   Subjects were placed in one of three conditions: 3-phases diagram, animation, or diagram-only.

The subjects in the 3-phases diagram and animation conditions were give two minutes to view a single diagram of a toilet tank in resting position. Subjects then saw either three diagrams of a toilet tank in different phases of its cycle or a computer controlled animation, which played the entire cycle twice. There was no time limit for these learning activities. After viewing the diagrams or animation, subjects were asked to write step-by-step the events that happen during a single cycle of a toilet's flush. To test spatial ability, the Paper Folding Test [6] was administered.

The subjects in the diagram-only condition were allowed to see a single diagram of the toilet tank in resting position. Instead of seeing this diagram for two minutes, as the other groups did, the diagram-only group was allowed an unlimited amount of time to view the diagram.   After viewing, subjects proceeded immediately to the written questions.

Subjects' written descriptions of the events that comprise a cycle of the toilet's flush were analyzed. To asses how subjects understood the temporal relationships between events, their descriptions were scored for the number of temporal conjunctions that signaled simultaneity, such as *meanwhile*, *as*, and *while*. The number of correct events mentioned in the description were counted in order to evaluate causal understanding. An event was considered to be a discrete movement of a single part, and the descriptions were scored for mention of 25 relevant events, determined by the experimenter prior to data collection. For instance, "As the *float rises* it brings the *float arm up* and gradually *pushes the inlet stopper back* in place," was counted as three relevant events (shown in italics).

3 Results

Subjects in all three conditions described roughly the same number of simultaneous relationships in their descriptions of a flush cycle. The number of simultaneous events did not differ significantly across conditions.

For causal understanding, there were no significant differences between the 3-phases and animation conditions. Out of a possible 25 events that occur during a single flush, subjects in the 3-phases condition mentioned a mean of 10.2 events. The mean number of events mentioned by the animation group was 9.7. The subjects in both of these conditions reported significantly more events than the diagram-only condition, which had a mean of 6.8 events $[F_{(2.55)} = 7.431, p < .05]$.

4 Discussion

This study supports previous claims that animation does not provide a significant benefit over learning the same material through static diagrams. The results indicate that the format of similar visual material does not affect how subjects constructed their mental models of the system in terms of temporal and causal structuring. That the subjects in the animation and 3-phases conditions were able to report more causal events than the diagram-only condition suggests that both the 3-phases diagram and the animation provide more causal information than what can be inferred from a single static diagram. However, the quality and quantity of this information does not seem to differ between those who studied the two displays.

While computer-controlled animation provides a real-time display of the events that occur within a mechanical system, the time constraint may eliminate any benefits that animation has over static diagrams. It may be the case that the speed of the animation prohibits subjects from fully benefiting from the additional temporal and causal information that is encoded in animations. Future studies could compare system-controlled animation to subject-controlled animation in order to determine how different forms of multimedia affect the causal and temporal understanding of a particular mechanical system. This may provide insight into what features of visual instruction are necessary for people to construct accurate mental models of the information that is presented via visual displays.

Acknowledgments. This research was supported by grant number N00014-97-1-0601 to Mary Hegarty from the Office of Naval Research.

References

1. Morrison, J.B., Tversky, B., Betrancourt, M.: Animation: Does it facilitate learning? In: Butz, A., Kruger, A., Olivier, P. (eds.): Smart Graphics: Papers from the AAAI Spring Symposium, Technical Report SS-00-04. AAAI Press, Menlo Park, California (2000) 53-60.
2. Park, O., Gittelman, S.S.: Selective use of animation and feedback in computer-based instruction. Educational Technology, Research, and Development, Vol. 40 (1992) 27-38.
3. Hegarty, M., Narayanan, N.H., Freitas, P.: Understanding Machines from Multimedia and Hypermedia Presentations. In: Otero, J., Graesser, A.C., Leon, J. (eds.): The Psychology of Science Text Comprehension. Erlbaum, Hillsdale, NJ. (in press).
4. Lowe, R.K.: Extracting information from an animation during complex visual learning. European Journal of Psychology of Education, Vol. 14(2). (1999), 225-244.
5. Reiber, L.P., Boyce, M.J., Assad, C.: The effects of computer animation on adult learning and retrieval tasks. Journal of Computer-Based Instruction, Vol. 17(2). (1990), 46-52.
6. Ekstrom, R.B., French, J.W., Harman, H.H., Derman, D.: Kit of factor-referenced cognitive tests. Educational Testing Service, Princeton, NJ. (1976).

External Representations Contribute to the Dynamic Construction of Ideas

Masaki Suwa[1] and Barbara Tversky[2]

[1] Information and Human Activity, PRESTO, JST
School of Computer and Cognitive Sciences, Chukyo University,
101 Tokodachi, Kaizu, Toyota, 470-0393, Japan
suwa@sccs.chukyo-u.ac.jp

[2] Department of Psychology, Stanford University, Stanford, CA 94305-2130, USA
bt@psych.stanford.edu

Extended Abstract

External representations such as diagrams, sketches, charts, graphs and scribbles on napkins play facilitatory roles in inference, problem-solving and understanding (e.g. [1],[2],[3],[4],[5],[6],[7],[8],[9]). How does the externality and visibility of representations facilitate inference and problem-solving? One benefit of external representations is on memory. They reduce working memory load by providing external tokens for the elements that must otherwise be kept in mind. This frees working memory to perform mental calculations on the elements rather than both keeping elements in mind and operating on them [2],[9]. External representations also serve as visuo-spatial retrieval cues for long term memory, evoking relevant information that might not otherwise be retrieved. Another benefit of external representations is to promote discovery and inference, both visuo-spatial and metaphorical. Perceptual judgements about size, distance, and direction are easily made from external representations (e.g.[4]). In a Venn diagram, set relations such as inclusion are abstractly mapped onto visuo-spatial diagrammatic features, enabling direct perceptual calculation. Visuo-spatial features such as proximity, connectivity, and alignment provide useful hints to selection of appropriate inference paths (e.g.[1],[6],[8]) and to proper understanding of the structure of a target system (e.g.[5]). Calculations requiring counting, sorting, or ordering are easily made by rearranging external spaces (e.g. [7]).

To serve these functions of memory, inference, and calculation, the interpretation of the external representation is static; it must stay the same in order not to introduce error in the operations performed from the external representation. External representations, however, are visuo-spatial displays, and it is known from research on perception that such displays, especially vague and ambiguous ones, can be interpreted and reinterpreted. Are there situations where the very instability of visuo-spatial displays can be used to advantage?

One situation where the instability of interpretations of external representations can be beneficial is design. Among the earliest of commentators on this was Schon [10], who proposed that freehand sketches serve as a medium for the dynamic generation of

M. Hegarty, B. Meyer, and N. Hari Narayanan (Eds.): Diagrams 2002, LNAI 2317, pp. 341–343, 2002.

new design ideas. In developing ideas for new projects, designers do not draw sketches to externally represent ideas that are already consolidated in their minds. Rather, they draw sketches to try out ideas, usually vague and uncertain ones. By examining the externalizations, designers can spot problems they may not have anticipated. More than that, they can see new features and relations among elements that they have drawn, ones not intended in the original sketch. These unintended discoveries promote new ideas and refine current ones. This process is iterative as design progresses. Seeing unintended relations and features in sketches requires release from previous interpretations. Previous interpretations can have a strong hold on observers, so preventing fixation and encouraging new interpretations are perceptual processes desirable for designers to acquire.

In recent years we have explored ways that designers use external representations to discover and develop design ideas. The project has used both naturalistic and experimental methods. We present data from both these projects that are relevant to the current analysis. In a large naturalistic study, novice and experienced architects were filmed as they designed a museum. Later, they watched their design session, and commented on every stroke of the pen. These protocols have been analyzed in detail [11],[12],[13]. In the experimental study, designers and novices were shown ambiguous sketches and asked to produce as many interpretations of them as possible.

The protocols from the design sessions showed clearly that new design ideas were likely to be generated immediately after discovering new relations and features in one's own sketches. Notably these new relations and features were unintentional byproducts of the aspects of the sketches drawn for other reasons. We call this process *detection of unintended relations and features* [12]. Detection of unintended relations and features is a significant impetus for the generation of new ideas. The generation of new ideas, in turn, was likely to become an impetus for further detection of unintended relations and features, so that each component process drives the other [12]. The joint occurrence of generation of new ideas and detection of unintended relations and features constituted the core cognitive processes of designers as they worked. As is common in discovery problems, the designers themselves were not able to predict what kind of unintended relations and features they were going to detect and what kinds of ideas they were going to generate. Productive design is situated in the physical setting of sketches. They are not merely a static medium for externalizing internal visions, but rather a physical environment from which ideas are generated on the fly.

The key to creative design, then, is the cyclic pattern of generating new ideas and detecting unintended relations and features in external representations. More generally, it can be regarded as a coordinated co-generation of new conceptual thoughts and perceptual discoveries in external representations. This appears to be a general phenomenon occurring in a broader context involving any kind of creation, not just in design processes [13]. An important issue is how to facilitate the co-generation of reinterpretations and novel ideas in inspecting external representations, given that it is by no means automatic. We argued that the cognitive skill we have called *constructive perception* promotes the discovery of new interpretations in external representations [14]. By constructive perception, we mean self-awareness of the ways that perception underlies interpretations of external representations. The self-awareness allows searching for other ways to perceive, enabling reorganization of the external representation to promote novel interpretations. Experienced designers are superior to laypeople in this skill [14]. This finding raises two issues, one cognitive, the other didactic. What constitutes the ability of constructive perception? How can people be trained to use it? Research on these will promote successful use of external representations for creative purposes.

References

1. Gelernter, H.: Realization of a Geometry-Theorem Proving Machine. In E.A. Feigenbaum and J. Feldman (eds.) Computer and Thought. MacGraw-Hill. (1963)
2. Newell, A., Simon, H. A.: Human Problem Solving. Prentice-Hall, N. J. (1972)
3. Koedinger, K. R., Anderson, J. R.: Abstract Planning and Perceptual Chunks: Elements of Expertise in Geometry. Cognitive Science. 14 (1990) 511-550
4. Clement, J.: Use of Physical Intuition and Imagistic Simulation in Expert Problem Solving. In D.Tirosh (ed.) Implicit and Explicit Knowledge. Ablex Publishing Co. (1994)
5. Petre M.: Why Looking isn't Always Seeing: Readership Skills and Graphical Programming. Communications of the ACM. 38(6) (1995) 33-44.
6. Larkin, J., Simon, H. A.: Why a Diagram is (sometimes) Worth Ten Thousand Words. Cognitive Science. 11 (1987) 65-99
7. Kirsh, D.: The Intelligent Use of Space. Artificial Intelligence. 73(1-2) (1993) 31-68
8. Narayanan, N. H., Suwa, M., Motoda, H.: Diagram-based Problem-Solving: The Case of an Impossible Problem. Proceedings of the 17th annual conference of the cognitive science society. Lawrence Erlbaum Associates, N. J. (1995) 206-211.
9. Tversky, B.: Spatial Schemas in Depictions. In M. Gattis (ed.), Spatial Schemas and Abstract Thought. Cambridge: MIT Press. (2001) 79-111.
10. Schon, D. A.: The Reflective Practitioner. Basic Books, New York. (1983)
11. Suwa, M., Tversky, B.: What Do Architects and Students Perceive in Their Design Sketches?: A protocol analysis. Design Studies. 18 (1997) 385-403.
12. Suwa, M., Gero, J.,Purcell, T.: Unexpected discoveries and S-invention of design requirements: Important vehicles for a design process. Design Studies. 21 (2000) 539-567.
13. Suwa, M., Tversky, B., Gero, J., Purcell, T.: Regrouping Parts of an External Representation as a Source of Insight. Proceedings of the third International Conference on Cognitive Science. Press of USTC, Beijing, China (2001) 692-696.
14. Suwa, M., Tversky, B.: Constructive Perception in Design, *in* J.S.Gero and M.L.Maher (eds.), Computational and Cognitive Models of Creative Design V. Key Centre of Design Computing and Cognition, University of Sydney, Australia (2001) 227-239

One Small Step for a Diagram, One Giant Leap for Meaning

Robert R. Hoffman, Ph. D.
John W. Coffey, Ed.D.
Patrick J. Hayes, Ph.D.
Alberto J. Cañas, Ph.D.
Kenneth M. Ford, Ph.D.
Mary J. Carnot, M.A.
Institute for Human and Machine Cognition
University of West Florida
40 Alcaniz St.
Pensacola, FL 32501 USA
rhoffman@ai.uwf.edu

Abstract. Concept-Mapping was used to create models of the knowledge of expert weather forecasters. STORM-LK (System To Organize Representations in Meteorology-Local Knowledge) demonstrates the use of CMap diagrams to generate large-scale multi-media knowledge models. This merger of interactive graphical communication with knowledge management provides a counterpoint for discussion of logical and psychological issues in diagrammatic reasoning.

1 Introduction

This project utilized Concept-Maps (CMaps) as a knowledge representation scheme in the domain of meteorology. CMaps are graphs that use nodes and labeled links to depict concepts and their relations [1]. CMaps have been widely used in education to promote meaningful learning and distance collaboration [2]. CMaps encode meaning in ways that are not yet well understood and are not yet fully useable by computational tools. The work described here is part of an ongoing program to investigate these aspects of CMaps, and use CMaps to build human-machine interfaces of increasing power.

2 Method

Participants (n = 22) were expert, journeyman, and apprentice forecasters at the Pensacola Naval Air Station. We conducted over 60 hours of CMapping interviews. To make a CMap, a knowledge engineer guided the domain expert while the another knowledge engineer constructed the CMap using the CMapTools software. In a second session, the initial CMap was refined to make sure that that each proposition was clear and precise. Figure 1 presents an example CMap.

3 Results

The rate of gain was about 2 mappable propositions per knowledge elicitation session minute. Other knowledge methods (e.g., protocol analysis, documentation analysis, the Critical Decision Method) were utilized in this project, and compared on the yield metric. One can quite safely conclude that CMapping is more efficient at generating

M. Hegarty, B. Meyer, and N. Hari Narayanan (Eds.): Diagrams 2002, LNAI 2317, pp. 344–346, 2002.
© Springer-Verlag Berlin Heidelberg 2002

knowledge models than other methods of knowledge elicitation or Cognitive Task Analysis [see 3, 4].

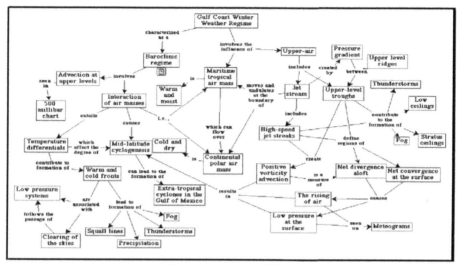

Fig. 1. An example CMap.

Figure 2 presents a screen shot of one of the final CMaps. Appended to concept-nodes are CMap icons that take one to the CMap indicated by the node to which the icon is attached. At the top node in every CMap is a CMap icon that takes one back to the Top Map and to all other CMaps that are associated to the given CMap. Hence, one can get from anywhere in the knowledge model to anywhere else in two clicks, at most.

An independent expert went over each CMap, providing changes and clarifications. Comparison of the verification data to results from a survey on how forecasters and apprentices learn from the "Local Forecasting Handbook" suggested that the propositions in CMap form can be analyzed at a rate much faster than in a traditional text format—as much as an order of magnitude faster.

4 Cmaps as Diagrams?

CMaps have considerable utility from the perspective of cognitive psychology. For instance, STORM-LK contains all of the information that is in the Local Forecasting Handbook, but also includes digital video-tutorials presented by the domain experts themselves. Since the CMaps are web-enabled, they allow real-time access to data (radar satellite, computer forecasts, etc.)—always within a context that provides the explanatory glue for the forecasting process. CMaps have great expressive power: Nodes do not have a restricted ontology—and the link labels do not have a restricted set of possible meanings, in contrast with Conceptual Graphs, for example [5].

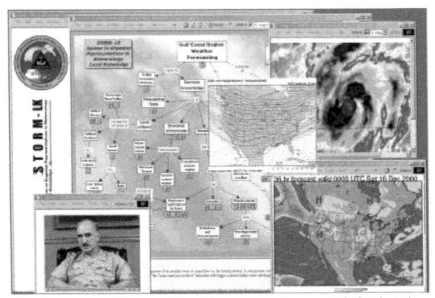

Fig. 2. A screen shot illustrating the variety of resources that are attached to the nodes.

This raises the issue of whether CMaps should be thought of as diagrams. CMaps are not just labeled graphs; they are diagrammatic in that their spatial layout is critical to their functionality. As people create, revise, and learn from CMaps, the "where" of the concept nodes is important. Our experience at CMapping, and teaching others to CMap, has shown repeatedly that people rely on locations to remember concepts and their inter-relations, and to navigate among them. Location in two-dimensions is not part of the "logic" of finished CMaps but is definitely part of the CMap diagramming process.

The great semantic freedom of CMaps is, however, a weakness from a computational point of view. STORM-LK represented an attempt to compose CMaps that were "propositionally coherent"—every node-link-node is a proposition. The possibility that knowledge models possessing great expressive power can also be propositionally coherent suggests additional capabilities and applications that are currently being explored. For instance, node-link-node sequences could be used to form search queries. It may be possible to introduce a mechanism for scoping.

References

1. Novak, J. D.Learning, creating, and using knowledge. Erlbaum, Mahwah NJ (1998).
2. Novak. J. D., & Wandersee, J. H. (Eds.) Special Issue: Perspectives on concept-Mapping. Journal of Research in Science Teaching **27** (1990)
3. Hoffman, R. R. The problem of extracting the knowledge of experts from the perspective of experimental psychology. The AI Magazine **8** (1987, Summer), 53-67.
4. Hoffman, R. R., Shadbolt, N., Burton, A. M., & Klein, G. A. Eliciting knowledge from experts: A methodological analysis. Organizational Behavior and Human Decision Processes **62** (1995) 129–158.
5. Sowa, J. F. (1984). Conceptual structures: Information processing in mind and machine. Addison-Wesley, Reading, MA (1984).

Understanding Static and Dynamic Visualizations

Sally Bogacz and J. Gregory Trafton

Naval Research Laboratory, Code 5513, Washington DC 20375-5337, USA
{bogacz,trafton}@itd.nrl.navy.mil

Abstract. Data from expert forecasters making weather reports (using the talk aloud method) were coded for dynamic comments as well as whether the visualization itself was static or dynamic. Preliminary results strongly suggest that meteorologists build dynamic mental models from static images.

1 Introduction

How and why do people use dynamic animations? On one hand, dynamic images allow one to see action as it unfolds, which can be extremely valuable in some domains (computational fluid dynamics, weather forecasting). In addition, it takes a great deal of effort to mentally animate static images (Hegarty & Sims, 1994). On the other hand, dynamic animations can impose a high workload because of the increased complexity of the image (Lowe, 1999) and it is often easier to see all the details of individual images if each image can be observed by itself. This question is further complicated because research comparing static and dynamic images is often quite contradictory, leading to suggestions to use animations only in limited situations (Betrancourt & Tversky, 2000).

What happens, then, when practitioners have a choice of whether to use static or dynamic visualizations? This study will examine expert meteorologists as they were making a weather forecast. The forecaster looked at many weather visualizations of different types (weather models, satellite imagery *etc.*) and each type could be examined either as an individual image (static) or as a dynamic animation (looping many static images together to create an animation). We examined the type of visualization the forecaster chose to examine as well as the kind of information (static or dynamic) that was extracted from each kind of visualization.

2 Method

2.1 Task

The task was to prepare a written brief for an airplane flown from an aircraft carrier to Whidbey Island, Washington State 12 hours in the future. In order to do this, forecasters had to determine detailed qualitative and quantitative information about weather conditions. This task took about 2 hours.

2.2 Participants, Apparatus, and Procedure

The data for this analysis were taken from four experienced forecasters, who had an average of 10 years of experience. Each forecaster talked aloud (Ericsson & Simon, 1993) as they worked on their forecasting task.

Forecasters looked at three types of weather visualizations: static images or text (static), dynamic animations (dynamic), and sets of visualizations (set). Sets

M. Hegarty, B. Meyer, and N. Hari Narayanan (Eds.): Diagrams 2002, LNAI 2317, pp. 347–349, 2002.
© Springer-Verlag Berlin Heidelberg 2002

were conceptually related weather charts (across time or pressure (height), typically). These weather charts could be displayed as an individual, set, or a dynamic animation. Dynamic animations were simply an animated "set" of non-text images.

In addition to coding the type of visualization, we coded the forecasters' extraction utterances as either static (simple extractions like "The temperature at Whidbey is 17 degrees") or dynamic (extractions that dealt with change over space or time, like "24 hours from now, definitely a lot of precip. moving into the area.").

A typical weather map is shown in Figure 1.

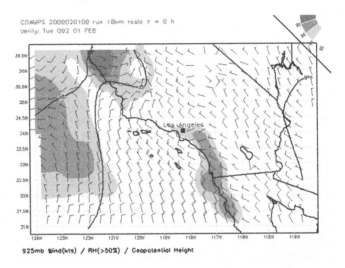

Fig 1. A visualization of a COAMPS model run showing wind speed, wind direction, temperature, and pressure at the surface level.

3   Results

Forecasters looked at an average of 26.5 visualizations. They extracted information from these visualizations an average of 4.6 times per visualization. Forecasters had the opportunity to look at dynamic images either by choosing animation loops or by scrolling through sets of images. However, they chose to look at static images most of the time: 57% of the visualizations were static images. But this reliance on static images did not prevent forecasters from making dynamic extractions about the weather for a substantial portion of the time: 37% of the extractions in this study were dynamic comments. Table 1 shows the percentage of dynamic extractions made for each type of visualization.

We expected the percentage of dynamic extractions to be greater for visualizations that had a dynamic quality to them either because they were animated images ("dynamic"), or because they formed sets of images ("set"). However, Table 1 demonstrates that forecasters made most of their dynamic comments to static images, $\chi^2(2) = 95.7$, p < .001, Bonferroni adjusted chi-square significant at p < .01. This pattern of results holds true even when the number of different types of visualizations is taken into account, $\chi^2(2) = 107.2$, $p < .001$, Bonferroni adjusted chi-squares significant at $p < .01$.

Table 1. *Percentage of dynamic extractions for each type of visualization*

Visualization	Dynamic extractions
Dynamic	10
Set	11
Static	79

4   Discussion

We showed that forecasting is a domain with a great deal of dynamic content; indeed, over a third of the weather related comments dealt with dynamics. Given this finding, one would expect forecasters to examine primarily (or at least many) dynamic animations. What we found, however, was forecasters examining primarily static visualizations, even though they had the choice to view either static or dynamic representations. What's more, the vast majority of dynamic comments occurred when forecasters were examining static images. This strongly suggests that forecasters were mentally animating static weather maps to build a qualitative mental model (Trafton *et al.*, 2000).

This research was supported in part by grants N00014-00-WX-20844 and N00014-00-WX-40002 to the second author from the Office of Naval Research.

References

1. BETRANCOURT, M. & TVERSKY, B. (2000). Effect of computer animation on users' performance: A review. *Travail Humain*, **63**(4), 311-329.
2. ERICSSON, K. A. & SIMON, H. A. (1993). *Protocol Analysis: Verbal reports as data.* (Revised edition). Cambridge MA: MIT Press.
3. HEGARTY, M. & SIMS, V. K. (1994). Individual differences in mental animation during mechanical reasoning. *Memory & Cognition*, **22** (4), 411-430.
4. LOWE, R. K. (1999). Extracting information from an animation during complex visual learning. *European Journal of Psychology of Education*, **14**(2), 225-244.
5. TRAFTON, J.G., KIRSCHENBAUM, S. S., TSUI, T. L., MIYAMOTO, R. T., BALLAS, J. A.. & RAYMOND, P. D. (2000). Turning pictures into numbers: extracting and generating information from complex visualizations. *International Journal of Human-Computer Studies*, **53** (5), 827-850.

Teaching Science Teachers Electricity Using AVOW Diagrams

Peter C-H. Cheng and Nigel G. Pitt

ESRC Centre for Research in Development, Instruction and Training
School of Psychology, University of Nottingham,
University Park, Nottingham NG7 2RD, UK.
peter.cheng@nottingham.ac.uk
ngp@psychology.nottingham.ac.uk

Abstract. An approach to the teaching of electricity is described which uses AVOW diagrams, a novel diagrammatic representation to visualize the laws of electricity. AVOW diagrams can help learners develop useful concepts and a more integrated understanding of electric circuit behavior than alternative teaching methods. In this study the practical potential of the approach to teach trainee science teachers was examined. In addition to the overall learning effect, trainee teachers with little previous experience of electric circuit behavior advanced after brief instruction to a level of understanding approaching the more knowledgeable trainees before instruction. For those trainees with prior knowledge the approach consolidated their understanding and they were able to quickly adopt the technique in the solution to more complex electric circuit problems.

1 Introduction

AVOW diagrams (Amps, Volts, Ohms, Watts) have been shown in classroom and laboratory studies to be effective in supporting learning about electric circuits [1-6]. Compared to conventional approaches the technique can help overcome the familiar difficulties and common misunderstandings that arise with this topic [7-10]. The diagrams simultaneously portray current (I), voltage (V), resistance (R) and power (P) of loads in a circuit, encapsulate the laws of electricity (V=IR and P=IV) and model series and parallel circuits and internal resistance, as shown in Figure 1.

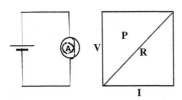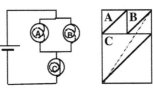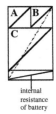

internal
resistance
of battery

Fig. 1. AVOW diagram for single load (left); AVOW diagram for a network with an ideal or a real battery

M. Hegarty, B. Meyer, and N. Hari Narayanan (Eds.): Diagrams 2002, LNAI 2317, pp. 350–352, 2002.
© Springer-Verlag Berlin Heidelberg 2002

This research examined the feasibility of using the approach for instructing teachers with varied prior knowledge in electricity given the limitations on available time and the demands made on them to teach beyond their own specialist areas.

2 Experiment and Results

Twenty-seven trainee science teachers were instructed for two hours in the AVOW diagram technique. They were classified into two groups PC and B depending on their university degree subject, physics/chemistry or biology, and hence into high or low prior knowledge of the topic, respectively. Participants completed a pre-test and post-test consisting of problem solving questions and a classification task requiring the components to be identified according to their insulator-like or conductor-like properties. The post-test also included a challenging transfer problem. On the problem solving questions both groups improved from pre-test to post-test but the PC group had a marginally significant increase whilst the increase for the B group was not significant. With the classification task both groups had significant improvements. Given the difficult circumstances for the trainees under which the study was conducted these results are encouraging.

The examination of the solutions produced by the trainees in the post-test showed that many in both groups were using the approach effectively. An example form the B group is shown in Figure 2. This was B2's diagrams for the electric circuit represented in Figure 1 before and after Bulb A is unscrewed. In addition to the diagrams the participant confirms her understanding when the bulb is disconnected in her written statement on the left-hand side.

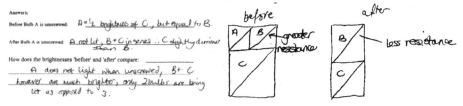

Fig. 2. Trainee B2's solution to a conceptually difficult problem.

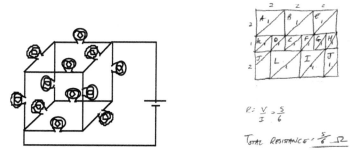

Fig. 3. "Cube" of light bulbs. What is its resistance?   **Fig. 4.** PC9's AVOW diagram solution

For those trainee teachers already with high prior knowledge, the AVOW approach helped some make good attempts at a complex task included in the post-test. Participants were asked to find the overall resistance between two connected corners of a cube of $1\,\Omega$ light bulbs as illustrated in Figure 3. Figure 4 shows the response of PC9 to the problem completed in a matter of minutes. Solving this problem the conventional algebraic way is particularly time consuming and difficult.

3   Conclusion

AVOW diagrams were being used by the trainees as problem solving tools and not merely as fixed visualizations of circuit properties. There was evidence that the more knowledgeable group's understanding of electricity grew. For the less knowledgeable it was not so clear but it appeared they were beginning to approach the standard of the more knowledgeable trainees before instruction. The experiment was conducted in a real teacher-training context and the results are encouraging as to the potential of the approach for supporting trainee teachers' acquisition of the knowledge they would need to teach this topic. A full version of this paper is available as a technical report from the ESRC Centre for Instruction, Training and Development (http://www.psychology.nottingham.ac.uk/research/credit/).

References

1.   Cheng, P. C-H.: AVOW Diagrams: A Novel Representational System for Understanding Electricity. In: Anderson, M., Meyer, B., Olivier, P., (eds.): Diagrammatic Representation and Reasoning. Berlin Springer (2001)
2.   Cheng, P. C-H.: AVOW Diagrams Enhance Conceptual Understanding of Electricity: Principles of Effective Representational Systems. *Cognitive Science*. (forthcoming).
3.   Cheng, P. C-H., & Shipstone, D. M.: Supporting learning and promoting conceptual change with box and AVOW diagrams. Part 1: Representational Design and Instructional Approaches. *International Journal of Science Education*. (forthcoming).
4.   Cheng, P. C-H., & Shipstone, D. M.: Supporting learning and promoting conceptual change with box and AVOW diagrams. Part 2: Their impact on student learning at A-level. *International Journal of Science Education*. (forthcoming).
5.   Shipstone, D. M., & Cheng, P. C-H.: The electric circuit: a new approach, Part 1. *School Science Review*. 83(303) (2001) 55-63.
6.   Shipstone, D. M., & Cheng, P. C-H.: The electric circuit: a new approach, Part 2. *School Science Review*. (in press 2002).
7.   Cohen, R., Eylon, B., Ganiel, U.: Potential difference and current in simple electric circuits. *American Journal of Physics*. 51 (1983) 407 - 412
8.   Shipstone, D. M.: A study of children's understanding of electricity in simple DC circuits. *European Journal of Science Education*. 6 (2) (1984) 185 – 198
9.   McDermott, L. C., Shaffer, P. S.: Research as a guide for curriculum development: An example from introductory electricity. Part 1: Investigation of student understanding. *American Journal of Physics*. 60 (11) (1992) 994 – 1003
10.   Millar, R., King, T.: Students' understanding of voltage in simple series electric circuits. *International Journal of Science Education*. 15 (3) (1993) 339 – 349

Conceptual Diagrams: Representing Ideas in Design

Fehmi Dogan[1] and Nancy J. Nersessian[2]

[1]College of Architecture, Georgia Institute of Technology
fehmi.dogan@arch.gatech.edu
[2]Program in Cognitive Science, Georgia Institute of Technology
nancyn@cc.gatech.edu

Studies in cognition [1, 2] have investigated the role of external visual representations in different domains in supporting reasoning, problem solving, and communication. These studies often are confined to domains that pose relatively well-defined problems [3], such as geometry [4] and physics [2], with fewer studies in domains where the problems are ill-defined [3], such as meteorology [5] and architecture [6].

In many studies of well-defined problems, diagrammatic representations illustrate either causal or temporal relationships between parts of entities and phenomena that the diagram represents. In architecture, diagrams are used to represent causal relationships, such as with orientation diagrams, or temporal relationships, such as with circulation diagrams. There is, however, another kind of diagram that is used to represent the main idea or the core of a design. We call these diagrams *conceptual diagrams*. They differ, potentially, from other diagrammatic representations studied thus far in that they represent an abstract conceptualization of a potential problem solution.

Diagrams in other fields can be interpreted as conceptual diagrams as well, such as a diagram that shows the electron orbiting around a nucleus in atomic physics, or the supply-demand diagram in economics. In the domain of scientific discovery, Nersessian [7] has shown how the use of conceptual diagrams helped Maxwell to construct and communicate his representation of the electromagnetic field concept.

Conceptual diagrams are abstract representations that embed the core of a conceptualization of a problem solution. They are concise, yet powerful aids in problem solving in that they provide high-level commitments constraining solutions. In architecture, they embed the core of a design solution encapsulating its *generic* characteristics and constraints and conveying the form of possible *specific* solutions. That they are not detailed prevents early commitment to a specific design solution and, thus, they facilitate exploratory reasoning. At the same time they are not ambiguous in the way sketches are in that they fix meaning and define a set of related solutions.

This latter is important because design problems are ill-defined in that either the initial state, the goal state, or the operators--or all of them--require further specification. With ill-defined problems there exists a set of potential goal states instead of one goal state. One way that architects delimit the range of alternatives is by analogy. Conceptual diagrams function in a way similar to analogies in that they provide constraints that restrict the set of specified goal states.

We propose salient characteristics of conceptual diagrams that are significant for

M. Hegarty, B. Meyer, and N. Hari Narayanan (Eds.): Diagrams 2002, LNAI 2317, pp. 353–355, 2002.

design cognition. First, as abstractions, conceptual diagrams provide idealizations that represent complex ideas in a simple and easily communicable and retainable form. Second, they can represent spatial, functional, or formal relationships. These two characteristics make them important in collaborative design problem solving and communication with the client. Third, they are easier to transform into final design solutions than conceptualizations represented verbally, in part, because conceptual diagrams are in the same representational mode as the end product of designers. We will now consider two examples illustrating the use of conceptual diagrams in design.

The first example is from an expert designer, Louis I. Kahn. Kahn formulated a design theory that held that a designer first finds a "conceptual idea" and then implement the idea in different schemes. The conceptual idea remains constant whereas the implementation changes. In one of his projects, the First Unitarian Church of Rochester, Kahn drew a conceptual diagram at his first meeting with the client (Figure 1).

This diagram conveys in generic form Kahn's conceptualization of the problem and of potential solutions. It consists of concentric circles that represent how different spatial components of the church building relate to each other. The main idea here is that all the components of the design are to be in a concentric relationship, i.e., they are defined in reference to the center that Kahn in another speech defined as the locus of contemplation. He described his concept as follows:

Fig. 1. Kahn's concept diagram

> A chapel, to me, is a space that one can be in, but must have excess of space around it, so that you don't have to go in. That means, it must have an ambulatory, so that you don't have to go into the chapel; and the ambulatory must have an arcade outside, so that you don't have to go into the ambulatory; and the object outside is a garden, so that you don't have to go into the arcade; and the garden has a wall, so that you can be outside of it or inside of it. [8, p. 86]

Kahn's conceptual diagram was crucial in communicating his main idea to the client. The client in turn used this concept to assess different schemes proposed by Kahn. It supported collaboration in that it assisted the client to see the problem and to contribute to the definition of the problem.

The second example is from a student's design project. The project asked students to integrate a leftover space on a university campus into its surrounding. The final presentation included a conceptual diagram and also two photographs that embed the conceptualization of possible design solutions (Figure 2).

Fig. 2. The diagram on the right represents the design concept.

The top photograph shows a group sitting around a circle, illustrating that the student views the problem as a problem of gathering of people around a common object. The bottom photograph shows an outdoor theater representing a precedent solution. His conceptual diagram is of a central circle and several tangent lines to the circle. The circle represents the central area of focus that gathers people around whereas the tangent lines represent the lines of flow that connect the area of focus to its immediate surroundings. The diagram helped the student to convey his main idea to the jury members and facilitated their comments and contributions to his design.

In conclusion, we have presented two examples of conceptual diagrams. In these examples the conceptual diagrams embed generic conceptualizations of solutions that constrain the possible allowable designs. Designers in both cases used the conceptual diagram to convey an idea, as well as employing it in their further problem solving. In Kahn's case, it ensured a commitment to the generic concentric composition, while allowing further explorations. In the second case, it enabled the student and the reviewers to form a shared understanding of the conceptualization behind the solution.

The study of conceptual diagrams has implications for understanding collaborative problem solving and for enhancing design education. In collaborative problem solving, conceptual diagrams make it easy to communicate and grasp the central ideas constraining the range of problem solutions under consideration. If used more extensively in architectural education, conceptual diagrams could serve to encourage students' self-reflections on high-level commitments in their design processes. Research into conceptual diagrams promises new insights about the role of diagrams in different contexts and, more generally, of external representations in cognition.

References

1. Zhang, J. (1994). "Representations in distributed cognitive tasks." Cognitive Science Journal 18(1): 87-122.
2. Larkin, J. H. and H. A. Simon (1987). "Why a diagram is (sometimes) worth ten thousand words." Cognitive Science 11: 65-99.
3. Reitman, W. (1964). Heuristic decision procedures, open constraints, and the structure of ill-defined problem. Human Judgements and Optimality. M. Shelly and G. Bryan (Eds.). New York, Wiley: 282-315.
4. Koedinger, K. R. and J. R. Anderson (1995). Abstract planning and perceptual chunks. Diagrammatic reasoning: cognitive and computational perspectives. B. Chandrasekaran, J. Glasgow and N. H. Narayanan (Eds.). Cambridge, MA, MIT Press.
5. Lowe, R. (2000). Animation of diagrams: an aid to learning? Theory and application of diagrams. M. Anderson, P. Cheng and V. Haarslev (Eds.). Berlin, Springer.
6. Do, E. Y.-L. and M. D. Gross (2001). "Thinking with diagrams in architectural design." Artificial Intelligence Review 15(1-2): 135-149.
7. Nersessian, N. (2002). Maxwell and "the Method of Physical Analogy": model-based reasoning, generic abstraction, and conceptual change. Reading natural history: essays in the history and philosophy of science and mathematics to honor Howard Stein on his 70th birthday. D. Malamet (Ed.). Lasalle, IL, Open Court.
8. Latour, A. (1991). Louis I. Kahn: writings, lectures, interviews. New York, Rizzoli.

A Survey of Drawing in Cross-Linguistic Communication*

Charlotte R. Peters and Patrick G.T. Healey

Information, Media and Communication Research Group
Department of Computer Science
Queen Mary, University of London
London E1 4NS, UK
{cp, ph}@dcs.qmul.ac.uk

Abstract. This paper presents a small survey of the use of drawing in cross-linguistic communication. The findings indicate that drawing is an infrequent but valuable element of cross-linguistic interactions. It is used opportunistically, for a wide variety of functions, predominantly in dyadic interactions. It is used more frequently to support the drawer in contributing to an interaction conducted in their 'second' language, than to support an addressee in understanding a contribution in the drawer's 'first' language. Also, two broad categories of drawing, 'cross-cultural' topics and language-use are more frequent in cross-linguistic interactions than in other contexts.

1 Introduction

Although it is natural to think of drawing as a solitary activity, in practice it is often integrated into communicative interactions. van Sommers [1] provides evidence that approximately half of all drawing takes place for or with an audience. The uses to which drawing is put in these communicative contexts are different from those to which it is put in solitary contexts. It is known that provides a valuable means of communication when language is compromised e.g. adults with Aphasia [2].

Anecdotally, drawing also appears to be of value in cross-linguistic interactions. Breakdowns in cross-linguistic communication, for example; in shops, banks, restaurants, hotels, tourist information offices often provoke the use of sketches and diagrams. This paper presents an initial, questionnaire-based investigation of the use of drawing in support of cross-linguistic interactions.

2 Method: Cross-Linguistic Drawing Questionnaire

A 'critical incident' questionnaire was used to elicit episodes of drawing in cross-linguistic interaction. This questionnaire required subjects to recall the last time

---

* We gratefully acknowledge the support of ATR Media Information Science Laboratories and the ERSC/EPSRC grant MAGIC: Multimodality and Graphics in Interactive Communication (L328253003).

M. Hegarty, B. Meyer, and N. Hari Narayanan (Eds.): Diagrams 2002, LNAI 2317, pp. 356–358, 2002.

that drawing occurred in an interaction with someone whom they do not share a first language. They were asked a series of questions with regard to the conversation, the drawing(s) and the shared language skills of the interactants.

Subjects. Nineteen females and 16 males (1 unspecified) with an average age of 34 took part. All were employees at an international research laboratory in Japan. Half were Japanese and half a mixture of other nationalities.

Materials. Most questions were open-ended. Blank spaces were allowed for subjects to use in answering. Questions with a limited range of answers utilised a tickbox method to select the appropriate answer(s). All tickboxes had an 'other' option, for alternative answers. Finally, for answers that could range from 'very ... ' to 'never' or 'none' a scale ranging from 1-7 was used ("1" being the strongest to "7" for the weakest).

Procedure. A small sample of fifty questionnaires were distributed among subjects who were known to have had cross-linguistic interactions and a good level of English, although this was not tested in any formal manner. Subjects had approximately one week to complete and return the questionnaire. 41 (of 50) questionnaires were returned. Those that were at least 30% incomplete or misunderstood were rejected, leaving 36 valid for analysis.

3   Findings

67% of the reported episodes occurred at work, the remainder were spread more or less evenly between situations such as; at home, in a bar, café or shop, on a boat, on the street and in a language class. Interactions mostly occurred with people that the respondents knew moderately well or better; consisting primarily of friends (31%) and colleagues (47%). Despite the workplace based sample, the interactions frequently related to a social or personal topic (15/24) for example, drawings of people, lifestyles, local customs, food, and language usage. By contrast, representations of technical objects and problem exploration occurred in only 27% of cases. This corresponds to 20% of comparable categories in van Sommers (1984) data, further indicating that the incidence even of ostensibly work based topics of topics is not entirely due to the composition of the sample. The main findings from the questionnaire data are:

1. Drawing Is Primarily Opportunistic. The work environment where most of the reported interactions took place provides a variety of graphical media. Whiteboards and more sophisticated graphical tools are also widely available. However, 81% of drawing took place on paper usually taken from the immediate environment of the interaction.

2. Drawing Is Relatively Infrequent. Approximately two thirds of those surveyed reported engaging in cross-linguistic interactions on a "frequent" basis [1] 39% of respondents cited the last episode of drawing in cross-linguistic interaction as occurring within the last week, for 22% it had taken place within the last six months and for the remainder it had taken place between six months and seven years ago.

3. Drawing Is Most Useful for Non-native Language Production. Respondents most frequently used drawing when interacting in their second or weaker language. 70% of drawings were produced in interactions in a shared language rather than the respondent's first language.

4. Drawing of Cross-cultural Topics and Second-Language Usage Is Predominant in This Context. The questionnaire data shows the presence of two categories of drawing, relevant to cross-linguistic communication, that are not present in van Sommers (1984) data. Firstly, the most frequent example of drawing activity is of cross-cultural topics (31%). This category is, of course, heterogeneous and includes cartoons, festivals, food, lifestyle, local customs and people. Secondly, 17% of drawing activity was with regard to second language usage. Examples include distinguishing between similar sounding nouns and adverbs, and words for specific items. Noticeably, van Sommers' (1984) categories: 'sketching hair and clothing' and 'defacing pictures' are not reported in the questionnaire data. This suggests, intuitively, that drawing is useful to communicate new or unfamiliar concepts to an uninformed participant.

Because the analysis is retrospective it is not possible to verify reports against what was drawn. However, the results are comparable to van Sommers (1984). They show that drawing in cross-linguistic communication is a spontaneous and social activity and has important communicative functions. Drawing appears to be of most value to the participant in a cross-linguistic interaction who is communicating in their non-native language, about concepts which at least one participant is uncertain, such as cross-cultural nuances or second language usage.

References

1. van Sommers, P.: Drawing and Cognition: Descriptive and Experimental Studies of Graphic Production Processes. Cambridge: Cambridge University Press (1984)
2. Lyon, J.: Drawing: its value as a communication aid for adults with aphasia. Aphasiology **9** (1995) 34–50

---

[1] The 'critical incident' technique used in this questionnaire does not provide direct data about frequency of occurrence of events.

Informal Tools for Designing Anywhere, Anytime, Anydevice User Interfaces

James A. Landay

Computer Science Division,
University of California,
Berkeley, CA 94720 USA
landay@cs.berkeley.edu

Abstract. We are now entering the era of pervasive computing, an era where people will access information and services anywhere, anytime, and from a wide variety of devices. The challenge for researchers and practitioners is how to support the design of user interfaces that will empower people to engage in these interactions easily and efficiently. Our work has been in creating design tools that support the best practices of user-centered design. Such practices include the informal techniques used during the early stages of design, such as sketching and "faking" interactions using Wizard of Oz techniques to test early designs. In this talk we will argue that tools with informal user interfaces best support these practices. Informal user interfaces support natural human input, such as speech and writing, while minimizing recognition and transformation of the in put. These interfaces that document, rather than transform, better support a designer's flow state. Unrecognized input embraces nuanced expression and suggests a malleability of form that is critical for activities such as early-stage design. We will illustrate this by examining informal tools we have created for designing information architectures and web sites, speech-based user interfaces, and eventually anytime, anywhere user interfaces that take advantage of a variety of modes of input and output on a range of devices.

Biography. James Landay is an Assistant Professor of Computer Science at the University of California, Berkeley. He received his Ph.D. from Carnegie Mellon University in 1996. His Ph.D. dissertation was the first to demonstrate the use of sketching in user interface design tools. He has published extensively in the area of user interfaces, including articles on user interface design tools, web site evaluation tools, gesture recognition, pen-based user interfaces, mobile computing, and visual languages.

M. Hegarty, B. Meyer, and N. Hari Narayanan (Eds.): Diagrams 2002, LNAI 2317, p. 359, 2002.
© Springer-Verlag Berlin Heidelberg 2002

Author Index

Lecture Notes in Artificial Intelligence (LNAI)

Lecture Notes in Computer Science